DESIGN THEORY
OF FLUIDIC COMPONENTS

H2 £21-60 M.L.

DESIGN THEORY OF FLUIDIC COMPONENTS

Joseph M. Kirshner

Fluidic Systems Research Branch
Harry Diamond Laboratories
Washington, D.C.
and
Department of Mechanical Engineering
George Washington University
Washington, D.C.

Silas Katz

Mechanical Engineering Department
Concordia University
Montreal, Quebec

ACADEMIC PRESS   New York   San Francisco   London   1975
A Subsidiary of Harcourt Brace Jovanovich, Publishers

COPYRIGHT © 1975 BY ACADEMIC PRESS, INC.
ALL RIGHTS RESERVED.
NO PART OF THIS PUBLICATION MAY BE REPRODUCED OR TRANSMITTED IN ANY FORM OR BY ANY MEANS, ELECTRONIC OR MECHANICAL, INCLUDING PHOTOCOPY, RECORDING, OR ANY INFORMATION STORAGE AND RETRIEVAL SYSTEM, WITHOUT PERMISSION IN WRITING FROM THE PUBLISHER. REPRODUCTION IN WHOLE OR IN PART FOR ANY PURPOSE OF THE UNITED STATES GOVERNMENT IS PERMITTED.

ACADEMIC PRESS, INC.
111 Fifth Avenue, New York, New York 10003

United Kingdom Edition published by
ACADEMIC PRESS, INC. (LONDON) LTD.
24/28 Oval Road, London NW1

Library of Congress Cataloging in Publication Data

Kirshner, Joseph M
  Design theory of fluidic components.

  Includes bibliographies.
  1. Fluidic devices.   I. Katz, Silas, joint author.
II. Title.
TJ853.K57   629.8'04'2   73-18940
ISBN 0-12-410250-6

PRINTED IN THE UNITED STATES OF AMERICA

CONTENTS

Preface ix
Acknowledgments xi

Chapter 1 Introduction to Fluidics

1.1 Introduction 1
1.2 The Basic Components 3
1.3 Circuit Considerations 10

Chapter 2 Passive Components

2.1 Resistance 14
2.2 General Fluidic Resistance 16
2.3 Entrance Resistance 18
2.4 Wall Shear Resistance 19
2.5 Exit Resistance 26
2.6 A Typical Fluidic Resistance 30
2.7 Fluid Diodes 31
2.8 Resistance Networks 37
2.9 Capacitance 43
2.10 Inertance or Inductance 56
Problems 56
Nomenclature 59
References 60

Chapter 3 Distributed Fluid Passive Components

3.1 Introduction to Transmission Lines 63
3.2 Distributed Parameter Theory 65
3.3 Lumped Parameter Approximations 69
3.4 Propagation Models for Circular Fluid Lines 75

v

3.5 Propagation Models for Rectangular Fluid Lines	87
3.6 Lumped Modeling of Fluidic Circuits	93
3.7 Circuit Theory	97
3.8 Frequency Response of Fluid Line Circuits	99
3.9 Impulse and Step Response of Fluid Lines	105
3.10 The Method of Characteristics	106
3.11 The Method of Characteristics for Small-Amplitude Signals	107
3.12 The Quasi Method of Characteristics for Small-Amplitude Signals	118
3.13 The Method of Characteristics for Large-Amplitude Signals	125
3.14 Matching	130
3.15 The Method of Characteristics for Fluidic Line Circuits	136
Problems	140
Nomenclature	142
References	144

Chapter 4 Jet Flows

4.1 Introduction	146
4.2 Laminar-Free Jets from Infinitesimal Apertures	147
4.3 Turbulent-Free Jets from Infinitesimal Apertures	161
4.4 Turbulent-Free Jets from Finite Apertures	173
4.5 Experimental Results on Plane Turbulent Jets	183
Problems	189
Nomenclature	190
References	191

Chapter 5 Jet Dynamics

5.1 General Development of Jet Dynamics	192
5.2 Response of Jet to an Impulse Function	195
5.3 Constant Pressure Gradient	197
5.4 Oscillating Pressure Gradient	199
5.5 Propagating Pressure Gradient	203
5.6 Transverse Impedance of a Jet	205
5.7 The Effects of Feedback (Edgetones)	207
5.8 Velocity Profile of Oscillating Jet	220
5.9 Effect of the Jet on the Pressure Field	222
Problems	224
Nomenclature	225
References	225

Chapter 6 Static Characteristic Curves

6.1 Introduction	227
6.2 Concept of Source and Load	227
6.3 The Two-Terminal Pair	234
6.4 Proportional Amplifier Characteristics	235
6.5 Bistable Switch	240

Contents vii

6.6	NOR Elements	247
6.7	Passive Logic Elements	249
	Problems	252
	Nomenclature	253
	References	254

Chapter 7 The Impact Modulator

7.1	Introduction	255
7.2	Centerline Total Pressure Decay of Free and Impinging Jets	259
7.3	The Effects of Control Flows on the Plate Decay Factor	262
7.4	Source Flow Modulation	268
7.5	Impact Modulator Pressure Gain	269
	Problems	274
	Nomenclature	275
	References	275

Chapter 8 The Vortex Triode

8.1	Historical Introduction	276
8.2	Basic Description of Vortex Triode	278
8.3	Analyses of the Vortex Triode	284
8.4	Vortex Triode Design Chart	300
8.5	The Vortex Triode as a Proportional Amplifier	303
	Problems	310
	Suggested Term Papers	311
	Nomenclature	311
	References	313

Chapter 9 The Beam Deflection Amplifier

9.1	Historical Introduction	315
9.2	Basic Operating Principles	320
9.3	Introduction to Amplifier Static Analysis	324
9.4	Analysis of Input Region	326
9.5	Effect of the Vents	348
9.6	Analysis of the Output Region	352
9.7	The Output Characteristics	358
9.8	Aspect Ratio	361
9.9	Introduction to Dynamic Analysis	361
9.10	The Input Impedance	362
9.11	The Transfer Function	364
9.12	Evaluation of the Impedances and of k_s	368
9.13	Staging of Amplifiers	370
	Problems	376
	Suggested Term Papers	377
	Suggested Research Projects	377
	Nomenclature	378
	References	379

Chapter 10 The Bistable Switch

10.1	Early History	382
10.2	Principles of Operation	383
10.3	Wall Attachment Theories	389
10.4	Theory of the Bistable Switch	408
	Problems	421
	Nomenclature	421
	References	422

Chapter 11 The Transition NOR

11.1	Introduction	425
11.2	Performance Criteria	426
11.3	Supply Characteristics	430
11.4	Transfer Characteristics	435
11.5	Input and Output Characteristics	437
11.6	Performance Optimization	439
11.7	Modified Verhelst Diagram	440
11.8	Dynamic Response	441
11.9	Alternate Transition Element Configurations	445
	Problems	447
	Suggested Research Projects	447
	Nomenclature	448
	References	448

Appendix A **Circular Transmission Line Characteristics for Air**   451

Appendix B **Rectangular Transmission Line Characteristics for Air**   453

Appendix C **Weighting Factors for Circular Sections**   458

Appendix D **Weighting Factors for Rectangular Sections**   462

Appendix E **Computer Programs**   465

Index   473

PREFACE

The field of fluidics is now almost 15 years old. From its beginnings, fluidics captured the imagination of the scientific and engineering community. Unquestionably, the concept of "no moving parts" heightened interest and was a most attractive feature.

After the initial stimulation, fluidics proved to be a difficult and sometimes frustrating subject. Part of the difficulty stemmed from the fact that fluidics is a hybrid field. Fluidics requires a knowledge of control systems to define useful functions and a knowledge of fluid mechanics to implement these functions. Beyond this, however, remained the complexity of fluidic components themselves. As a result, the early years of fluidic research consisted of a patchwork of separate investigations.

In the past few years the bits and pieces of theoretical and experimental fluidics research have begun to coalesce into an integrated whole. As the accumulated portions of the complex and difficult jigsaw puzzle have fallen into place, it has at last become possible to say with assurance not only that we understand most of the basic mechanisms involved in fluidics, but also that we understand them sufficiently well so that fluidic components and circuits can be analyzed with engineering accuracy.

A significant reason for the success that has been achieved was the recognition that for control systems low-power devices are adequate and desirable not only because of their lower operating costs but because they are considerably less noisy. Low power and the associated small signal transients permit linearization of the fundamental fluid equations. Thus impedance concepts become useful and dynamic analysis is simplified.

The computer, too, has been a factor in obtaining solutions for those cases in which many (often nonlinear) equations are necessary to describe a fluidic component, such as the vortex amplifier and the bistable switch.

To be sure, there are still unsolved problems and we have attempted to point out some of them. The basic purpose of this book, however, is to present a solid foundation of fluidics as a science rather than an art. In this connection we have tried to tie together the most pertinent results of research in fluidics and in closely allied fields.

Although present theory gives results sufficiently accurate for engineering design, it is not possible to justify all the assumptions used. Thus, in a scientific sense the theory is not always satisfying, but in an engineering design sense the theory does seem to be satisfactory. For this reason we have designated this book as "*Design* Theory of Fluidic Components."

We do not mean to imply that the book is a designer's guide to fluidics. To the contrary, it is intended primarily as a reference work for fluidic research engineers. The text has considerable detail that would not interest a hardware-oriented engineer. The book is also suitable for a graduate course in fluidics. Indeed, a preliminary version of the material contained herein has already been used for a graduate mechanical engineering course at George Washington University.

The book is divided roughly into the introductory chapter plus two parts. The first part (Chapters 2–5) considers passive fluid components and the theory of jets. These are the essential building blocks of fluidics.

After discussing in Chapter 2 the factors that give rise to lumped resistance, capacitance, and inductance, we then in Chapter 3 review fluid transients in lines. The various models of the transmission line are examined and compared as are the relations between impedance concepts and the method of characteristics. Various jet velocity distributions are covered in Chapter 4, and in Chapter 5 a simplified theory of the motion of jets in a pressure gradient is developed.

The second part (Chapters 6–11) deals with the active fluidic components and begins with a consideration of the characteristic curves that are necessary to describe the performance of the active components. Following this we show the extent to which the performance of specific fluidic components can be predicted analytically. The current literature has made it possible to relate the static and dynamic characteristics of most of the more important fluidic devices to their geometric and physical properties. Sufficient detail is presented for most of the analyses so that they can be followed and used by the reader in designing devices. In a few cases the mathematics necessary is so involved that the results can be obtained only by the use of a computer, and in those cases we have somewhat reluctantly omitted the details of the analysis.

We hope this book will prove as informative to readers as compiling it has been for us.

ACKNOWLEDGMENTS

A great many individuals from many countries have contributed to the emergence of fluidics as a science. We have tried in all cases to identify the originators of the concepts involved. We have also endeavored to include all important material that bears directly on the subjects we discuss. Nevertheless, because there is so much work published in the area of fluidics, we undoubtedly have missed some papers that should have been covered.

Most of the material in Chapter 4 originally appeared in *Fluidics Quarterly*, Volume 1, Issue 3 (April 1968). We thank the publishers of *Fluidics Quarterly* for permitting us to use this material. The work on this monograph was done while we were both with the Fluidic Systems Research Branch of the Harry Diamond Laboratories. We would like to thank the Harry Diamond Laboratories for permission to use material originating in HDL reports and for preparing the many drawings presented in this book. We also appreciate the excellent typing done by Mrs. Ercelle Janifer, Miss Billie Richardson, and Miss Dawn Perry. We are sincerely grateful to Mr. Robert D. Hatcher, formerly Chief of the Systems Research Laboratory at HDL, for reading the original manuscripts and making many valuable suggestions.

Chapter 1

INTRODUCTION TO FLUIDICS

1.1 INTRODUCTION

Fluidics as a technology dates from 1959 when intensive efforts on no-moving-part fluid devices began at the Harry Diamond Laboratories (then the Diamond Ordnance Fuze Laboratories) as a result of a search for methods and devices for increasing the reliability of systems.

Today, there exists a literature of over 4000 papers and over 700 U.S. patents on fluidics. Scientists from all over the world are solving the problems involved and engineers are incorporating fluidic devices in many types of systems.

The original components demonstrated at HDL in March 1960 were very difficult to stage, had low gains, and were quite noisy; very little was known of the physical mechanisms involved.

The newer commercial components not only have much better characteristics, less noise, and are easier to stage but also are being built in modular form to minimize tubing; several types allow snap-on or plug-in construction. As a result, hundreds of applications have now been found for fluidics and many more are being considered.

After wrestling for a number of years with extremely difficult and complex problems, many of the major problems have been solved (at least to a first approximation), and we know with reasonable accuracy the theoretical relationships between the geometry of many of our fluidic devices and their performance characteristics.

1.2 THE BASIC COMPONENTS

1.2.1 The Beam Deflection Proportional Amplifier

Several types of beam deflection amplifiers are illustrated in Figs. 1.1a and 1.1b. Like many fluidic devices the amplifier has the type of construction known as two-dimensional; that is, the planform as illustrated in Fig. 1.1a has some depth perpendicular to the paper and is covered from above and below by cover plates. The power jet nozzle is then a rectangular slit. The height of this slit (the distance between the top and bottom plates) divided by the slit width (the nozzle width) is called the aspect ratio.

In all of the illustrations shown, the power or supply jet is at the bottom. One of the simpler models is that shown at top left of Fig. 1.1b. The two outlets face the power nozzle so that when the power jet is undeflected it splits equally and the flow and pressure differences at these two outlets are zero. Just below the outlets are the vents which dump excess fluid from the power jet that does not enter the receivers (outlets). The lowest pair of openings are the controls. The pressure signal is applied across these controls to cause the power jet to deflect.

The amplifier in Fig. 1.1a and the one at the top right of Fig. 1.1b are similar to the first except that they have a center dump between the two outlets.

The other three amplifiers in Fig. 1.1b have two sets of nozzles that can affect

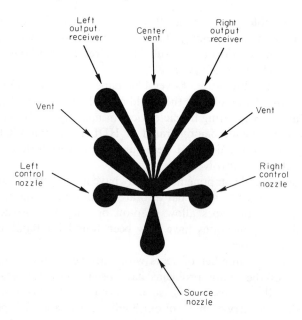

Fig. 1.1a A beam deflection amplifier.

1.2 The Basic Components

Fig. 1.1b Beam deflection amplifiers.

the interaction region. Both sets can be used as inputs or one set can be used for the input signal and the other as vents.

A typical output pressure difference versus input (control) pressure difference is shown in Fig. 1.2. The slope of this curve at any point is the pressure gain at the point.

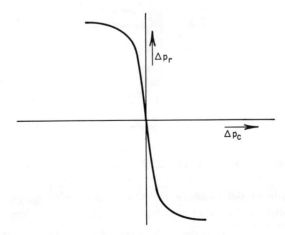

Fig. 1.2 Output pressure difference as a function of control pressure.

1.2.2 The Wall Attachment Bistable Switch

The wall attachment switch (Fig. 1.3) is a bistable device. Since the jet entrains fluid on both sides, the presence of the walls causes molecules of the fluid to be evacuated between the jet and each wall causing a low-pressure region on each side. This condition is unstable and the jet therefore attaches to one wall or the other. The jet can be detached from one wall and caused to attach to the opposite wall by a suitable positive pressure signal into the control on the attached side. If the outlet is open, flow issues from the outlet on the attached side. If the output on the attached side is blocked, the flow will issue from the vents but the jet will remain attached to the wall, in which condition a pressure exists within the outlet region that may be as much as 80% of the power jet pressure. The percent of power jet pressure appearing at the outlet is called the recovery pressure.

A figure of merit for a bistable device is the fan-out, which is defined as the number of similar devices that can be switched by a single device.

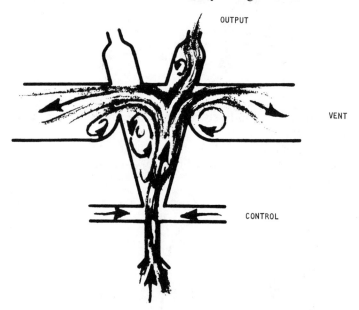

Fig. 1.3 Bistable switch.

1.2.3 Beam Deflection Inverter and NOR Elements

A monostable beam deflection device may be used as an inverter. In its stable state with no input the power jet exits from the signal outlet. While deflected by a control jet, however, the power jet is discharged into a dump. Wall attachment may or may not be used to provide the stable state.

1.2 The Basic Components

An inverter gives an output when there is no input and no output when there is an input. If more than one inlet is used, an output is obtained only if all the inputs are zero. Such a device is called a NOR element. Figure 1.4 illustrates one type of NOR element that utilizes beam deflection.

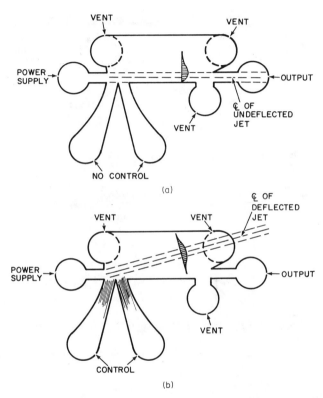

Fig. 1.4 NOR element (deflection type): (a) undeflected jet, (b) deflected jet with control.

1.2.4 Transition and the Turbulence NOR Element

Jets of low Reynolds numbers may remain laminar for an appreciable distance after leaving a nozzle whereas high Reynolds number jets become turbulent shortly after leaving the nozzle. There is a Reynolds number region where the jet, although normally laminar, is very sensitive to disturbances and easily becomes turbulent. In particular, it can become turbulent by disturbing it with a second jet of much lower flow.

This property may be used to produce an inverter. A laminar jet issuing from a small pipe is allowed to enter another small pipe in line with the first one but an appreciable distance downstream. If the jet is disturbed by another jet

(Fig. 1.5), it will become turbulent and very little of the flow will reach the downstream (receiver) pipe.

If there are two (or more) control jets, this device is a **NOR** unit. Because there is almost no effect on the controls when the outputs are loaded, this device is very easy to stage.

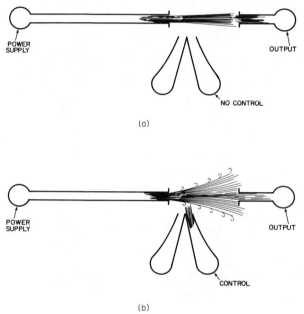

Fig. 1.5 NOR element (transition type): (a) laminar jet (high recovery), (b) turbulent jet (low recovery).

1.2.5 Impact Devices

If two round jets of the same diameter are directed head-on at each other and both lie on the same axis, the following phenomena are observed:

(1) If the jets have the same amount of momentum, the fluid will fan out in all directions in the plane perpendicular to the axis from the region of collision of the two jets.

(2) If the momentum of one jet is changed, the impact region will move toward the lower-momentum jet nozzle.

(3) If the impacting jets are disturbed by a third (control) jet applied at right angles to one of the impacting jets, the jet with which it impacts will increase its turbulence and consequently it will spread more than previously.

In an impact device the power jet from the receiver side impacts with the other power jet within a small chamber to which the outlet of the device is

1.2 The Basic Components

coupled (Fig. 1.6, top). A control jet perpendicular to the impacting jets causes spreading as well as some misalignment of one of the jets, thus allowing less of its momentum to enter the chamber. As a result the impact point moves out of the chamber and the flow pattern radiating from the impact point tilts

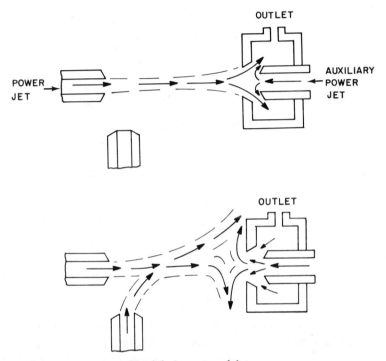

Fig. 1.6 Impact modulator.

(Fig. 1.6, bottom). This lowers the pressure in the chamber and decreases the flow into the outlet. The change in outlet pressure is thus of the opposite sense to the change in control pressure. The effect is used to produce both proportional and digital elements.

1.2.6 Vortex Devices

Consider fluid entering a cylinder tangentially and leaving through an axial drain, as in Fig. 1.7. For simplicity, we consider an incompressible fluid. Because of the conservation of angular momentum, the tangential velocity of the fluid will increase as it spirals in toward the drain. For an inviscid fluid it is easy to show that the tangential velocity u_θ, is given by

$$u_\theta = k/r$$

where k is a constant and r is the radial distance from the center.

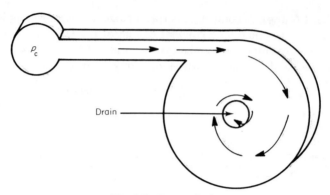

Fig. 1.7 Vortex diode.

If the total pressure is p_c, then for an incompressible, inviscid fluid, we have

$$p_c = p_1 + \tfrac{1}{2}\rho u^2$$

where p_1 is the static pressure, u is the fluid velocity, and ρ is the density. Now since $u_\theta = k/r$ and $|u_\theta| \le |u|$, it follows that, for a small enough value of r, the maximum possible value of the term $\tfrac{1}{2}\rho u^2$ occurs when $p_1 = -p_a$ (where p_a is atmospheric pressure) As a result

$$p_c + p_a = \tfrac{1}{2}\rho u_{\max}^2$$

Because u cannot take on values greater than $u_{\max}$ there exists a minimum value of r that defines a limit circle within which the fluid cannot penetrate.

Another way of thinking about this phenomenon is to consider the fact that the centrifugal force increases as the fluid spirals inward until eventually the centrifugal force of the inner layers of rotating fluid is sufficient to balance the total pressure forces.

1.2.7 Vortex Diode

It follows that if the radius of the drain is less than that of the limit circle an inviscid fluid would be trapped and merely continue whirling around without ever leaving the drain. Flow in the other direction (from the drain toward the tangential arm), however, would not "see" such a phenomenon, so that the geometry of Fig. 1.7 results in a fluid diode. Unfortunately, viscosity effects in any real fluid are quite important so that actual vortex diodes cannot really completely cut off the flow.

Fluid diodicity may be defined in several different ways. In most of the technical literature the diodicity is the ratio of pressure drops in the two directions for the same volume flow. With this definition diodicity of the order of 200

1.2 The Basic Components

has been obtained. However, it is also possible to use the flow ratio at the same pressure drop as the diodicity. In this case the maximum diodicity would be the square root of 200 or about 14.

1.2.8 Vortex Triode

If we add another source of flow as in Fig. 1.8, a basic vortex triode is obtained. Fluid from the power source p_s, ordinarily proceeds straight in toward the drain. The addition of flow from the control p_c causes the resultant flow to spiral in toward the center, resulting in a decreased output because of the centrifugal force effect. Since the change in net flow is greater than the control flow causing it, the device has flow gain.

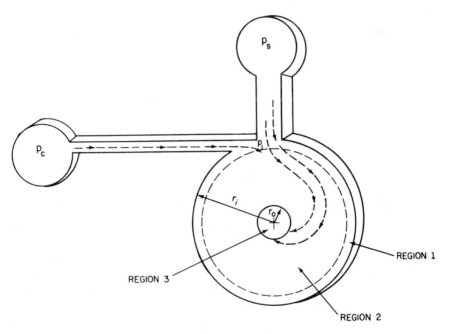

Fig. 1.8 Vortex triode.

The vortex triode is a throttling device rather than a diverter (as the beam deflection amplifier is); that is, the control cuts down the power and flow output rather than throwing it away. This is a decided advantage when appreciable power is to be controlled.

High pressure gains (changes in output pressure versus changes in input pressure) may also be obtained from a vortex amplifier, but the control pressure level must be greater than the supply pressure. This property makes the devices difficult to stage.

1.3 CIRCUIT CONSIDERATIONS

At present there is no generalized fluid circuit theory that compares to electrical circuit theory. Fluid circuits have several special features which make them difficult to analyze. Among the difficulties are:

(1) *Fluid Density*. Density and density changes substantially affect the circuit performance. As a consequence of density, the fluid particles carry momentum and this in turn may produce circuit components with directional characteristics. Density changes, furthermore, necessitate the use of specific thermodynamic processes (i.e., isothermal, adiabatic) in component analysis and the designation of flow as "incompressible" and "compressible."

(2) *Fluid Viscosity*. The viscous forces acting on fluid circuit components depend on the type of flow. There are generally two main types. If the flow is well ordered, fluid layers slide over one another and we call this "laminar flow," The other type, "turbulent flow," occurs when large-scale fluctuations are superimposed on the mean flow. The viscous forces are much larger in turbulent flow than in laminar flow. Thus circuit components of identical geometry have different characteristics in laminar and in turbulent flow.

With these difficulties as a background let us now attempt to define the fluid analogs of electric current and voltage.

1.3.1 Fluid Circuit Signal Variables

From the continuity equation of fluid mechanics, the fluid analog of electrical current is mass flow $\dot{m}$. The fluid analog of voltage, however, is more difficult to obtain. In electricity the voltage concept may be derived from the conservation of energy. The derivation is not straightforward for fluids since one has the option of considering all the available energy or only the mechanical energy.

If we assume that for most cases only the mechanical energy needs to be considered and that the fluid is homogeneous, the fluid voltage analog ε at a point is given by

$$\varepsilon = \int (1/\rho) \, dp_\sigma + u^2/2 \tag{1.1}$$

where p_σ represents the static pressure.

The particular value of the integral $\int (1/\rho) \, dp_\sigma$ depends on the thermodynamic process. In addition to the theoretical difficulties in calculating ε, there are practical difficulties even in measuring it at a single point. The problem becomes completely unmanageable when we realize that we must average the distribution of ε across a section.

Fortunately for most fluidic control circuits and devices the Mach number is usually less than 0.3, so that the change in density (throughout the circuit at a

1.3 Circuit Considerations

specific time) is small compared to the density. Under these conditions Eq. (1.1) can be written as

$$\varepsilon \cong (p_\sigma/\bar{\rho}) + u^2/2 \qquad (1.2)$$

and the current analog as

$$\dot{m} = \bar{\rho} q \qquad (1.3)$$

where $\bar{\rho}$ is the average density throughout the circuit, and q is the volume flow.

The product of the voltage and current analogs, the power E, may be expressed as

$$E = \varepsilon \dot{m} = (p_\sigma + \bar{\rho} u^2/2) q \qquad (1.4)$$

Now, note that when $\bar{\rho}$ is constant Eqs. (1.3) and (1.4) are also satisfied for the voltage analog p and the current analog q in the form

$$\text{voltage analog} = p = p_\sigma + \bar{\rho} u^2/2 \qquad (1.5a)$$
$$\text{current analog} = q \qquad (1.5b)$$

where p is the total pressure at a point.

The signal variables described in Eq. (1.5) are easier to measure than the more accurate analogs (ε and $\dot{m}$).

It is possible to obtain an additional simplification of signal variables by recognizing that at the low Mach numbers for which Eq. (1.5a) holds, the velocity portion of the total pressure in Eq. (1.5a) is much less than the static pressure portion. For example, when the Mach number is less than 0.3 the dynamic pressure term ($\bar{\rho} u^2/2$) of Eq. (1.5a) is less than 5% of the static term and the dynamic term is less than 1% of the static term for Mach numbers below 0.1. Since most fluidic circuits operate with small amplitude signals, the static pressure is sometimes an acceptable approximation of the voltage analog. However, it is always necessary to remember that the correct voltage analog is really the total pressure. Failure to use total pressure can lead to erroneous results in the analysis of circuit components such as at sudden enlargements and branches.

1.3.2 Fluid Circuit Regimes

The models that are used for fluid circuit components depend on the amplitude and frequency of the signal variables. The chart shown in Fig. 1.9 separates the amplitude–frequency map into discrete regimes for a fluid line. Each regime represents a particular amplitude and frequency division. Actually the divisions are not as clearly defined as shown in Fig. 1.9, but occur over ranges of amplitude and frequency. The major amplitude divisions are shown at Mach numbers of 0.3 and 1.0. As the Mach number decreases below 0.3 the accuracy of the incompressible assumption improves; however, the circuit model remains the same.

The major frequency divisions depend upon the viscous characteristic frequency ω_v. For a circular line, $\omega_v = 32\mu/\rho d^2$ (where μ is the absolute viscosity and d is the line diameter). At low Mach numbers (less than 0.3), when the frequency is less than $0.1\omega_v$ the line is primarily resistive and capacitive (isothermal); when the frequency is greater than $10\omega_v$ the line is inertive and capacitive (adiabatic). This latter condition is sometimes referred to as "lossless." Between these frequency values ($0.1\omega_v < \omega < 10\omega_v$) the appropriate line model is resistive, inertive, and capacitive. This regime is more difficult to work with and to analyze than the other low Mach number regimes ($\omega < 0.1\omega_v$ and $\omega > 10\omega_v$). When the frequency becomes of the order of a/d (where a is the acoustic velocity and d is the duct diameter), the length of a wave becomes comparable to or less than the duct diameter and the duct must be treated as a waveguide.

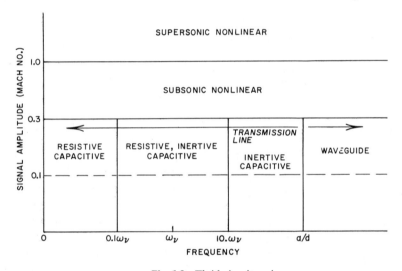

Fig. 1.9 Fluid circuit regimes.

The chart shown in Fig. 1.9 does not indicate the frequency conditions under which the line may be treated as lumped or as distributed. This does not depend on the viscosity and diameter, as the characteristic frequency does, but rather on the line length. Lines may be considered as lumped when their lengths are short compared to a wavelength in the frequency regime of operation.

Whenever it is possible, the system components should be designed to operate in the lumped and in the resistive regimes. If data transmission rather than control is desired the lossless regime should be used. The intermediate region should be avoided to the extent feasible.

Except for our discussion of the method of characteristics and an occasional other comment, this text will be confined to the regimes for which the Mach number is less than 0.3. This corresponds to air pressure changes or differences of roughly 7 kN/m² or roughly 1 psi.

1.3 Circuit Considerations

We will completely ignore the waveguide region and will thus in general restrict ourselves to wavelengths appreciably larger than the diameter. Indeed the major portion of our text will concern itself with wavelengths appreciably larger than the lengths of the various elements of the circuit so that lumped circuit analysis holds.

Chapter 2

PASSIVE COMPONENTS

2.1 RESISTANCE

Fluidic resistance is the passive dissipative component of fluidic circuits. Resistance, in general, is defined as the ratio of across variable to through variable [1]. If we choose the fluidic across variable as the total pressure drop Δp, and the fluidic through variable as the volume flow q, the fluidic resistance R is

$$R \equiv \Delta p/q \qquad (2.1)$$

Equation (2.1) defines only the dc resistance. When there is a linear relation between total pressure drop and volume flow the dc resistance is constant. However, the relation is often nonlinear for fluid components (Fig. 2.1). In this case the dc resistance varies with operating point. We then use a small signal linear approximation about the operating point to define an ac or variational resistance, R_{ac}, for one-dimensional fluid flow as

$$R_{ac} \equiv d(\Delta p)/dq \qquad (2.2)$$

The total pressure–volume flow relation in fluidic resistance is usually of the form $\Delta p = aq + bq^2$, where a and b are dimensional constants. Thus we may express the ratio of variational resistance to dc resistance as

$$R_{ac}/R = (a + 2bq)/(a + bq) \qquad (2.3)$$

Figure 2.2 shows the resistance ratio R_{ac}/R plotted against the parameter bq/a. For small values bq/a the resistance ratio equals unity.

The equivalence of variational and dc resistance is commonly referred to as

2.1 Resistance

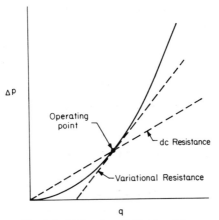

Fig. 2.1 DC and variational resistance.

linear resistance. For large values of bq/a the variational resistance approaches twice the value of the dc resistance, and we often call this a *square law resistance* $(p \sim q^2)$.

Since most fluidic resistive configurations produce square law resistances it is more convenient to use a coefficient of resistance or loss coefficient K_L to provide a measure of the dissipative process. The loss coefficient is defined as

$$K_L = \Delta p/(\rho \bar{U}^2/2) \tag{2.4}$$

where ρ is the fluid density and $\bar{U}$ is some characteristic velocity (usually the average velocity at the smallest cross section). In addition, as a practical matter, we now introduce the apparent loss coefficient K_A, which is

$$K_A = \Delta p_a/(\rho \bar{U}^2/2) \tag{2.5}$$

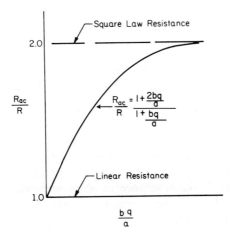

Fig. 2.2 Ratio of variational to dc resistance.

where Δp_σ is the static pressure drop. In the following section we will discuss the relation between the total loss coefficient and the apparent loss coefficient.

From Eqs. (2.2) and (2.4) the relation between variational resistance and loss coefficient is

$$R_{ac} = \rho K_L q / A^2 \tag{2.6}$$

where $q = \bar{U}A$, and A is the area at the minimum cross section.

2.2 GENERAL FLUIDIC RESISTANCE

Passages or ducts provide resistance in a fluidic circuit. The resistance depends on the shape of the passage and is the result of a combination of the following factors: (a) wall shear, (b) separation, (c) directional changes.

In each particular shape a different factor may predominate. For a long constant-area tube, wall shear produces most of the resistance. In a diffuser or a passage with a sudden area change, separation effects predominate. On the other hand, directional changes are significant in a curved pipe or bend.

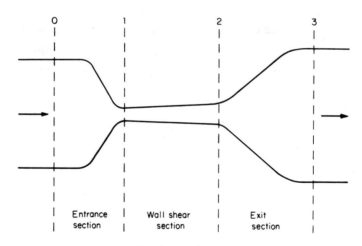

Fig. 2.3 General fluidic resistance.

Figure 2.3 shows the configuration of a general fluidic resistance between terminals 0 and 3. There are three distinct sections in this resistance: an entrance section (01), a wall shear section (12), and an exit section (23). Thus we may consider the general fluidic resistance as three separate resistances in series. Since, in this special case, the characteristic velocity is the same for each resistance, we may write that

$$(K_L)_{03} = (K_L)_{01} + (K_L)_{12} + (K_L)_{23} \tag{2.7}$$

2.2 General Fluidic Resistance

The loss coefficients of entrance and exit sections [$(K_L)_{01}$ and $(K_L)_{23}$] are termed *minor losses* because they are usually much smaller than the *major loss* due to the wall shear $(K_L)_{12}$. Fluidic circuits, however, often use short constant-area sections, and it is then possible for the minor losses to exceed the major loss.

Since the total pressure cannot rise in the flow direction, the resistance of any passive configuration must be positive. Static pressure, on the other hand, can rise, as in a subsonic diffuser, for example. Thus static pressure measurements alone may give a distorted representation of the circuit. Nevertheless, because of the ease of measurement, test data are usually presented in terms of static pressure. In Eqs. (2.4) and (2.5) we defined the total loss coefficient and the apparent loss coefficient. To relate these coefficients, consider the energy equation for incompressible flow between two sections, i.e., 1 and 2 of a variable area passage [2, p. 51] (see Fig. 2.3),

$$\int_{A_1}\left(\frac{\rho u_1^2}{2}+p_1\right)u_1\,dA_1 = \int_{A_2}\left(\frac{\rho u_2^2}{2}+p_2\right)u_2\,dA_2 + (K_L)_{12}\frac{\rho \bar{U}^2}{2}\bar{U}A \quad (2.8)$$

where p_1 and p_2 are the static pressures and u_1 and u_2 are the velocities at sections 1 and 2, in the passage. Now, we define an energy distribution factor β_e such that

$$\beta_e = (\textstyle\int_A u^3\,dA)/\bar{U}^3 A \quad (2.9)$$

where $\bar{U} = (1/A)\int u\,dA$. The factor β_e is equal to unity when the velocity distribution in the passage is uniform.

For all other cases β_e is greater than unity. For example, in the case of fully developed laminar flow in a round passage

$$u(r) = 2\bar{U}(1 - 4r^2/d^2) \quad (2.10a)$$

where $u(r)$ is the axial velocity at the radius r, $\bar{U}$ is the average velocity, and d is the diameter. The factor β_e is easily found to be 2.00.

For fully developed laminar flow between parallel plates

$$u(y) = 1.5\bar{U}(1 - 4y^2/h^2) \quad (2.10b)$$

where $u(y)$ is the axial velocity at a distance y from the midplane and h is the distance between the plates. In this case $\beta_e = 1.54$.

When the flow is turbulent the velocity profile is rather uniform and β_e generally lies between 1 and 1.10.

If the static pressure is uniform over each cross section

$$\int\left(\frac{\rho u^2}{2}+p_\sigma\right)u\,dA = \beta_e\frac{\rho \bar{U}^3 A}{2} + p_\sigma \bar{U}A$$

and therefore Eq. (2.8) reduces to

$$(K_L)_{12} = (K_A)_{12} + \beta_{e1}(A/A_1)^2 - \beta_{e2}(A/A_2)^2 \quad (2.11)$$

where A is the area of the smallest cross section between sections 1 and 2. There are two conditions under which the total loss coefficient and the apparent loss coefficient are equal. In one condition the cross sections at the measuring terminals A_1 and A_2 are much larger than the smallest cross section A. The other condition occurs in a length of constant-area passage where the energy distribution factors at the beginning and end of the test section are equal. This is the familiar condition of fully developed flow.

The general fluidic resistance (Fig. 2.3) may now also be expressed in terms of apparent loss coefficients. We obtain from application of Eqs. (2.7) and (2.11)

$$(K_L)_{03} = (K_A)_{01} + (K_A)_{12} + (K_A)_{23} + \beta_{e0}(A_1/A_0)^2 - \beta_{e3}(A_2/A_3)^2 \quad (2.12)$$

In Sections 2.3–2.5 which follow we present the resistance coefficients for the entrance, wall shear, and exit portions of the general fluid resistance. Then in Section 2.6 we demonstrate a specific example of the general fluidic resistance.

2.3 ENTRANCE RESISTANCE

In the entrance section of the general fluid resistance there is an area decrease and a velocity increase. The resulting conversion of potential energy to kinetic energy is very efficient and the total loss coefficient is quite small. However, the apparent loss coefficient can be appreciable.

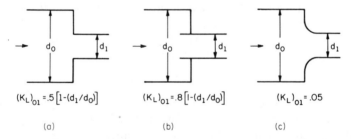

Fig. 2.4 Entrance sections: (a) square edge, (b) re-entrant, (c) bell mouth.

Figure 2.4 shows three sudden contractions: The total loss coefficients for these configurations with circular cross sections are given in many fluid mechanics textbooks and engineering handbooks [3–5] and are repeated on Fig. 2.4. In the square edge and re-entrant sections the fluid stream contracts (vena contracta) to a diameter less than d_1 as it enters the smaller area. An energy loss then results from the subsequent expansion back to the diameter d_1. For the bell mouth section any radius greater than $0.14d_1$ prevents formation of a vena contracta. The total loss coefficient, for this case, is small and is due to wall shear.

There are less data available for the loss coefficients of rectangular sections. When the section is square the losses are approximately equal to those of the

2.4 Wall Shear Resistance

round section. For planar fluidic type sections (constant depth) there would be additional wall shear effects and thus the sudden contraction would produce larger loss coefficients. In most practical cases, however, the losses in the entrance section (01) will be small compared to the losses in the wall shear and exit sections.

The velocity distributions throughout the sudden contraction usually remain close to uniform. The energy distributions factors β_{e0} and β_{e1} are, therefore, approximately equal to 1, and the relation between apparent and total loss of the sudden contraction is, from Eq. (2.11),

$$(K_A)_{01} = (K_L)_{01} + 1 - (A_1/A_0)^2 \qquad (2.13)$$

2.4 WALL SHEAR RESISTANCE

The wall shear portion of the general fluidic resistance (Fig. 2.3) always has an entrance length in which the velocity distribution is changing. Fully developed flow is seldom reached in fluidic passages.

In the developing or entrance region of a constant-area passage the boundary layer grows in much the same way that it does on a flat plate. The boundary layer growth is somewhat slower than that on a flat plate because the fluid outside the boundary layer must accelerate to maintain flow continuity. As a consequence the flow in fluidic passages is almost always laminar. To demonstrate this we make the conservative assumption that the critical flat-plate Reynolds number, $N_{Rl} = 5(10)^5$, also applies to distances l along the passage. Then we may write

$$N_{Rl} = (l/d_e)N_{Rd_e} = 5 \times 10^5$$

where d_e is the equivalent diameter (4 times the cross-sectional area divided by the perimeter) and N_{Rd_e} is the Reynolds number based on the equivalent diameter ($\bar{U}d_e/\nu$). Figure 2.5 shows the relation between critical Reynolds number N_{Rd_e}

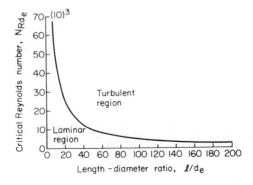

Fig. 2.5 Critical Reynolds number in the inlet region of a constant-area passage.

and the l/d_e ratio of the passage. The region beneath the curve represents laminar flow conditions throughout the entire tube length. Thus, for example, if $l/d_e = 20$, the flow is laminar throughout if $N_{Rd_e} \leq 25{,}000$. The region above the curve means that there is a transition from laminar to turbulent flow in the passage. If $l/d_e = 60$ and $N_{Rd_e} = 20{,}000$ the flow is laminar for the first 25 equivalent diameters of length and is thereafter turbulent.

There are two analytical approaches to the solutions of the boundary layer equations in the entrance length [2, p. 102]. One approach assumes an appropriate polynomial form for the changing velocity profile [6] and the other linearizes the equations [7, 8]. The profile method is more accurate near the tube entrance where the boundary layer is small. The linearizing method is preferable farther downstream where the profile is almost fully developed. Since neither method yields a solution that is accurate throughout the entire entrance region we often rely on curves developed from experimental data.

In addition to these two analytical approaches, there is a semiempirical method [9] that assumes a solution as a linear combination of the frictionless flow solution and the fully developed viscous solution. We shall demonstrate this method in Section 2.4.3 when we consider the loss coefficient of the nozzle–baffle resistance.

2.4.1 Resistance of Straight Passages

The resistance of straight constant-area passages with fully developed flow is well known. In this case the apparent loss coefficient and the total loss coefficient are equal. They are usually expressed in the form

$$(K_A)_s = (K_L)_s = \frac{c_s(l/d_e)}{N_{Rd_e}} \tag{2.14}$$

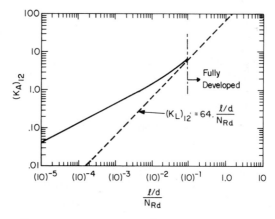

Fig. 2.6 Apparent loss coefficient for circular tube.

2.4 Wall Shear Resistance

Here c_s is a constant that depends on the shape of the cross section. For round tubes $c_s = 64$, and for parallel plates $c_s = 96$ [8]. In rectangular passages c_s is a function of the aspect ratio (short length to long length) and has values in the range from 64 to 96. Square sections have approximately the same values of c_s as round sections. A rectangular cross section of low aspect ratio has about the same value of c_s as the infinite parallel plates. Between these extremes, c_s has been measured as 62 and 76 for aspect ratios of 0.5 and 0.2 [10].

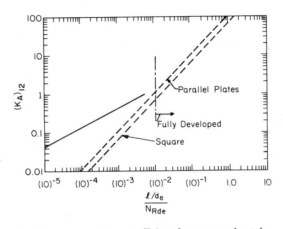

Fig. 2.7 Apparent loss coefficient for rectangular tube.

The form of the loss coefficient given in Eq. (2.14) produces a linear resistance because N_{Rd_e} is proportional to the velocity. When the entrance region is taken into effect and the flow is not fully developed throughout the length, the resistance is nonlinear. The solid line in Figs. 2.6 and 2.7 shows the apparent loss coefficient in round tubes [6] and in rectangular passsages [11]. These apparent coefficients approach the fully developed coefficients (dashed lines) when the length becomes appreciable or the Reynolds number becomes small. In the developing region of tubes the apparent coefficient may be expressed empirically [6] as

$$(K_A)_{12} = 13.74 \left(\frac{l/d_e}{N_{Rd_e}} \right)^{1/2} \qquad (2.15)$$

where $(l/d_e)/N_{Rd_e} < (10)^{-3}$.

Equation (2.15) applies equally well to rectangular passages [11].

Although the flow in the developing region is laminar the resistance of components that operate in this region are not necessarily linear. To insure linearity the apparent loss coefficient must follow the relation shown by the dashed line in Fig. 2.6. The junction at which the measured data (solid line) follow the dashed line is not well defined since the approach is asymptotic. However, when $(l/d_e)/N_{Rd_e} \geq 0.2$ the two lines become very close to each other. Thus good linearity is obtained when the Reynolds number is less than $5l/d_e$. In some cases where less precise linearity is satisfactory, the Reynolds number limitation may be raised, for example, to $10l/d_e$.

2.4.2 Resistance of Curved Passages

In curved passages a helical secondary flow is superimposed on the primary axial flow. As a result the maximum axial velocity is displaced from the center of the cross section toward the outer wall. The increased motion gives a curved passage a larger resistance than the same length of straight passage. Since there is no equivalent in curved passages to the Hagen–Poiseuille law for straight passages we rely entirely on experiments to provide information on curved passage resistance.

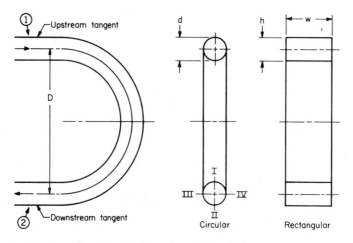

Fig. 2.8 Configuration of curved channels.

Figure 2.8 shows a curved channel of either circular or rectangular cross section. The straight section upstream of the curved channel is called the upstream tangent. Similarly, the straight section following the curved channel is called the downstream tangent. Although not part of the curved passsage, these straight sections influence the loss coefficient that is attributable to curvature because they affect the energy distribution factors at the entrance and exit of the curved channel (sections 1 and 2).

The measurement of apparent loss coefficient in curved flow is difficult for several reasons. First, the static pressure distribution is not uniform across each cross section. To obtain a measure of the static pressure at section 2 (Fig. 2.8), for example, requires the average of static pressure readings at I and II, or III and IV, or at all four locations. In addition β_{e1} and β_{e2} are not easily specified as in the straight passages. The energy distribution factor (β_{e1}) depends on the

2.4 Wall Shear Resistance

length of the upstream tangent. The exit factor (β_{e2}) is a function of the entrance factor, the geometry of the curved passage, the Reynolds number, and to some extent of the downstream tangent. Thus, an inordinately large number of tests would be required to completely specify the loss coefficients for curved channels. As a practical matter the experiments usually refer only to fully developed curved flow. To accomplish this the terminals are selected at a considerable distance downstream of the entrance (section 1) and upstream of the exit (section 2). Test results provide the following empirical relation between the loss coefficients for fully developed curved flow and fully developed straight flow [12]:

$$\frac{K_{Lc}}{K_{Ls}} = \frac{1}{1 - [1 - (11.6/N_D)^{0.45}]^{2.2}} \tag{2.16}$$

where N_D is the Dean number, which equals $N_{Rd_e}(d_e/D)^{1/2}$, and D is the diameter of curvature. Equation (2.16) is shown plotted in Fig. 2.9. When the Dean number

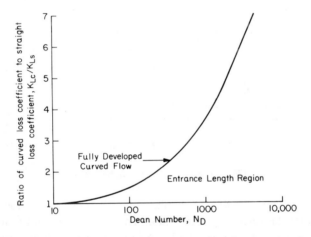

Fig. 2.9 Relation between fully developed curved flow and fully developed straight flow.

is small (small curvature) the loss coefficient for curved flow is essentially equal to the straight flow loss coefficient. As the Dean number increases the ratio K_{Lc}/K_{Ls} increases monotonically. At large Dean numbers the curved loss coefficient is five to seven times the straight loss coefficient.

Apparent loss coefficients have also been measured in the entrance region of curved channels with various lengths of upstream tangent [13]. However, there are not nearly enough data to cover all the possibilities. Unlike straight channels, the apparent loss in the entrance of curved channels is less than the fully developed curved loss and therefore lies in the region below the curve shown in Fig. 2.9. For example, a curved channel with $l/d_e = 100$ has about a 10% smaller ratio than that given in Eq. (2.16) for the same Reynolds number.

2.4.3 Resistance of Nozzle–Baffle

Figure 2.10 shows an axisymmetric nozzle–baffle type of fluid resistance. In this configuration, the flow depends on the distance d_b between nozzle face and baffle. The direction of flow may be radially in towards the nozzle or radially out away from the nozzle. However, outflow is the more common. When the nozzle–baffle operates with outflow, the cross-sectional area through which the flow passes increases in the flow direction. In this respect the nozzle–baffle is similar to a diffuser. However, there are also viscous effects due to the wall shear on the baffle and nozzle face.

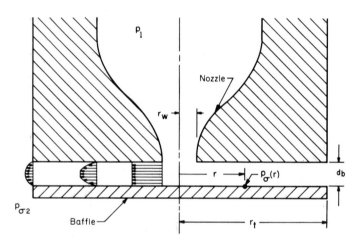

Fig. 2.10 Nozzle–baffle configuration.

Let us first consider the flow in the flat region between nozzle face and baffle. We assume that the velocity profile is uniform when $r = r_w$ and that it develops as r increases. As a consequence of the increasing area the average velocity decreases and the boundary layer grows at a greater rate than it would in a straight tube. The profile, therefore, becomes fully developed in a shorter distance than in a tube. The static pressure $p_\sigma(r)$ in the region between nozzle face and baffle is assumed to be a linear combination of the effects of viscous wall shear and inviscid ideal diffusion [9, 14], and is expressed as

$$\frac{p_\sigma(r) - p_{\sigma 2}}{\rho \overline{U}^2/2} = 24 \frac{r_t/d_b}{N_{Rd_b}} \left(\frac{r_w}{r_t}\right) \ln\left(\frac{r_t}{r}\right) + \beta_{et}\left(\frac{r_w}{r_t}\right)^2 - \beta_e\left(\frac{r_w}{r}\right)^2 \quad (2.17)$$

where the subscripts w and t refer to the radial positions at the nozzle and at the outer edge of the nozzle–baffle, $\overline{U}$ is the average velocity at $r = r_w$ (minimum area), $p_{\sigma 2}$ is the static pressure at $r = r_t$ and $N_{Rd_b} = \overline{U}d_b/\nu$. The formulation for static pressure drop given in Eq. (2.17) can also be obtained by series approximation to the Navier–Stokes equation [15]. The first term on the right-hand side

2.4 Wall Shear Resistance

of Eq. (2.17) represents wall shear effects and the last two terms are the ideal diffuser effects. Since the velocity profile develops so rapidly in the flat region, the energy distribution factor for fully developed flow ($\beta_e = 1.543$) is a good approximation throughout the entire region ($r_w \leq r \leq r_t$). In the case of inflow, however, the profile develops much more slowly and the variation in energy distribution factor must be considered [15].

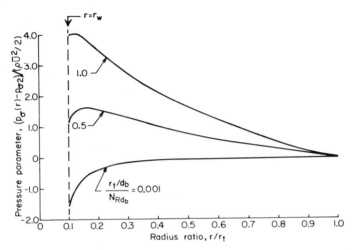

Fig. 2.11 Static pressure distribution between nozzle and baffle ($r_t/r_w = 10$).

Equation (2.17) is plotted in Fig. 2.11 for an outer radius ratio (r_t/r_w) equal to 10. When the Reynolds number parameter $r_t/(d_b N_{Rd_b})$ is small the static pressure increases monotonically with radial distance. In this region diffuser effects are large and wall shear effects are practically negligible. As the Reynolds number parameter increases, wall shear increases and diffuser effects decrease. Thus we obtain a pressure distribution that has a positive pressure gradient near the nozzle and a negative pressure gradient farther downstream. For example, at $r_t/(d_b N_{Rd_b}) = 0.5$ the pressure gradient is positive for $r/r_t < 0.16$ and negative for $r/r_t > 0.16$. If the parameter exceeds 1.285 (for $r_t/r_w = 10$) the pressure gradient is negative throughout the entire region between the nozzle face and baffle. This is an indication that wall shear effects predominate and the diffuser effects are very small.

It is generally preferable to operate the nozzle–baffle with a negative pressure gradient throughout the entire region. In this way the erratic effects associated with separation can be avoided. To determine the Reynolds number parameter for which the pressure gradient is always negative, the derivative of Eq. (2.17) is set equal to zero and evaluated at $r = r_w$ [14]. The result is

$$\frac{r_t/d_b}{N_{Rd_b}} \geq \frac{\beta_e}{12} \frac{r_t}{r_w} \tag{2.18}$$

Thus with $\beta_e = 1.543$ the parameter must exceed 1.285 when $r_t/r_w = 10$. Figure 2.11 shows that for the parameter equal to unity there is still a small positive slope (pressure gradient) near the nozzle (r_w).

The apparent loss coefficient of the nozzle–baffle can be obtained from Eq. (2.17) and the energy equation. When we assume that the energy loss from section 1 to the entrance of the flat region is small, the result is

$$(K_A)_{12} = 24 \frac{(r_t/d_b)}{N_{Rd_b}} \left(\frac{r_w}{r_t}\right) \ln \frac{r_t}{r_w} + \beta_{et}\left(\frac{r_w}{r_t}\right)^2 \qquad (2.19)$$

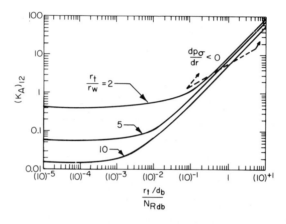

Fig. 2.12 Apparent loss coefficient for nozzle–baffle.

Figure 2.12 shows the apparent loss coefficient plotted against the Reynolds number parameter for various values of r_t/r_w. In the region above the dashed line that intersects the loss curves, the pressure gradient is always negative.

2.5 EXIT RESISTANCE

In the entrance section of the general fluid resistance (01 of Fig. 2.3) the fluid velocity increases. We have found that the conversion of potential energy to kinetic energy produces only a small loss. In the exit section (23) of the general fluid resistance the situation is quite different. This process develops larger losses because the fluid separates from the passage walls and forms turbulent eddies.

Figure 2.13 shows two very common axisymmetric exit sections. In the diffuser the loss coefficient depends on the included angle θ and the area ratio A_2/A_3. When the included angle is small there is a long wall between sections 2 and 3 and wall shear losses contribute significantly to the total loss. As the angle increases wall shear effects decrease and separation losses increase. There is an optimum angle for minimum energy loss. The nozzle–baffle, previously discussed,

2.5 Exit Resistance

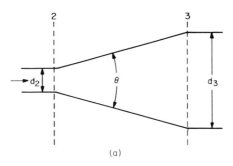

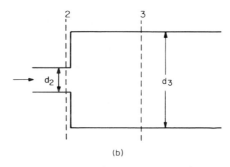

Fig. 2.13 Exit sections: (a) diffuser, (b) sudden enlargement.

is a type of diffuser, although it is generally used to provide resistance while the diffuser is used to reduce resistance. The sudden enlargement is a type of diffuser with included angle equal to 180 degrees. The energy loss in the sudden enlargement is due entirely to separation.

The uniformity of the inlet profile at section 2 affects the total loss coefficients as well as the apparent loss coefficient. A large nonuniformity at the inlet increases the total loss coefficient by hastening the onset of separation. A nonuniform exit profile at section 3 does not affect the total loss but has a considerable affect on the apparent loss. When the exit energy distribution factor β_{e3} is larger than 1 the static pressure continues to rise in the constant-area section downstream of the exit and maximizes about 6 diameters downstream from the diffuser exit [16, 17].

Figure 2.14 shows the total loss coefficients $(K_L)_{23}$ of a subsonic diffuser plotted against the included angle for various values of the area ratio [17, 18]. The loss is minimum when the included angle is between 5 and 10 degrees. The loss reaches a maximum at about 60 degrees for area ratios of 2.34 and 4.00. The total loss of the sudden enlargement ($\theta = 180$ degrees) is very close to the Borda–Carnot relation

$$(K_L)_{23} = [(A_2/A_3) - 1]^2 \tag{2.20}$$

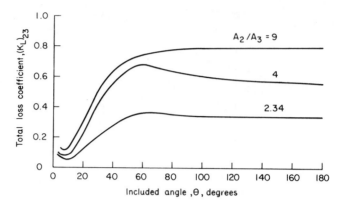

Fig. 2.14 Total loss coefficients for diffusers.

If we assume that the profiles at the inlet and outlet of the sudden enlargement are uniform, the apparent loss coefficient can be obtained from Eqs. (2.11) and (2.20) and is

$$(K_A)_{23} = 2(A_2/A_3)[(A_2/A_3) - 1] \tag{2.21}$$

Figure 2.15 shows the loss coefficients from the sudden enlargement. The total loss coefficient decreases as the area ratio increases and is always positive. This merely confirms the fact that the total pressure must always decrease in the flow direction. It also indicates that the energy loss decreases as the area ratio increases toward 1. The apparent loss coefficient, on the other hand, is always

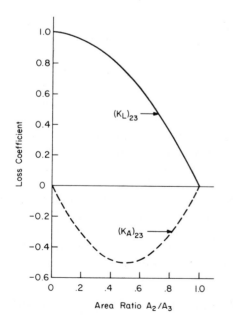

Fig. 2.15 Loss coefficients for sudden enlargements.

2.5 Exit Resistance

negative. Moreover it reaches a minimum value when the area ratio (A_2/A_3) is 0.5. How are we to interpret this minimum in apparent loss coefficient when the minimum in the total loss coefficient occurs at $A_2/A_3 = 1$? On the surface it appears that we can design a sudden enlargement of minimum resistance by making the area ratio equal to 0.5.

To understand the reason for the discrepancy consider the configuration shown in Fig. 2.16. We have a large pressurized reservoir at pressure p_s that

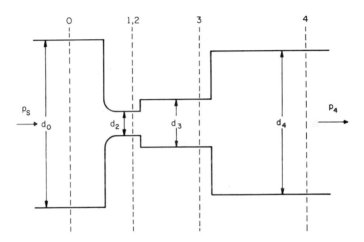

Fig. 2.16 Sudden enlargement to maximize flow.

supplies flow to a nozzle at section 2 (the designation of this as section 1,2 on Fig. 2.16 infers that the nozzle is a passage of negligible length and permits us to retain the numbering previously used for the exit section) and the flow ultimately discharges to atmospheric pressure. We want to design a sudden enlargement from section 2 to 3 such that the flow through the nozzle is a maximum. However, if the static pressure at section 3 is above atmospheric pressure we have another sudden enlargement from section 3 to the atmosphere at section 4. Thus the total loss coefficient from the reservoir to atmosphere depends on $(K_L)_{01}$, $(K_L)_{12}$, $(K_L)_{23}$, and $(K_L)_{34}$. If we assume that $(K_L)_{01}$ and $(K_L)_{12}$ are negligible, the total pressure drop from p_s to p_4 is

$$p_s - p_4 = (K_L)_{23}(\rho \overline{U}_2^2/2) + (K_L)_{34}(\rho \overline{U}_3^2/2) \tag{2.22}$$

Now, from continuity, $A_2 \overline{U}_2 = A_3 \overline{U}_3$ and the total loss coefficient for the sudden enlargement given in Eq. (2.20) and Eq. (2.22) becomes

$$(K_L)_{04} = [(A_2/A_3) - 1]^2 + [(A_3/A_4) - 1]^2 (A_2/A_3)^2 \tag{2.23}$$

The enlargement into atmosphere A_3/A_4 is very small so that Eq. (2.23) reduces to

$$(K_L)_{04} = [(A_2/A_3) - 1]^2 + (A_2/A_3)^2 \tag{2.24}$$

The total loss coefficient expressed in Eq. (2.24) is also minimum when $A_2/A_3 = 0.5$. We see, therefore, that the minimum resistance can be determined from either loss coefficient if the whole circuit is considered.

2.6 A TYPICAL FLUIDIC RESISTANCE

In the previous sections we considered separately the entrance, wall shear, and exit resistances of the general fluidic resistance. Let us now apply this information to a specific example.

Figure 2.17 shows a common type of fluidic resistance that consists of a well-shaped entrance, a constant-area section of length l, and a sudden enlargement

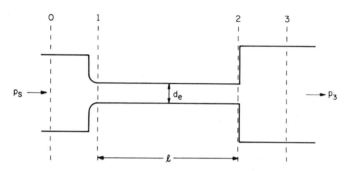

Fig. 2.17 Typical fluidic resistance with bell mouth entrance, constant-area straight section, and sudden enlargement exit.

at the exit. The entrance and exit area (A_0, A_3) are large compared to the constant-area section ($A_1 = A_2$). The total pressure drop across the composite resistance, as given by Eq. (2.12), is

$$(p_s - p_3)/(\rho \bar{U}^2/2) = (K_A)_{01} + (K_A)_{12} + (K_A)_{23} \qquad (2.25)$$

since the last two terms of Eq. (2.12) are negligible.

The apparent loss coefficients given in Eq. (2.25) are:

(a) From Eq. (2.13), for a well-shaped entrance the total loss is small and $(K_A)_{01} = 1$.

(b) The straight constant-area section has an apparent loss coefficient that is a function of the Reynolds number parameter (Fig. 2.6); thus, $(K_A)_{12} = g((l/d_e)/N_{Rd_e})$.

(c) From Eq. (2.21), a sudden enlargement with $A_2/A_3 \ll 1$ has $(K_A)_{23} = 0$.

When these values of the apparent loss coefficient are substituted into Eq. (2.25), the result is

$$(p_s - p_3)/(\rho \bar{U}^2/2) = 1 + g((l/d_e)/N_{Rd_e}) \qquad (2.26)$$

2.7 Fluid Diodes

We arrive at a more convenient form if we express Eq. (2.26) in terms of the volume flow, $q = (\pi d_e^2/4)\bar{U}$, so that

$$(p_s - p_3) d_e^2 = \left\{1 + g\left[\frac{(l/d_e)\pi v/4}{q/d_e}\right]\right\}\left(\frac{8\rho}{\pi^2}\right)\left(\frac{q}{d_e}\right)^2 \qquad (2.27)$$

Equation (2.27) is plotted in Fig. 2.18 when the fluid is air at standard conditions. The pressure drop parameter $(p_s - p_3)d_e^2$ is plotted against the flow

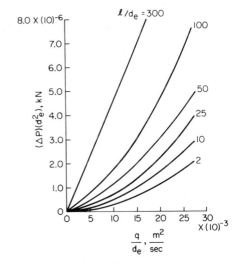

Fig. 2.18 Pressure–flow relation for typical fluidic resistance shown in Fig. 2.17.

parameter q/d_e for various values of the length-to-diameter ratio. As l/d_e increases the relation between pressure drop and volume flow becomes more linear. When the Reynolds number parameter exceeds 0.2, $g = 64(l/d_e)/(N_R)$, and Eq. (2.27) becomes

$$p_s - p_3 = (8\rho q^2/\pi^2 d_e^4) + (128\mu l q/\pi d_e^4) \qquad (2.28)$$

The second term on the right-hand side of Eq. (2.27) is the familiar Hagen–Poiseuille expression for fully developed flow and the first term represents the energy loss of the sudden enlargement.

2.7 FLUID DIODES

A diode is a two-terminal component that has a large resistance in one flow direction and a small resistance in the other. At sufficient signal levels, electronic diodes may have a resistance ratio or diodicity of 10^6 or more. In fluid terminology, performance ratio D_i is the measure of diodicity and is defined as

$$D_i = K_{LR}/K_{LF} \qquad (2.29)$$

where K_{LF} is the loss coefficient in the forward direction (easy) and K_{LR} is the loss coefficient in the reverse direction (hard). Moving-part fluid components, such as check valves, have performance ratios comparable to electronic diodes but no-moving-part fluid diodes have relatively small ratios. At the present time, the largest reported value is about 180. One of the difficulties in using a diode with a performance ratio of this magnitude is that the resistance of the rest of the circuit degrades the overall diodicity.

Two totally different, no-moving-part components, are called diodes. One is a two-terminal component [19–22] in the same sense as described above. The vortex diode as shown in Fig. 2.19a is an example of this type. The other fluid component [23–28] has more than two terminals and usually is similar in principle

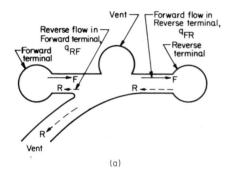

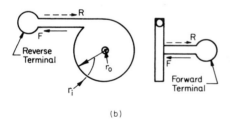

Fig. 2.19 (a) Vented and (b) nonvented fluidic diodes.

to the component shown schematically in Fig. 2.19b. In this latter type fluid flow in the forward direction passes from the forward terminal to the reverse terminal, but flow from the reverse terminal vents to a terminal at a reference pressure. The vented component is sometimes called a *flow diode* to distinguish it from the nonvented *pressure diode*. However, the vented component represents a fluid network with diodicity rather than a diode. TEE and WYE connections are examples of other fluid networks that exhibit diodicity. We confine our discussion at this time to the nonvented two-terminal diode. In Section 2.8.3 we consider the vented diodes in more detail.

Diodicity occurs in fluid components when the flow pattern changes with the flow direction. Thus, in effect, each diode configuration represents two distinct

2.7 Fluid Diodes

resistances, one of which is larger than the other. The forward flow resistance is usually the result of wall shear or separation and should be kept as small as possible. Diodes are most often classified, however, by the principle through which the reverse flow resistance is obtained. There are two distinct mechanisms that are presently being used to produce a large reverse resistance: (a) rotational flow and (b) momentum interaction. The following sections show that the rotational flow diodes are significantly superior to the momentum interaction diodes.

2.7.1 Rotational Flow Diodes

One type of rotational flow diode is the Thoma counterflow brake [20, 29], which is better known today as the vortex diode (Fig. 2.19a). In the reverse direction fluid enters a chamber tangentially and forms a vortex. As the fluid rotates toward a central exit port the radial and tangential velocities increase. As a result there is a centrifugal force which opposes the total pressure forces and causes an increase in resistance. In the forward direction the fluid passes into the chamber through a sudden enlargement and exits through a sudden contraction.

To predict the performance ratio of the vortex diode we calculate the reverse flow coefficient and the forward flow coefficient separately. The total reverse flow coefficient of the vortex diode consists of an entrance loss, a vortex chamber loss, and an exit loss. In this respect the diode takes on the form of the general fluidic resistance with sections 01, 12, and 23. The cross sections before and after the diode (0 and 3) have relatively large areas and negligible velocity. Thus from Eq. (2.12) the total reverse flow coefficient is

$$K_{LR} = (K_A)_{01} + (K_A)_{12} + (K_A)_{23} \tag{2.30}$$

From Eqs. (2.13) and (2.21), $(K_A)_{01} \cong 1$ and $(K_A)_{23} \cong 0$. As a result Eq. (2.30) becomes

$$K_{LR} = 1 + (K_A)_{12} \tag{2.31}$$

The apparent loss coefficient $(K_A)_{12}$ may be derived from the energy, continuity, and momentum equations as follows. From the energy equation for inviscid flow,

$$(\rho/2)(u_{\theta i}^2 + u_{ri}^2) + p_1 = (\rho/2)(u_{\theta 0}^2 + u_{r0}^2) + p_0 \tag{2.32}$$

where p_1 is the static pressure inside the chamber at $r = r_i$, p_0 the static pressure at $r = r_0$, u_{ri} the radial velocity at $r = r_i$, $u_{\theta i}$ the tangential velocity at $r = r_i$, u_{r0} the radial velocity at $r = r_0$, $u_{\theta 0}$ the tangential velocity at $r = r_0$, r_i the radius of the chamber, and r_0 the radius of the outlet.

But from continuity

$$u_{r0} = (r_i/r_0)u_{ri} \tag{2.33}$$

and from the conservation of angular momentum

$$u_{\theta 0} = (r_i/r_0)u_{\theta i} \tag{2.34}$$

The substitution of Eqs. (2.33) and (2.34) into Eq. (2.32) yields

$$p_1 - p_0 = (\rho/2)[u_{ri}^2 + u_{\theta i}^2][(r_i^2/r_0^2) - 1] \tag{2.35}$$

Since u_{ri} is much smaller than $u_{\theta i}$ we may select the reference velocity $U = u_{\theta i}$ and thus Eq. (2.35) becomes

$$(K_A)_{12} = \frac{p_1 - p_0}{\rho U^2/2} = \left(\frac{r_i}{r_0}\right)^2 - 1 \tag{2.36}$$

From Eqs. (2.31) and (2.36) the total reverse flow coefficient for the vortex diode becomes

$$K_{LR} = (r_i/r_0)^2 \tag{2.37}$$

Equation (2.37) indicates that we can achieve large reverse flow coefficients by simply making the radius ratio large. However, because of viscosity the loss coefficients will be much less than predicted by Eq. (2.37).

In the forward flow direction the loss coefficient is essentially the result of two sudden enlargements. Thus we may approximate the forward loss coefficient as $K_{LF} = 2$.

The performance ratio of the vortex diode can now be estimated. For example, if we assume incompressible flow and a vortex chamber with a radius ratio of 5, the reverse loss coefficient is 25 and the forward loss coefficient is 2. Thus the performance ratio is 12.5. This estimate agrees remarkably well with test data [30]. Vortex diodes with larger radius ratios and therefore higher performance ratios ($D_i = 43$) have been reported [29]. These devices produced less diodicity than the present theory predicts. However the circuits contained other diode components in addition to the vortex chamber so that a direct comparison is not possible. We will discuss the vortex diode as a special case of the vortex triode in Chapter 8.

The cascade diode is another fluid diode based on rotational flow in one direction and separated type flow in the other [30, 31]. Figure 2.20 shows a schematic drawing of a cascade diode that has four cascades, each with six blades, mounted inside a tube. Forward flow (left to right on Fig. 2.20) passes in the axial direction between the cascades. Reverse flow (right to left), on the other hand, is deflected by the blades so that the flow rotates from one cascade to another. For reverse flow the first blade in each cascade makes a 45-degree angle with the axis. The following blades have progressively larger angles until they reach 90 degrees. These latter blades are the first blades contacted in the forward flow. Since they are set at 90 degrees they impart no spin to the fluid. The diode shown in Fig. 2.20 is a straight cascade because the blades are mounted directly behind each other. However, the rotation of the fluid is enhanced when

2.7 Fluid Diodes

each cascade follows a helical pattern. Although forward resistance is also increased by this means, there is a larger increase in reverse resistance and diodicity increases.

The performance of the cascade diode depends mainly on the number of cascades, the number of blades per cascade, and the helix angle. Performance ratios up to 181 have been obtained for a four-cascade, 53 blades per cascade, diode with a helix angle of 90 degrees [31]. The calculations of Jacobs and Baker [31] indicate that improvement of the performance to about 350 is possible if the number of blades per cascade is increased.

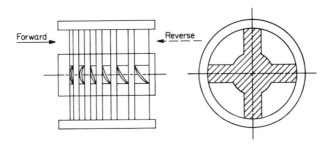

Fig. 2.20 Cascade diode.

A disadvantage of the cascade diode is that it is geometrically complex and, therefore, expensive to fabricate. In addition there is the possibility of deterioration of the performance through mechanical damage to the blades.

2.7.2 Momentum Interaction Diodes

The scroll diode [30] and the Tesla diode [19, 32] use momentum interaction to achieve reverse flow resistance.

Figure 2.21a shows the configuration of the scroll diode. This diode is essentially a converging nozzle with a sudden enlargement that has an axial annular cup. In the reverse flow direction fluid separates at the nozzle throat and enters the annular cup. The cup directs the fluid back toward the oncoming flow. Finally the flow turns again to exit through the annular passage between the cup and the sudden enlargement. In the forward direction the fluid does not enter the cup. The fluid flows through the throat and then into a diffuser section to minimize forward losses.

Measurements on scroll diodes of various geometries have yielded performance ratios in the range from 3 to 7 [30]. With careful redesign the ratio may increase to about 10. This, however, seems to be the upper limit of performance for the scroll diode.

The Tesla diode (Fig. 2.21b) is a series of connected branches and flow loops. In the reverse flow direction fluid enters the branches at a small angle and loops

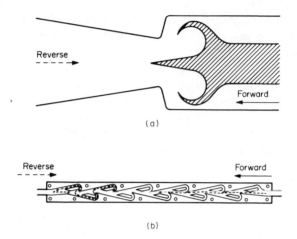

Fig. 2.21 Momentum interaction diodes: (a) scroll diode, (b) Tesla diode.

around to oppose the main flow. Forward flow is predominantly axial in direction and incurs an enlargement loss when it passes over the branching sections. Figure 2.22a shows a schematic model of a single stage Tesla diode [32]. In

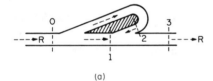

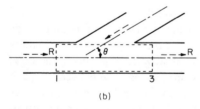

Fig. 2.22 Tesla diode model for reverse flow: (a) single-stage Tesla diode, (b) momentum interaction model.

reverse flow, fluid passing section 0 divides between the in-line section 1 and the branch section 2. The branch flow then loops around and is directed at the in-line flow. The two flows then recombine to enter section 3.

The control volume shown in Fig. 2.22b provides a method of determining the reverse loss coefficient [32]. Application of the momentum equation between sections 1 and 3 leads to the expression

$$p_1 - p_3 = \rho \bar{U}_3^2 - \rho \bar{U}_1^2 + (A_2/A_3)\bar{U}_2^2 \cos \theta \tag{2.38}$$

2.8 Resistance Networks

where the areas are A_1, A_2, and A_3 ($A_1 = A_3$) and θ is the angle that the return loop makes with the in-line direction. From Eq. (2.38), after some rearrangement, the reverse loss coefficient becomes

$$K_{LR} = 1 - \bar{U}_1^2/\bar{U}_3^2 + 2(A_2/A_3)(\bar{U}_2^2/\bar{U}_3^2) \cos\theta \qquad (2.39)$$

If we now assume that the static and total pressures are equal at sections 1 and 2, then the velocities must also be equal ($\bar{U}_1 = \bar{U}_2$). With this relation and the continuity equation, Eq. (2.39) reduces to

$$K_{LR} = \frac{(A_2/A_3)^2 + 2(A_2/A_3)(1 + \cos\theta)}{(1 + A_2/A_3)^2} \qquad (2.40)$$

The maximum reverse loss coefficient occurs when $\theta = 0$ degrees and $A_2/A_3 = 2$ and has a magnitude of 1.33. Of course it is not possible practically to make the angle zero. Nevertheless this value is useful as an indication of the maximum possible magnitude. In forward flow this configuration is a type of sudden enlargement, and therefore we would expect the forward total loss coefficient to fall between 0 and 1 (Fig. 2.15). Test data on typical geometries of single-stage Tesla diode [32] result in reverse coefficients of about 1 and forward coefficients of about 0.25. Thus the performance ratio of the diode is in the vicinity of 4. Series or parallel connection of a number of single-stage Tesla diodes should not alter performance to any large extent since both forward and reverse resistances change in the same proportion.

2.8 RESISTANCE NETWORKS

2.8.1 Branch Resistance

A fluidic resistive circuit that occurs frequently is the branch or junction. Figure 2.23 shows a schematic drawing of the equal-area TEE branch. For dividing flow (Fig. 2.23a), fluid enters the in-line section 1 and leaves through in-line section 2 and branch-line section 3. For mixing flow (Fig. 2.23b), fluid flows in the opposite direction in the branch line. Thus fluid enters at sections 1 and 3, and leaves at section 2. The flow pattern at the junction is different in mixing and dividing flows. As a result the loss coefficients for the two cases also differ.

Let us consider the dividing flow condition as a variable area passage with entrance at section 1 and exits at sections 2 and 3. If we assume that the sections are far enough from the junction to insure uniform velocity and pressure distributions the energy equation [Eq. (2.8)] for the branch is

$$p_1 \bar{U}_1 = p_2 \bar{U}_2 + p_3 \bar{U}_3 + (K_L)_{12}(\rho \bar{U}_1^2/2)\bar{U}_2 + (K_L)_{13}(\rho \bar{U}_1^2/2)\bar{U}_3 \qquad (2.41)$$

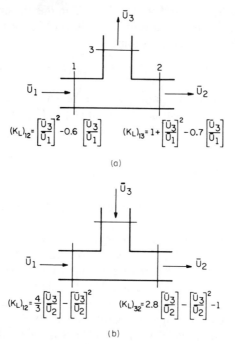

Fig. 2.23 Branch resistance: (a) dividing flow ($\bar{U}_1 \bar{U}_2 = \bar{U}_3$), +(b) mixing flow ($\bar{U}_2 = \bar{U}_1 + \bar{U}_3$).

where $(K_L)_{12}$ is the total loss coefficient for that portion of the in-line fluid at section 1 that remains in-line and $(K_L)_{13}$ is the total loss coefficient for the portion of fluid that turns into the branch line. The pressures, of course, refer to the total pressures at sections 1, 2, and 3. In dividing flow at an equal-area branch with incompressible flow, the continuity equation yields $\bar{U}_1 = \bar{U}_2 + \bar{U}_3$. Thus Eq. (2.41) may be written as

$$\left(p_1 - p_2 - (K_L)_{12} \frac{\rho \bar{U}_1^2}{2}\right) \bar{U}_2 + \left(p_1 - p_3 - (K_L)_{13} \frac{\rho \bar{U}_1^2}{2}\right) \bar{U}_3 = 0 \quad (2.42)$$

Since the terms in parentheses in Eq. (2.42) can never be negative and the velocities are generally not zero, we conclude that

$$(K_L)_{12} = (p_1 - p_2)/(\rho \bar{U}_1^2/2) \quad (2.43a)$$

$$(K_L)_{13} = (p_1 - p_3)/(\rho \bar{U}_1^2/2) \quad (2.43b)$$

The total loss coefficients for an equal-area TEE [33] based on experimental data [34] for dividing flow are expressed as

$$(K_L)_{12} = (\bar{U}_3/\bar{U}_1)^2 - 0.6(\bar{U}_3/\bar{U}_1) \quad (2.44a)$$

$$(K_L)_{13} = 1 + (\bar{U}_3/\bar{U}_1)^2 - 0.7(\bar{U}_3/\bar{U}_1) \quad (2.44b)$$

2.8 Resistance Networks

We observe that the loss coefficients depend on the portion of in-line flow that leaves through the branch.

In the case of mixing flow the loss coefficients ($K_L = \Delta p/(\rho \bar{U}_2^2/2)$) are [33]

$$(K_L)_{12} = \tfrac{4}{3}(\bar{U}_3/\bar{U}_2) - (\bar{U}_3/\bar{U}_2)^2 \qquad (2.45a)$$

$$(K_L)_{32} = 2.8(\bar{U}_3/\bar{U}_2) - (\bar{U}_3/\bar{U}_2)^2 - 1 \qquad (2.45b)$$

Once again the coefficients depend on the fraction of the total flow that passes through the branch line. For unequal-area TEE's or WYE junctions there are experimental data [34, 35] on the loss coefficients, but in these cases the coefficients have not been expressed analytically.

2.8.2 A Typical Branch Circuit

Figure 2.24 shows a typical branch circuit for dividing flow. The upstream in-line section (1) receives flow from a stagnation chamber at pressure p_s. The flow divides at the equal-area TEE junction into a branch line 3 and a downstream in-line section 2. Orifices of area A_5 and A_4 ($A_5, A_4 < A$) at the ends of the downstream and branch lines (sections 4 and 5) connect the lines to atmospheric pressure through large sudden enlargements. We seek to find the portion of the total flow that passes through the branch and downstream lines as a function of the orifice sizes.

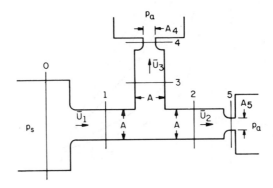

Fig. 2.24 A typical branch circuit.

The losses from sections 2 and 3 to atmosphere are due to the sudden enlargements and are equal to unity. Thus, $p_3 = \rho \bar{U}_4^2/2$ and, since $A_4 \bar{U}_4 = A \bar{U}_3$, we obtain

$$p_3 = (A/A_4)^2 \rho \bar{U}_3^2/2 \qquad (2.46a)$$

and, similarly for section 2,

$$p_2 = (A/A_5)^2 \rho \bar{U}_2^2/2 \qquad (2.46b)$$

Now from Eqs. (2.43) and (2.44)

$$p_1 - p_3 = [1 + (\bar{U}_3/\bar{U}_1)^2 - 0.7(\bar{U}_3/\bar{U}_1)](\rho \bar{U}_1^2/2) \quad (2.47\text{a})$$

and

$$p_1 - p_2 = [(\bar{U}_3/\bar{U}_2)^2 - 0.6(\bar{U}_3/\bar{U}_1)](\rho \bar{U}_1^2/2) \quad (2.47\text{b})$$

The combination of Eqs. (2.46), (2.47), and the continuity equation ($\bar{U}_1 = \bar{U}_2 + \bar{U}_3$) leads to

$$\left(\frac{\bar{U}_3}{\bar{U}_1}\right)^2 \left[\left(\frac{A}{A_4}\right)^2 - \left(\frac{A}{A_5}\right)^2\right] + \left(\frac{\bar{U}_3}{\bar{U}_1}\right)\left[2\left(\frac{A}{A_5}\right)^2 - 0.1\right] + \left[1 - \left(\frac{A}{A_5}\right)^2\right] = 0 \quad (2.48)$$

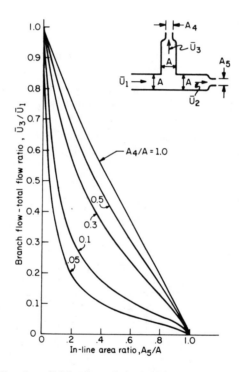

Fig. 2.25 Flow in a dividing branch for various branch area ratios, A_4/A.

The information contained in Eq. (2.48) is given in Fig. 2.25, which shows the portion of the total flow in the branch line versus the in-line orifice ratio A_5/A for various values of the branch-line orifice ratio A_4/A. When the in-line orifice area equals the junction area ($A_5/A = 1$) there is no branch flow at any branch-line orifice ratio. At the other extreme, when $A_5/A = 0$ all the flow passes out the branch line. In between the flow division depends on both area ratios. For example, at $A_5/A = 0.4$, and $A_4/A = 0.05$, 10.5% of the fluid leaves through the branch and 89.5% leaves through the downstream line.

2.8 Resistance Networks

2.8.3 Vented Diodes

Information about vented type fluid diodes appears almost exclusively in the patent literature [23–28]. Perhaps the reason for this is that the devices are valuable but are difficult to analyze. Thus there are very little technical data on these diodes. In addition, even the performance criteria for the vented diodes have not been clearly defined.

Let us begin therefore by defining the performance (D_i) of the vented diode as the square of the ratio of the forward flow in the reverse terminal q_{FR} (Fig. 2.19b) to the reverse flow in the forward terminal q_{RF} at the same pressure difference between terminals. That is

$$D_i = (q_{FR}/q_{RF})^2 \qquad (2.49)$$

Unlike the nonvented diodes, the performance of vented diodes can be increased by connection of several stages in series. This is a consequence of the increase in the number of vents. Actually the overall performance is approximately the product of the performance of the stages.

In this section we describe two types of vented diodes: (a) the vortex vent diode [24], and (b) the wall-effect diode [36]. Figure 2.26a shows a schematic

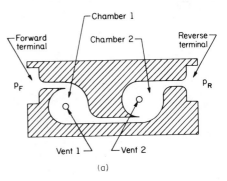

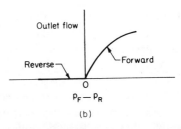

Fig. 2.26 (a) Vortex vent fluid diode, and (b) typical characterist.

drawing of a vortex vent diode [24]. In the forward direction fluid enters chamber 1 and passes into chamber 2 with only a small vortex flow out through vent 1. Similarly the flow in chamber 2 passes to the reverse terminal with another small vortex flow going out at vent 2. As a result most of the forward flow passes to the reverse terminal. In the reverse direction on the other hand, the fluid first enters into chamber 2 where a strong vortex is formed. This increases the reverse resistance to flow. At the same time the largest portion of the reverse flow spirals out through vent 2 leaving only a small amount left to enter chamber 1. The fluid which does reach chamber 1 is subjected to another strong vortex, and most of this fluid flows out through vent 1. The remaining small quantity of reverse flow then passes to the forward terminal. We recognize, then, that this device is two vented vortex diode stages in series. Figure 2.26b shows the flow rate through the downstream terminal (forward terminal for reverse flow and reverse terminal for forward flow) as a function of pressure difference between the terminals. There is essentially no reverse flow through to the forward terminal, whereas the forward flow through the reverse terminal is appreciable. The performance index of this device is very high and it increases with an increase in pressure difference.

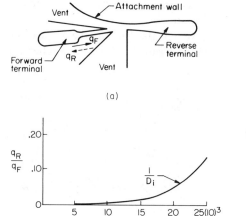

Fig. 2.27 (a) Wall-effect vent diode, and (b) performance.

Figure 2.27a shows a wall-effect diode [36]. Except for a small loss, forward flow is received at the in-line reverse terminal. Reverse flow, however, attaches to the curved wall and mainly passes out the vent. Figure 2.27b shows the ratio of reverse to forward flow in the forward terminal plotted against the Reynolds number of the flow in the reverse terminal. Although this ratio squared is not quite the inverse of the performance [Eq. (2.49)] it is close enough for practical purposes. To determine the performance of the wall-effect diode consider, for example, that at a Reynolds number of 10,000 the flow ratio is only 1%. This is equivalent to a performance of 10^4. If two wall-effect diodes (operating at the same Reynolds number) were placed in series, the overall performance would be about 10^8. As the Reynolds number increases the performance decreases. In this case more flow from the attached wall spills over into the forward terminal.

2.9 CAPACITANCE

We define capacitance as the measure of the change in pressure due to the storage of volume flow,

$$C \equiv \frac{\int q\, dt}{\Delta p} \qquad (2.50)$$

Consider fluid entering a chamber. The fluid is compressible so that the amount of fluid within the chamber may be increased by compressing it and the volume of the chamber may be allowed to vary. Then from the continuity equation

$$\rho q = \frac{d}{dt}(V\rho) = \rho \frac{dV}{dt} + V \frac{d\rho}{dt}$$

where V is the chamber volume and ρ is the fluid density,

$$q = dV/dt + (V/\rho)\, d\rho/dt \qquad (2.51)$$

We assume small changes in pressure and therefore small changes in volume so that

$$V = \bar{V} + V', \qquad p = \bar{p} + p', \qquad \rho = \bar{\rho} + \rho'$$

where $V'/V \ll 1$, so that Eq. (2.51) may be written

$$q = dV/dt + (\bar{V}/\bar{\rho})\, d\rho/dt \qquad (2.52)$$

Now

$$d\rho/\bar{\rho} = (1/\beta_M)\, dp \qquad (2.53)$$

where β_M is the bulk modulus.

Inserting this into Eq. (2.52),

$$q = dV/dt + (\bar{V}/\beta_M)\, dp/dt$$

$$\int q\, dt = \left(\frac{dV}{dp} + \frac{\bar{V}}{\beta_M}\right) \int dp \qquad (2.54)$$

$$C = \frac{\int q\, dt}{\Delta p} = \frac{dV}{dp} + \frac{\bar{V}}{\beta_M} \qquad (2.55a)$$

dV/dp is often called the compliance.

If the volume is fixed, then

$$C = \bar{V}/\beta_M \qquad (2.55b)$$

2.9.1 Spring-Loaded Piston

As a particular example consider fluid pushing against a spring-loaded massless piston that is vented to ambient on the opposite side (Fig. 2.28a).

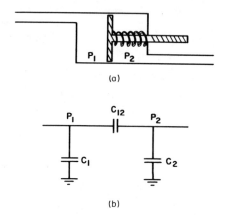

Fig. 2.28 Capacitance: (a) point-to-point capacitance formed by spring-loaded massless piston in a chamber; (b) equivalent circuit.

We first concern ourselves only with that side of the chamber for which the pressure is p_1.

If the spring constant is k_s and the area of the piston is A_p, then

$$A_p\, dp = k_s\, dx = k_s\, dV/A_p$$

where dx is the compression of the spring due to the applied force $A_p\, dp$, whence

$$dV/dp = A_p^2/k_s \tag{2.56}$$

so that from Eq. (2.55) the capacitance of the spring-loaded piston for small changes of pressure is

$$C = (A_p^2/k_s) + (\bar{V}/\beta_M) \tag{2.57}$$

If we assume a gas having a polytropic relation between pressure and density, i.e.,

$$p = c\rho^n \tag{2.58a}$$

$$dp = cn(\bar{\rho})^{n-1}\, d\rho = (n\bar{p}/\bar{\rho})\, d\rho \tag{2.58b}$$

Now from Eq. (2.53) and Eq. (2.58b)

$$\beta_M = \bar{\rho}\,\frac{dp}{d\rho}\, n\bar{p}$$

so that when the fluid is gas the capacitance is given from Eq. (2.55) as

$$C = (dV/dp) + (\bar{V}/n\bar{p}) \tag{2.59a}$$

2.9 Capacitance

or, when the volume is fixed,

$$C = \overline{V}/n\overline{p} \tag{2.59b}$$

and, for the particular case of the spring-loaded massless piston (acting single sided) when the fluid is gas,

$$C = (A_p^2/k_s) + (\overline{V}/n\overline{p}) \tag{2.60}$$

The equivalent circuit for this is a capacitance to ground.

In liquids the compressibility term $\overline{V}/\beta_M$ will usually be much smaller than the compliance term A_p^2/k_s and may be neglected.

If the opposite side of the piston is part of the circuit (Fig. 2.28a), then a similar result but with a different meaning is obtained.

In this case the force acting on the spring is $A_p\,d(p_2 - p_1)$ instead of $A_p\,dp$ and we define the capacitance across the piston due to compliance as

$$C_{12} = \frac{\int q\,dt}{\int d(p_1 - p_2)} = \frac{A_p^2}{k_s} \tag{2.61}$$

The compressibility capacitance associated with each side of the piston is as before

$$C_1 = \frac{\int q\,dt}{\int dp_1} = \frac{\overline{V}}{n\overline{p}_1}, \qquad C_2 = \frac{\int q\,dt}{\int dp_2} = \frac{\overline{V}}{n\overline{p}_2}$$

The equivalent circuit is shown in Fig. 2.28b.

Since the bulk modulus of liquids is very large, the fixed-volume capacitance of liquid systems is quite small and is often neglected. The fixed-volume capacitance of gas systems, on the other hand, depends on the polytropic exponent and the absolute pressure. In many fluidic systems the absolute pressure does not change appreciably and thus the variation in capacitance due to pressure changes is small. In addition the nonlinear effect of absolute pressure is defined even when there are large pressure changes. The effects of changes in the polytropic exponent, however, are more uncertain because the exponent depends on the heat transfer between the fluid and the enclosure. In general, slow changes in small enclosures follow an isothermal process, whereas fast changes in large enclosures follow an adiabatic process. Since slow and fast are relative terms we will present an analytical formulation for the polytropic exponent.

2.9.2 Derivation of Polytropic Exponent for Enclosures with Sinusoidal Inputs

To obtain a formulation for the polytropic exponent we use the equation of state

$$p = \rho R_g T \tag{2.62a}$$

or

$$dp = R_g T\,d\rho + \rho R_g\,dT \tag{2.62b}$$

where R_g is the gas constant and T is the absolute temperature.

Eliminating the density between Eqs. (2.58b) and (2.62b) and using the relation

$$R_g = c_p(\gamma - 1)/\gamma$$

yields

$$n = \left[1 - \rho c_p\left(\frac{\gamma - 1}{\gamma}\right)\left(\frac{\Delta T}{\Delta p}\right)\right]^{-1} \tag{2.63}$$

where γ is the ratio of specific heats and c_p is the specific heat at constant pressure. The value of the exponent n thus depends upon the relation between the average temperature change (ΔT) and the pressure change (Δp). This relation is obtained by solution of the thermal energy equation. Daniels [37] solved this equation for three enclosures (long cylinders, spheres, and narrow rectangular parallelepipeds) using harmonic functions for pressure and temperature. The solutions give the following formulations for the average temperature–pressure ratio:

Long Cylinder

$$\left(\frac{\Delta T}{\Delta p}\right)_c = \left(1 - \frac{2J_1(j^{3/2}\omega^{1/2}F)}{j^{3/2}\omega^{1/2}FJ_0(j^{3/2}\omega^{1/2}F)}\right)\frac{1}{\rho c_p} \tag{2.64a}$$

Sphere

$$\left(\frac{\Delta T}{\Delta p}\right)_s = \left(1 - \frac{3\coth(j^{1/2}\omega^{1/2}F)}{j^{1/2}\omega^{1/2}F} + \frac{3}{j\omega F^2}\right)\frac{1}{\rho c_p} \tag{2.64b}$$

Narrow Box

$$\left(\frac{\Delta T}{\Delta p}\right)_b = \left(1 - \frac{\tanh(j^{1/2}\omega^{1/2}F)}{j^{1/2}\omega^{1/2}F}\right)\frac{1}{\rho c_p} \tag{2.64c}$$

where $F = [N_p r_w^2/v]^{1/2}$, N_p is the Prandtl number,* r_w is the characteristic dimension of the enclosure (radius for cylinder and sphere, half-depth for narrow box), v is the kinematic viscosity, ω is the angular frequency, $j = \sqrt{-1}$, and J_0, J_1 are Bessel functions. The substitution of the average temperature–pressure ratio from Eq. (2.64) into Eq. (2.63) results in

Long Cylinder

$$n_c = \gamma\left[1 + \frac{2(\gamma - 1)}{j^{3/2}\omega^{1/2}F} \frac{J_1(j^{3/2}\omega^{1/2}F)}{J_0(j^{3/2}\omega^{1/2}F)}\right]^{-1} \tag{2.65a}$$

Sphere

$$n_s = \gamma\left\{1 + 3(\gamma - 1)\left[\frac{\coth(j^{1/2}\omega^{1/2}F)}{j^{1/2}\omega^{1/2}F} - \frac{1}{j\omega F^2}\right]\right\}^{-1} \tag{2.65b}$$

---

* N_p is the kinematic viscosity divided by the thermal diffusivity.

2.9 Capacitance

Narrow Box

$$n_b = \gamma \left\{ 1 + (\gamma - 1) \left[\frac{\tanh(j^{1/2}\omega^{1/2}F)}{j^{1/2}\omega^{1/2}F} \right] \right\}^{-1} \quad (2.65c)$$

Figure 2.29 shows the magnitude of the polytropic exponents expressed in Eq. (2.65) for $\gamma = 1.4$. There is a continuous transition from isothermal to adiabatic as the frequency-size parameter $\omega^{1/2}F$ increases. The polytropic exponent like the temperature distribution is a function of the ratio of surface

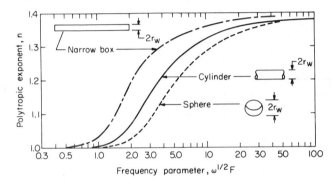

Fig. 2.29 The transition from isothermal to adiabatic capacitance.

area to volume. The narrow box enclosure has the smallest area–volume ratio $(1/r_w)$ and is therefore the enclosure that approaches adiabatic conditions at the lowest value of $\omega^{1/2}F$. The area–volume ratios for the long circular cylinder and the sphere are $2/r_w$ and $3/r_w$. Thus the sphere remains isothermal for larger values of $\omega^{1/2}F$ than the long cylinder with the same radius and has the slowest transition to adiabatic conditions. For finite-length cylindrical enclosures the surface area–volume ratio increases and the transition curve shifts toward the transition curve for the sphere. For cylinders with l/d ratios above 2.37 the polytropic exponents are within 2% of the long-cylinder values [38].

Figures 2.30–2.32 are log–log plots of frequency versus size with the polytropic exponent as a parameter. This is a nonnormalized representation of Eq. (2.65) for air at standard conditions. Each value of polytropic exponent is constant along a straight line that has a slope of -2 on the log–log plots. The regions beneath the $n = 1.05$ line and above the $n = 1.35$ line are essentially isothermal and adiabatic, respectively. For example, a long cylinder (Fig. 2.30) of 10 mm diameter at a signal frequency of 0.2 Hz is practically isothermal ($n < 1.05$), whereas the same chamber with a signal frequency of 1.8 Hz has a polytropic exponent of 1.2. On the other hand, a spherical chamber of the same diameter (10 mm) requires a frequency signal of 3.8 Hz to reach a polytropic exponent of 1.2.

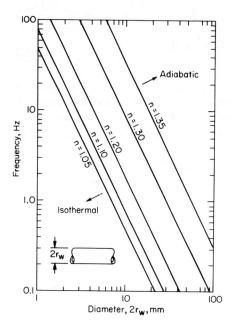

Fig. 2.30 Polytropic exponent for sinusoidal signals in air-filled cylindrical enclosures.

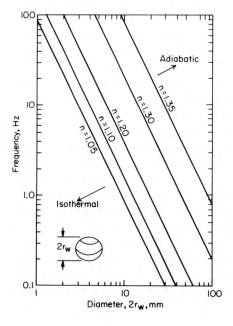

Fig. 2.31 Polytropic exponent for sinusoidal signals in air-filled spherical enclosures.

2.9 Capacitance

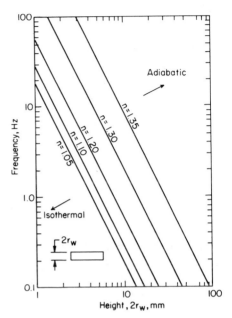

Fig. 2.32 Polytropic exponent for sinusoidal signals in air-filled narrow box enclosures.

Equation (2.59b) shows the relation between capacitance and polytropic exponent. Since the polytropic exponent [Eq. (2.65)] has real and imaginary parts, the capacitance is also a complex quantity [39–41]. This means that the enclosure actually represents a component that has resistance as well as capacitance. The resistance is a consequence of the transfer of thermal energy between the gas and the walls of the enclosure. In the isothermal region the resistance of the enclosure can exceed the capacitive reactance for cylindrical and spherical enclosures but can only equal the capacitive reactance for narrow box enclosures. Under adiabatic conditions there is no heat transfer and the enclosure is purely capacitive.

2.9.3 Response of an RC Circuit to a Step Input

The expressions derived for the polytropic exponent [Eq. (2.65)] for sinusoidal signals are also valid for Laplace transforms [42]. The response of time-dependent resistance–capacitance circuits to a step input may then be obtained by finding the inverse transform. For example, the substitution of $j\omega = s$ in Eq. (2.65b) gives the capacitance of a spherical enclosure as

$$C_s = \frac{V}{\gamma p}\left\{1 + 3(\gamma - 1)\left[\frac{\coth(s^{1/2}F)}{s^{1/2}F} - \frac{1}{sF^2}\right]\right\} \qquad (2.66)$$

The pressure in the enclosure when a step input in pressure is applied to the enclosure through a linear resistance R is

$$P(s) = \frac{1/\tau_a}{s\left\{s\left[1 + 3(\gamma - 1)\left(\frac{\coth(s^{1/2}F)}{s^{1/2}F} - \frac{1}{sF^2}\right)\right] + \frac{1}{\tau_a}\right\}} \quad (2.67)$$

where $P(s) = \mathscr{L}\{p(t)\}$ and τ_a is the adiabatic time constant $RV/\gamma p$. For values of $s^{1/2}F$ greater than 2 the hyperbolic cotangent is always very close to unity. With this approximation Eq. (2.67) becomes

$$P(s) = \frac{1/\tau_a}{s\left[s + \frac{3(\gamma - 1)s^{1/2}}{F} + \left(\frac{1}{\tau_a} - \frac{3(\gamma - 1)}{F^2}\right)\right]} \quad (2.68)$$

The approximation limits the validity of the inverse transform to times less than $0.35F\tau_a^{1/2}$ (see Problem 2.14). In most practical cases, however, this is the only period of interest. If Eq. (2.68) is now factored and separated into partial fractions we obtain

$$P(s) = \frac{1}{\tau_a(f - e)}\left[\frac{1}{s(s^{1/2} + e)} - \frac{1}{s(s^{1/2} + f)}\right] \quad (2.69)$$

where

$$e = \frac{3(\gamma - 1)}{2F} + \frac{1}{2}\left[\frac{3(\gamma - 1)(3\gamma + 1)}{F^2} - \frac{4}{\tau_a}\right]^{1/2}$$

$$f = \frac{3(\gamma - 1)}{2F} - \frac{1}{2}\left[\frac{3(\gamma - 1)(3\gamma + 1)}{F^2} - \frac{4}{\tau_a}\right]^{1/2}$$

Note that in the useful range ($\tau_a < 4/F^2$) the roots e and f are complex conjugates with small real part and large imaginary part.

The pressure–time relation is then

$$p(t) = \frac{1}{\tau_a}\left[\frac{1 - \exp(e^2 t)\,\text{erfc}(e\sqrt{t})}{e(f - e)} - \frac{1 - \exp(f^2 t)\,\text{erfc}(f\sqrt{t})}{f(f - e)}\right] \quad (2.70)$$

where it can be proved that $p(t)$ is real when e and f are complex conjugates.

Figure 2.33 shows the RC circuit response [Eq. (2.70)] when the enclosure is a sphere 50 mm in diameter, the adiabatic time constant is 0.200 sec, and the gas is air at standard conditions. The purely adiabatic and isothermal responses are also shown. At the beginning of the rise ($t/\tau_a < 0.5$) the theoretical response is close to adiabatic. As time increases the response is midway between adiabatic and isothermal. Eventually when $t/\tau_a > 4.0$ the response is close to isothermal. This representation agrees with experimental results [43, 44].

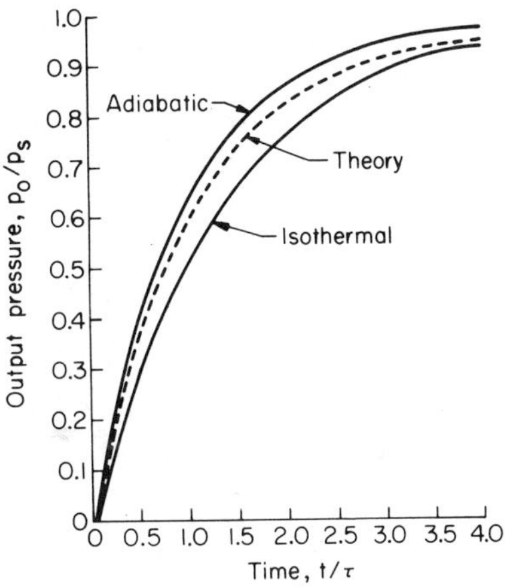

2.9 Capacitance

Fig. 2.33 Response of RC circuit to a step input (50-mm diameter spherical enclosure, $\tau = 0.200$ sec, fluid → air, $F = 5.43$ sec$^{1/2}$).

The form of the response given for a spherical enclosure in Eq. (2.70) also serves for the cylindrical and narrow box enclosures if the roots (e, f) are adjusted. Thus, for the cylinder,

$$e, f = \frac{\gamma - 1}{F} \pm \left[\frac{\gamma(\gamma - 1)}{F} - \frac{1}{\tau_a}\right]^{1/2}, \qquad t < 0.25 F \tau_a^{1/2} \qquad (2.71a)$$

and, for the narrow box,

$$e, f = \frac{\gamma - 1}{2F} \pm \frac{1}{2}\left[\frac{(\gamma - 1)^2}{F^2} - \frac{4}{\tau_a}\right]^{1/2}, \qquad t < 0.35 F \tau_a^{1/2} \qquad (2.71b)$$

2.9.4 Variable Volume Capacitance

The following sections discuss variable volume capacitance such as diaphragms, bellows, and accumulators. As a consequence of the solid barrier, diaphragm and bellows elements have two connecting terminals, one on each side of the barrier. They are thus the fluid analogs of electrical blocking capacitors. Accumulators, on the other hand, have only one connecting terminal in the medium of the working fluid. Thus, despite the variable volume, the accumulator represents a capacitance to ground.

Variable volume devices also have compressibility capacitance [Eq. (2.59a)]. In fluid circuits there is always a compressibility capacitance to ground at each terminal. Thus, for example, the accumulator consists of two grounded capacitors (one variable and one compressible) in parallel. The compressibility capacitances are not included in the following discussion but should not be overlooked when the fluid medium is a gas.

2.9.4.1 Diaphragm Capacitance

A typical diaphragm device (Fig. 2.34) consists of a circular sheet of flexible material clamped between two flanges [45]. The initial tension in the flexible sheet is small. The diaphragm separates the assembled device into an input

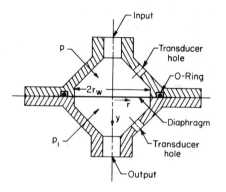

Fig. 2.34 Diaphragm configuration.

chamber and an output chamber. The capacitance of the device depends on the relation between the change in input chamber volume and the difference in chamber pressures. To formulate the capacitance we begin with the deflection $y(r)$ of a clamped circular diaphragm without flexural stiffness [46]:

$$y(r) = 0.662 r_w \left[\frac{(p - p_1) r_w}{E d_d} \right]^{1/3} \left[1.0 - 0.9 \left(\frac{r}{r_w} \right)^2 - 0.1 \left(\frac{r}{r_w} \right)^5 \right] \quad (2.72)$$

where d_d is the diaphragm thickness and E is the modulus of elasticity of the diaphragm material. The volume displaced by the deflected diaphragm is

$$V = 2\pi \int_0^{r_w} y(r) r \, dr = 1.085 r_w^3 \left[\frac{(p - p_1) r_w}{E d_d} \right]^{1/3} \quad (2.73)$$

Thus the relation [Eq. (2.73)] between displacement volume and pressure difference is nonlinear. In this case we define the variational capacitance of the diaphragm as

$$C \equiv dV/d(p - p_1) \quad (2.74)$$

and assume that the static relation [Eq. (2.73)] also applies to dynamic conditions at low frequencies. Differentiation of Eq. (2.73) then produces an expression for the variational capacitance in terms of the geometric and physical properties of the diaphragm. The result is

$$C = \frac{0.362 r_w^3}{(p - p_1)} \left[\frac{(p - p_1) r_w}{E d_d} \right]^{1/3} \quad (2.75)$$

2.9 Capacitance

This capacitance applies to small-amplitude signals Δp about an average signal $p - p_1$, where $\Delta p/(p - p_1) \ll 1$. Figure 2.35 shows the relation between the capacitance of the diaphragm and the pressure difference for various values of the geometric mechanical parameter $r_w^{10/3}/(Ed_d)^{1/3}$. The linearized capacitance is infinite when the chamber pressures are equal. Practically, however, any finite amplitude input signal creates a pressure difference which reduces the capacitance to a finite value. However, if $\Delta p/(p - p_1)$ is not small compared to unity, then the capacitance is nonlinear; that is, it is dependent on the signal amplitude Δp.

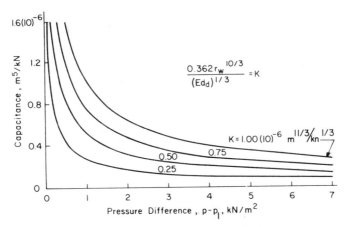

Fig. 2.35 Diaphragm capacitance.

2.9.4.2 Bellows

Figure 2.36 shows the configuration of a bellows with circular cross section [47]. The volume change is the result of the pressure difference across the free end. A force balance yields the relation between these variables as

$$V = (\pi^2 d^4/16k_b)(p - p_1) \tag{2.76}$$

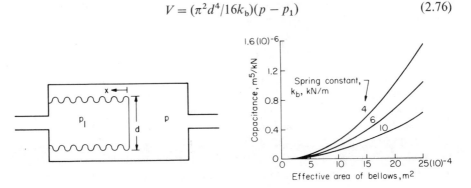

Fig. 2.36 Bellows configuration.

Fig. 2.37 Bellows capacitance.

where d is the diameter and k_b the spring constant of the bellows. The variable volume capacitance of the bellows is therefore

$$C = \pi^2 d^4/16 k_b \tag{2.77}$$

Bellows should not be extended beyond their free length. This alters the spring constant, thereby changing capacitance and also increasing wear. Thus ordinary bellows should be connected in circuits such that the pressure p is always greater than p_1. Differential pressure bellows and diaphragms, however, may be used in either direction. Figure 2.37 shows the capacitance of bellows for various values of spring constant given in units of kN/m.

2.9.4.3 *Accumulators*

Fluid systems that operate with liquids use accumulators to provide capacitance because the compressibility (fixed volume) capacitance is small. Figure 2.38

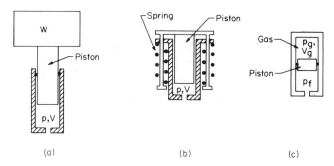

Fig. 2.38 Accumulators: (a) weight loaded, (b) spring loaded, (c) gas loaded.

shows schematic drawings of three basic types of accumulators [48]. In the weight-loaded accumulator the weight W on top of a piston maintains constant pressure in a fluid in a cylinder beneath the piston. The capacitance of the weight-loaded accumulator is

$$C = AV/W \tag{2.78}$$

where A is cross-sectional area of the piston and W is applied weight. The capacitance depends on the volume of the liquid in the cylinder.

The capacitance of the spring-loaded accumulator (Fig. 2.38b) has the same form as that given for the bellows in Eq. (2.77). All accumulators, however, are one-terminal capacitances to ground.

There are several different configurations for gas-loaded accumulators. In one device, the gas-loaded piston accumulator (Fig. 2.38c), a piston separates a closed volume of gas V_g from the low-compressibility working fluid. When the pressure in the working fluid changes the piston moves until the pressure in the

2.9 Capacitance

gas p_g equals the pressure in the working fluid. The variable volume capacitance of the gas-loaded piston accumulator is

$$C = V_g/np_g \tag{2.79}$$

2.9.4.4 Jet Barrier

When a differential pressure acts across a fluid jet, the jet bends and displaces a volume adjacent to it. Thus, there is a variable volume capacitance associated with the combination of a jet and side cavities [49, 50]. To relate the volume displaced by the jet and the pressure difference we must first define the jet boundaries. If the jet width is assumed small, the position of the jet centerline alone indicates the location of the jet barrier. Figure 2.39 shows the curvature

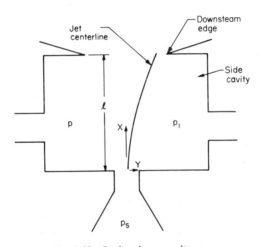

Fig. 2.39 Jet barrier capacitance.

of the jet centerline that results from the pressure difference $p - p_1$. The jet barrier extends from the power nozzle to the downstream edges of the side cavities. The interaction of the side pressures with the power jet momentum produces a centerline deflection y_0 that can be expressed as

$$y_0 = (x^2/4p_s b)(p - p_1) \tag{2.80}$$

where b is the nozzle width, p_s is the jet supply pressure, and x is the downstream distance. The volume displaced by the curved centerline is

$$V = h \int_0^l y_0 \, dx = (hl^3/12p_s b)(p - p_1) \tag{2.81}$$

where h is the jet depth and l is the distance from the nozzle to the downstream edges of the side cavities. The variable volume capacitance of the jet-cavity configuration is therefore

$$C = hl^3/12p_s b \tag{2.82}$$

We shall discuss the jet impedance again in Chapter 5.

2.10 INERTANCE OR INDUCTANCE

The analog of inductance in electricity is the inertance of a fluid. This property has been called both inductance and inertance when applied to fluids, and we shall use both terms indiscriminately.

The inertance is defined (by analogy) as

$$L = \Delta p/(dq/dt) \tag{2.83}$$

Tubing or pipe has inertance which can be found as follows.

Consider a pipe of cross-sectional area A and of length l. From Newton's second law,

$$A \Delta p = \rho A l \, du/dt = \rho l \, dq/dt$$

whence

$$L = \frac{\Delta p}{(dq/dt)} = \frac{\rho l}{A} \tag{2.84}$$

At frequencies for which the length of pipe is small compared to a wavelength, an open pipe has resistance and inertance. Its impedance Z is given by

$$Z = R + j\omega L \tag{2.85}$$

where $\omega = 2\pi f$ is the angular frequency and $j = \sqrt{-1}$.

There will, of course, also be inertance associated with moving-part devices used in fluidic circuits. Of major concern is the inertance associated with diaphragms or bellows used to obtain point-to-point capacitance. Although efforts are usually made to construct the devices so that the inertance is fairly small, the inertive reactance increases with frequency and may have to be taken into account.

PROBLEMS

2.1 The axial pressure gradient for laminar flow in a round pipe is given by

$$dp/dx = (2/r)\mu \, du/dr$$

where μ is the viscosity.

Show that for fully developed laminar flow the resistance of a duct of length L is given by

$$R = 128\mu L/\pi d^4$$

where d is the duct diameter.

2.2a The axial pressure gradient for flow between parallel plates is

$$dp/dx = (\mu/y) \, du/dy$$

Problems

Show that the resistance per unit width of a section of length L is given by

$$R_w = 12\mu L/h^3$$

2.2b Show by a plausibility argument that the resistance of a rectangular passage can be approximated by

$$R = 12\mu L(w^2 + h^2)/h^3 w^3$$

where w is width of the passage.

2.3 Show that in the inviscid vortex diode the radial velocity is small compared to the tangential velocity.

2.4 Find the energy distribution factor β_e for turbulent flow given that

$$u(r) = \bar{U}(1 - 2r/d)^{1/n}$$

where r is the distance from the axis, d is the diameter of the duct, and $n = 7$.

2.5 A typical fluidic resistance (Fig. 2.17) has a diameter d_e of 2.0 mm and a length l of 40.0 mm. The inlet and outlet diameters are much larger than d_e. The fluid medium is air

$$\nu = 14.85(10)^{-6} \text{ m}^2/\text{sec}, \qquad \rho = 1.2 \text{ kg/m}^3$$

(a) Find the flow rate through the resistor when the pressure difference $p_s - p_3$ is 10 kN/m². (Do not assume fully developed flow throughout.)

(b) Compare the flow rate from part (a) with the flow rate that would be calculated by assuming that the flow in the diameter d_e is always fully developed.

2.6 The fluidic resistance shown in Fig. 2.16 has an orifice d_2 and sudden enlargements to d_3 and d_4. If $d_0 = d_4 = 10.0$ mm, $d_2 = 1.0$ mm, and $p_s - p_4 = 20$ kN/m², find the flow rate through the resistor when

(a) $d_3 = 1.50$ mm,   (b) $d_3 = 2.00$ mm,   (c) $d_3 = 4.00$ mm

Assume wall shear is small and that the fluid medium is air ($\rho = 1.2$ kg/m³).

2.7 A control system requires a fluid component that has a pressure–flow relation $\Delta p = kq^3$. Can this function be achieved with some combination of passive fluid components? Discuss your answer.

2.8 A capillary tube has a diameter of 0.5 mm and a length of 20.0 mm. In a laboratory air experiment with uncalibrated equipment the following measurements are made: (a) static pressure difference from entrance to exit, 1.558 kN/m²; (b) flow rate, $q = 10$ cm³/sec; (c) entrance profile, uniform; and (d) exit profile, parabolic. Are these measurements consistent with what you know about fluid resistance? Explain your answer.

2.9 The nozzle–baffle configuration shown in Fig. 2.10 is to be used as a resistance to ground in a specific fluidic air system application. If $r_w = 1.0$ mm, $p_1 = p_{\sigma_2} = 8.0$ kN/m² and $q = 20$ cm³/sec must be maintained:

(a) What gap size d_b is required when $r_t = 10.0$ mm?
(b) What gap size d_b is required when $r_t = 20.0$ mm?

2.10 The flow entering the typical branch circuit shown in Fig. 2.24 is measured at section 1 as 250 cm³/sec. The TEE junction connects three circular lines, each with a diameter of 4.0 mm. The in-line leg is terminated with an orifice of diameter $d_5 = 2.5$ mm. Find the flow out of the in-line leg and the branch leg when (a) the branch-line termination $d_4 = 1.0$ mm, and when (b) the branch-line termination is changed to $d_4 = 4.0$ mm.

2.11 A capillary fluid resistance of $R = 0.5 \times 10^6$ kN-sec/m⁵ is placed at the entrance of a cylindrical tank of diameter 20 mm and length 100 mm. A sinusoidal pressure signal p_s is applied to the resistance–tank combination. What is the amplitude of the pressure in the tank when (a) $p_s = 10 \sin t$ (kN/m², t in seconds); (b) $p_s = 10 \sin 200\, t$ (kN/m²); (c) the input signals are those given in parts (a) and (b) but the capacitance is treated as isothermal.

2.12 A capacitance is required in a hydraulic fluidic circuit. Experiments show that a thin sheet of rubber of thickness 0.1 mm stretched across a chamber 2.0 cm in diameter is satisfactory for the application.

(a) If the modulus of elasticity of rubber is 1250 kN/m² and the operating pressure difference across the sheet is 1.0 kN/m², find the fluid capacitance used in the application.

(b) What length chamber of 2.0 cm diam would be required if the compressibility capacitance was used instead of the rubber sheet ($\beta_M = 2 \times 10^6$ kN/m²).

2.13 An orifice is placed at the entrance of a large tank of capacitance C. If the orifice characteristic is $(p_s - p) = kq^2$ (where p is the tank pressure, p_s is the upstream pressure, and q the flow through the orifice into the tank), show that for a step change in p_s the tank pressure is related to time by

$$(p/p_s) = (t^2/kC^2 p_s)^{1/2} - \tfrac{1}{4}(t^2/kC^2 p_s).$$

Discuss the response.

2.14 In Section 2.9.3 the step response of an RC fluidic circuit is derived. The derivation is based on an approximation of the Laplace transform which is within 1% when $s^{1/2}F > 2$ for a sphere and for a narrow box and $s^{1/2}F > 3$ for a cylinder.

Show that the estimate for the times for which the approximation is valid may be obtained by assuming that the step function response can be approximated by

$$p(t) = 1 - \exp(-t/\tau_a)$$

for which the transform is

$$sP(s) = 1/(\tau_a s + 1)$$

and then finding the value of t, namely t_1 such that values of $t < t_1$ correspond to values of $s > 4/F^2$.

To find this value, set

$$\lim_{s \to 4/F^2} sP(s) = \lim_{t \to t_1} p(t)$$

Using this equality plot the relation between t_1/τ_a and F^2/τ_a. Approximate this relation with a parabola (or transform the variables to a form that will result in approximately a straight line), then:

(a) Show that the restriction $s^{1/2}F > 2$ corresponds approximately to $t < 0.35F\sqrt{\tau_a}$.

(b) Show that the restriction $s^{1/2}F > 3$ corresponds approximately to $t < 0.25F\sqrt{\tau_a}$.

NOMENCLATURE

a	linear portion of resistance
A	Area
A_p	Piston area
c_p	Specific heat at constant pressure
c_s	Shape factor
c_v	Specific heat at constant volume
C	Capacitance
C_a	Adiabatic capacitance per unit length
d	Diameter
d_b	Distance between nozzle face and baffle
d_d	Diaphragm thickness
d_e	Equivalent diameter
D	Diameter of curvature
D_i	Diodicity
E	Modulus of elasticity
f	Frequency
h	Distance between plates
j	$\sqrt{-1}$
J_1, J_0	Bessel functions
k_b	Bellows spring constant
k_s	Spring constant
K_A	Apparent loss coefficient
K_L	Loss coefficient
l	Length
L	Inertance (inductance)
n	Polytropic coefficient
N_P	Prandtl number
N_R	Reynolds number

N_{Rl}	Reynolds number at distance l from beginning of passage
N_{Rd_b}	Reynolds number based on d_b
N_{Rd_e}	Reynolds number based on equivalent diameter d_e
p_0	Static pressure at $r = r_0$
p_s	Supply pressure
p_σ	Static pressure
p_1	Static pressure inside chamber at $r = r_i$
p'	Small change in pressure
q	Volume flow
q_{FR}	Forward flow in the reverse terminal
q_{RF}	Reverse flow in the forward terminal
r	Radius
r_i	Radius of vortex chamber
r_0	Radius of outlet hole
r_t	Radius to outer edge of nozzle-baffle
r_w	Nozzle radius, characteristic dimension of inclosure
R	dc resistance
R_{ac}	ac resistance
R_g	Gas constant
s	Laplace transform variable
T	Temperature
u	Axial velocity
u_r	Radial velocity
u_θ	Tangential velocity
$\bar{U}$	Characteristic velocity
V	Volume
V'	Small change in volume
W	Weight
y	y coordinate
y_0	Deflection of jet centerline
Z	Impedance
β_e	Energy distribution factor
β_M	Bulk modulus
γ	Ratio of specific heats
ν	Kinematic viscosity
Δp	Total pressure drop, or deviation of pressure
μ	Viscosity
ρ	Density
τ_a	$RV/\gamma p$, Adiabatic time constant (below Eq. (2.67))
ω	Angular frequency

REFERENCES

1. H. E. Koenig, Y. Tokad, and H. K. Kesavan, "Analysis of Discrete Physical Systems," p. 42. McGraw-Hill, New York, 1967.
2. J. M. Kirshner, "Fluid Amplifiers." McGraw-Hill, New York, 1966.
3. J. K. Vennard, "Elementary Fluid Mechanics," 4th ed., pp. 313–315. Wiley, New York, 1961.
4. T. Baumeister and L. Marks, Standard Handbook for Mechanical Engineers," 7th ed., pp. 3–63. McGraw-Hill, New York, 1967.
5. R. C. Binder, "Fluid Mechanics," 3rd ed., p. 128. Prentice-Hall, Englewood Cliffs, New Jersey, 1955.

References

6. A. H. Shapiro, R. Siegel, and S. J. Kline, Friction factors in the laminar entry region of a smooth tube. *Proc. U.S. Nat. Congr. Appl. Mech.*, 2nd, 1954 pp. 733–741.
7. H. L. Langhaar, Steady flow in the transition length of a straight tube. *J. Appl. Mech.* **9**; *Trans. ASME* **64**, A55–58 (1942).
8. L. S. Han, Hydrodynamic entrance lengths for incompressible laminar flow in rectangular ducts. *J. Appl. Mech.* **27**; *Trans. ASME Ser. E* **82**, 403–409 (1960).
9. H. Rouse, "Elementary Mechanics of Fluids," p. 166. Wiley, New York, 1946.
10. E. M. Sparrow, C. W. Hixon, and C. Shavitt, Experiments on laminar flow development in rectangular ducts. *J. Basic Eng. Trans. ASME Ser. D* 116–121 (1967).
11. K. N. Reid, Static characteristics of fluid amplifiers. *Proc. Fluid Power Res. Conf.*, Oklahoma State Univ. (1967).
12. C. M. White, Streamline flow through curved pipes. *Proc. Roy. Soc. London* **123**, 645–663 (1929).
13. G. H. Keulegan and K. H. Beij, Pressure losses for fluid flow in curved pipes. *J. Res. Nat. Bur. Stand.* **18**, 89–114 (1937).
14. J. J. McGinn, Observations on the radial flow of water between fixed parallel plates. *Appl. Sci. Res. Sect. A* **5**, 255–264 (1955).
15. L. A. Zalmanzon, "Components for Pneumatic Control Instruments," pp. 63–66. Pergamon, Oxford, 1965.
16. K. N. Reid, "Static and Dynamic Interation of a Fluid Jet and a Receiver Diffuser." Sc.D. Thesis, MIT, September 1964.
17. H. Peters, "Conversion of Energy in Cross-Sectional Divergences Under Different Conditions of Inflow," TM 737. NACA, March 1934.
18. A. H. Gibson, "Hydraulics and Its Applications," 4th ed., p. 93, D. Van Nostrand, 1930.
19. N. Tesla, "Valvular Conduit." U.S. Patent 1,329,559, 3 February 1920.
20. D. Thoma, "Fluid Lines." U.S. Patent 1,839,616, 5 January 1932.
21. J. H. Bertin *et al.*, "Aerodynamic Valve." U.S. Patent 2,670,011, 23 February 1954.
22. B. O. Ayers, "Variable Impedance Vortex Diode." U.S. Patent 3,521,657, 28 July 1970.
23. T. Reader, "Fluid Diode." U.S. Patent 3,375,842, 2 April 1968.
24. C. C. K. Kwok, "Vortex Vent Fluid Diode." U.S. Patent 3,461,897, 19 August 1969.
25. R. L. Blosser, "Fluidic Diode or Sensor Device." U.S. Patent 3,472,258, 14 October 1969.
26. R. B. Hartman, "Fluidic Diodes." U.S. Patent 3,472,256, 14 October 1969.
27. T. W. Bermel, "Fluidic Diode." U.S. Patent 3,480,030, 25 Nov. 69.
28. R. W. Hatch, "Fluid Diode." U.S. Patent 3,481,353, 2 Dec. 69.
29. R. Heim, "An Investigation of the Thoma Counterflow Brake." *Trans. Hyd. Inst. Munich Tech. Univ.*, Bull. 3, pp. 13–28.
30. P. J. Baker, A comparison of fluid diodes. *Cranfield Fluidics Conf.*, 2nd Paper D6 (1967).
31. B. E. A. Jacobs and P. J. Baker, The cascade diode. *Cranfield Fluidics Conf.*, 3rd Paper K5 (1968).
32. F. W. Paul, Fluid mechanics of the momentum flueric diode. *IFAC Symp. Fluidics* (1968).
33. S. Katz, Mechanical potential drops at a fluid branch. *Trans. ASME J. Basic Eng.* (1968).
34. G. Vogel, "Untersuchen Uber Den Verlust in Rechtwinkligen Rohrverzweigungen," Mitt. Hydraulischem Inst. Tech. Hochschule Munchen No. 2, 1928 [Investigations of the Losses in Right Angle Pipe Joints, Part II (Translated by Sverdup & Parcel Inc.), 1953].
35. F. Peterman, "Der Verlust in Schiefwinkligen Rohrverzeigingen." 1928 [Investigation of Losses in Oblique Angled Pipe Joints, (Translated by Sverdup & Parcel Inc.), 1953].
36. G. Belforte, On the behavior of air diodes. *Cranfield Fluidics Conf.*, 3rd Paper F6 (1968).
37. F. B. Daniels, Acoustical impedance of enclosures. *J. Acoust. Soc. Amer.* **19**, 569 (1947).
38. F. Biagi and R. K. Cook, Acoustic impedance of a right circular cylindrical enclosure. *J. Acoust. Soc. Amer.* **36**, 506 (1954).
39. H. Gerber, Acoustic properties of fluid-filled chambers at infrasonic frequencies in the absence of convection. *J. Acoust. Soc. Amer.* **36**, 1427 (1964).

40. H. Gerber, "Transient and Sinusoidal Thermal Diffusion, Convection, and Related Infrasonic Pressure Changes in Enclosed, Homogeneous Fluids." U.S. Naval Ord. Lab. Tech. Rep 62–94, 1963.
41. S. Katz and E. Hastie, "The Transition from Isothermal to Adiabatic Capacitance in Cylindrical Enclosures." Fluerics 31, HDL-TM-71-35, December 1971.
42. F. T. Brown, "Pneumatic Pulse Transmission with Bistable-Jet-Relay Reception and Amplification." Sc.D. Thesis, MIT, May 1962.
43. J. Dagan, "A Study of Pneumatic Capacitors." ME Thesis, Sir George Williams Univ., Montreal, Canada, March 1971.
44. J. Dagan and C. K. Kwok, "Study of Pneumatic Capacitors." ASME Paper 71-WA/Flcs-3, December 1971.
45. S. Katz and E. Hastie, "Pneumatic Passive Lead Network for Fluidic Systems." Paper 5-E1 Preprints of the 1971, JACC, August 1971.
46. R. J. Roark, "Formulas for Stress and Strain," 3rd ed., p. 233. McGraw-Hill, New York, 1954.
47. J. H. Howard, Metal bellows. *Mach. Des.* **26**, 137–148 (1954).
48. Fluid Power Handbook and Directory. Published by Hydraulic and Pneumatics Magazine, Sect. A, p. A 81. (1968: 1969).
49. F. M. Manion, "Tuned and Regenerative Flueric Amplifiers." U.S. Patent 3,513,867, 26 May 1970.
50. F. T. Brown and R. A. Humphrey, "Dynamics of a Proportional Amplifier: Part 2." ASME Paper No. 69-WA/Flcs-3; *Trans. Basic Eng.* **92**, 303–312 (1970).

Chapter 3

DISTRIBUTED FLUID PASSIVE COMPONENTS

3.1 INTRODUCTION TO TRANSMISSION LINES

The widespread use of electrical transmission lines in communication has led to a well-known transmission line theory. Textbooks, such as Ware and Reed [1] and King [2] present the basic theory in electrical terminology. The theory shows that a transmission line can be completely specified by two functions of a complex variable, (a) the propagation constant Γ and (b) the characteristic impedance Z_c, both of which depend on the series impedance and shunt admittance of the line. In the electrical case the complex functions are fundamental properties of the line. For fluid lines, however, the functions depend not only on the fluid properties but also on the flow regime, the signal level, the line shape, and the heat transfer. Thus the application of transmission line theory to fluidic lines requires the adaptation and modification of the equations of fluid motion and energy.

The earliest work on fluid transmission line models was done in the field of acoustic transmission. Helmholz [3] and Kirchhoff [3] provided a formulation for the high-frequency series impedance of the line assuming that there was no heat transfer. Rayleigh [3] showed that at low frequencies the series impedance of circular lines reduced to the Poiseuille dc resistance and that the flow was isothermal. Crandall [4] determined a series impedance that was valid for all frequencies, but he neglected heat transfer effects. Thus the propagation constant and characteristic impedance were accurate only for high frequencies. Daniels [5], on the other hand, did consider heat transfer and developed expressions for the shunt admittance of cylindrical enclosures for all frequencies. His concern, however, was not with transmission lines but with the acoustical impedance of enclosures and did not consider the series impedance. Iberall [6] provided the first complete solution which included the effects of both distributed friction and heat transfer. However, his original presentation is cumbersome and does not readily lend itself to use in practical applications. Iberall's discussion of the approximate solution obtained by Rohmann and Grogan [7] presents the complete solution in a more concise manner. Nichols [8] derived Iberall's solution and interpreted it in the form of an equivalent circuit. The work is performed in the frequency domain and includes both high- and low-frequency approximations. Karam and Franke [9] used Nichols' approach to obtain an improved high-frequency approximation. Brown [10, 11] provided a solution in the Laplace domain as well as a rather broad-band approximation. From this he obtained the impulse and step response for a matched line. Since the inversion process from the Laplace to the time domain is quite difficult, Kantola [12] introduced a Laplace domain series solution for the pressure and flow in a matched line. The advantage here is that the inversion to the time domain is easier and this facilitates the determination of step and impulse responses. Karam [13] presents a simplified approximation for the step response. The state of the art in circular fluid lines is summarized by Goodson and Leonard [14].

All the work cited above applies exclusively to circular lines. Since rectangular lines are more convenient for fabrication and packaging of fluidic circuits, there have been several recent papers on this subject. Schaedel [15] derived infinite series solutions for the rectangular line at any aspect ratio. Although the series converge rapidly they are rather unwieldy. Healey and Carlson [16] confirmed Schaedel's theory experimentally. They also show that the use of the hydraulic mean diameter of the rectangular section, in the existing circular model, often yields accurate results. As part of our discussion we include the results of another approach to the rectangular line. In this latter approach an approximation is made to reduce the problem from three dimensions to two dimensions. A comparison is then made between the three rectangular line models.

The transmission line portion of this chapter is divided into three main subject areas:

(1) Sections 3.2 and 3.3 deal with distributed transmission line theory and the lumped approximations of the theory. The material in this part is quite general and is not restricted to fluid lines. The results are given in terms of the propagation constant and the characteristic impedance.

(2) Sections 3.4 and 3.5 consider propagation models for circular and rectangular fluid lines. There are several different models based on specific assumptions. These models are compared with each other. In addition, approximations to the exact model are discussed and their range of validity is indicated.

(3) Sections 3.6–3.8 demonstrate the use of the fluid propagation models in the transmission line theory to predict the pressure and flow in fluid circuits.

3.2 Distributed Parameter Theory

Information on the exact models is presented in tabular form to enable the solution of a wide variety of problems. Special emphasis is given to branch circuits and lines with discontinuous change in cross section.

3.2 DISTRIBUTED PARAMETER THEORY

Figure 3.1 shows the circuit model for an infinitesimal length Δx of a uniform transmission line. A finite-length transmission line contains an infinite number of these infinitesimal elements in series. The signal variables E (for potential) and I (for flow) are general. With appropriate assumptions they apply to electrical, fluid, or any other continuous media. The signal variables E and I in

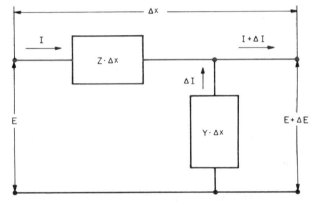

Fig. 3.1 Infinitesimal transmission line element.

this presentation may be either in the frequency ($j\omega$) or in the Laplace (s) domain. Each infinitesimal element in this model has a series impedance $Z \cdot \Delta x$, and a shunt admittance $Y \cdot \Delta x$ (where Z and Y are the impedance and admittance per unit length in terms of $j\omega$ or s).

The impedance and admittance cause the signal variables E and I at position x to change to $E + \Delta E$ and $I + \Delta I$ at position $x + \Delta x$. This is described as

$$\Delta E = -(Z \cdot \Delta x)I \quad (3.1\text{a})$$

$$\Delta I = -(Y \cdot \Delta x)E \quad (3.1\text{b})$$

In the limit as Δx approaches zero, Eq. (3.1) becomes

$$\partial E/\partial x = -ZI \quad (3.2\text{a})$$

$$\partial I/\partial x = -YE \quad (3.2\text{b})$$

and we assign a known boundary condition at the line entrance; i.e., $E = E_i$ and $I = I_i$ at $x = 0$. The combination of Eqs. (3.2a) and (3.2b) yields the wave equation

$$\partial^2 E/\partial x^2 = ZYE \tag{3.3}$$

which has the solution

$$E = A_1 \exp[(ZY)^{1/2}x] + A_2 \exp[-(ZY)^{1/2}x] \tag{3.4}$$

where A_1 and A_2 are constants of integration. From Eqs. (3.2a) and (3.4) the flow is

$$I = -A_1(Y/Z)^{1/2} \exp[(ZY)^{1/2}x] + A_2(Y/Z)^{1/2} \exp[-(ZY)^{1/2}x] \tag{3.5}$$

We may evaluate A_1 and A_2 from the boundary conditions at $x = 0$. Equations (3.4) and (3.5) may then be written as

$$E = \frac{[E_i + (Z/Y)^{1/2}I_i] \exp[-(ZY)^{1/2}x]}{2} + \frac{[E_i - (Z/Y)^{1/2}I_i] \exp[(ZY)^{1/2}x]}{2} \tag{3.6a}$$

$$I = \frac{[(Y/Z)^{1/2}E_i + I_i] \exp[-(ZY)^{1/2}x]}{2} + \frac{[-(Y/Z)^{1/2}E_i + I_i] \exp[(ZY)^{1/2}x]}{2} \tag{3.6b}$$

where E_i and I_i are the values of E and I at $x = 0$.

The exponential factor $(ZY)^{1/2}$ is called the propagation constant per unit length Γ, and is a complex variable in the frequency domain. Thus Γ is often given in the form

$$\Gamma \equiv (ZY)^{1/2} = \alpha + j\beta \tag{3.7}$$

where the real part α is called the attenuation constant because it causes a magnitude change in the signal variables. The imaginary part β is called the phase constant and represents the phase change per unit length in the signal passing along the line.

We are now in a position to interpret the meaning of Eq. (3.6) in terms of wave changes along the length of the line. The first term on the right-hand side of Eqs. (3.6a) and (3.6b) decays as x increases. This is the direct wave contribution to the potential and flow at any point on the line. The second term which decays with decreasing x provides the reflected wave contribution.

Another important property of the transmission line is the characteristic impedance Z_c. The characteristic impedance is defined as the input impedance of an infinite line. Since the potential at the load end of an infinite line is zero we may use Eq. (3.6a) to evaluate the characteristic impedance as

$$Z_c \equiv \lim_{x \to \infty} (E_i/I_i) = (Z/Y)^{1/2} \tag{3.8}$$

The characteristic impedance is a complex property of the transmission line. We shall shortly show that a finite line that is terminated by its characteristic impedance has the same input impedance as an infinite line.

3.2 Distributed Parameter Theory

If the definition of the characteristic impedance equation (3.8) is used in conjunction with Eq. (3.6), and the latter equation is rearranged, we may obtain four transmission matrix formulations for the transmission line. Two of the most useful matrices are

$$\begin{bmatrix} E \\ I \end{bmatrix} = \begin{bmatrix} \cosh \Gamma x & -Z_c \sinh \Gamma x \\ -Y_c \sinh \Gamma x & \cosh \Gamma x \end{bmatrix} \begin{bmatrix} E_i \\ I_i \end{bmatrix} \quad (3.9a)$$

$$\begin{bmatrix} E_i \\ I_i \end{bmatrix} = \begin{bmatrix} \cosh \Gamma x & Z_c \sinh \Gamma x \\ Y_c \sinh \Gamma x & \cosh \Gamma x \end{bmatrix} \begin{bmatrix} E \\ I \end{bmatrix} \quad (3.9b)$$

where Y_c is the characteristic admittance of the line $(1/Z_c)$.

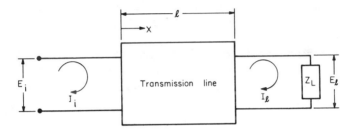

Fig. 3.2 Transmission line with load termination.

Figure 3.2 is a schematic drawing of a finite-length transmission line that has a terminating load impedance Z_L. The transmission matrix [Eq. (3.9b)] for a finite length l is

$$\begin{bmatrix} E_i \\ I_i \end{bmatrix} = \begin{bmatrix} \cosh \Gamma l & Z_c \sinh \Gamma l \\ Y_c \sinh \Gamma l & \cosh \Gamma l \end{bmatrix} \begin{bmatrix} E_l \\ I_l \end{bmatrix} \quad (3.10)$$

where E_l and I_l are the values of E and I at $x = l$.

Since the potential and flow at the end of the line are related to the load impedance $(E_l = Z_L I_l)$, the input impedance Z_{in} of a terminated line [from Eq. (3.10)] is

$$Z_{in} = \frac{E_i}{I_i} = Z_c \left(\frac{Z_L \cosh \Gamma l + Z_c \sinh \Gamma l}{Z_L \sinh \Gamma l + Z_c \cosh \Gamma l} \right) \quad (3.11)$$

Let us consider the input impedance when the line is terminated in three rather common ways: (a) short circuit $(Z_L = 0)$, (b) open circuit $(Z_L = \infty)$, and (c) matched $(Z_L = Z_c)$. For a short-circuited transmission line (open tube for fluid line) the potential at the end is always maintained at the reference or ground potential (zero in this case). Thus, the load impedance Z_L must also be zero and Eq. (3.11) gives the input impedance of the short-circuited line Z_{is} as

$$Z_{is} = Z_c \tanh \Gamma l \quad (3.12)$$

An open circuit transmission line (blocked tube for fluid line) has no flow at the load. Thus the load impedance Z_L must be infinite. In this case Eq. (3.11) reduces to the input impedance of the open circuit line Z_{io} where

$$Z_{io} = Z_c \coth \Gamma l \tag{3.13}$$

When the load impedance equals the characteristic impedance ($Z_L = Z_c$) the numerator and denominator of Eq. (3.11) have canceling factors and the input impedance equals the characteristic impedance ($Z_{in} = Z_c$). Thus we have now proved the previous statement that a finite line loaded with its characteristic impedance behaves as an infinite line. This is the condition of "matched impedance." The line has only a forward-moving direct wave. The amplitude of the reflected wave [second term on the right-hand side of Eq. (3.6)] is zero for the matched condition and does not contribute to the signal variables in the line.

On the other hand, the open and short circuit lines are "mismatched lines." In these cases there are reflections which affect the signal variables in the line. To demonstrate the effects of matching let us determine the potential along the line as a function of the load termination. The relation between the potential E at any point x and the entrance potential is $E/E_i = (E/E_l)(E_l/E_i)$. Both of the ratios on the right-hand side of this equation can be obtained from Eqs. (3.9) and (3.10) by considering the pertinent transmission line length to the end of the line. The result is

$$\frac{E}{E_i} = \frac{\sinh \Gamma l(1 - x/l) + (Z_L/Z_c) \cosh \Gamma l(1 - x/l)}{\sinh \Gamma l + (Z_L/Z_c) \cosh \Gamma l} \tag{3.14}$$

For the open circuit, short circuit, and matched load impedances Eq. (3.14) becomes

$$\left(\frac{E}{E_i}\right)_o = \frac{\cosh \Gamma l(1 - x/l)}{\cosh \Gamma l}, \quad \frac{Z_L}{Z_c} = \infty \quad \text{(open circuit)} \tag{3.15a}$$

$$\left(\frac{E}{E_i}\right)_s = \frac{\sinh \Gamma l(1 - x/l)}{\sinh \Gamma l}, \quad \frac{Z_L}{Z_c} = 0 \quad \text{(short circuit)} \tag{3.15b}$$

$$\left(\frac{E}{E_i}\right)_m = e^{-\Gamma l(x/l)}, \quad \frac{Z_L}{Z_c} = 1 \quad \text{(matched)} \tag{3.15c}$$

When the real and imaginary parts of the propagation factor (Γl) are known, the relation between potential and distance along the line is also known. Figure 3.3 shows the magnitude of the potential along the line when αl and βl are arbitrarily selected as 0.1 and 5.0 ($\Gamma l = 0.1 + j5.0$). For the mismatched cases there are standing wave patterns. The open circuit condition (blocked tube) has large resonant peaks. The magnitude of these peaks depends on the values of αl and βl. The matched line has no standing wave pattern. Instead we obtain a monotonically decreasing potential.

3.3 Lumped Parameter Approximations

The potential at the end of an open-circuited transmission line is [from Eq. (3.15a)]

$$\left(\frac{E_l}{E_i}\right)_o = \frac{1}{\cosh \Gamma l} \tag{3.16}$$

We shall often use this special case in the following sections to compare the distributed solution with lumped approximations and also to compare fluid line models for the transmission line.

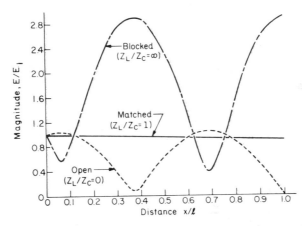

Fig. 3.3 The effect of loading on potential along line, where the propagation factor $\Gamma l = 0.10 + j5.00$.

3.3 LUMPED PARAMETER APPROXIMATIONS

Although the distributed parameter approach produces an accurate representation of transmission lines, the hyperbolic operators in the matrix forms [Eqs. (3.9a) and (3.9b)] are difficult to apply in practical cases. For this reason we often use a simplified transmission line model which consists of dividing the distributed impedances into a finite number of discrete impedances. The simplified model is called a lumped parameter model.

Figure 3.4 shows the schematic drawing of a general lumped parameter model. The line is divided into lumps, each of which contains a series impedance and a shunt admittance. The arrangement of the impedance and admittance in each lump may be split in the form of a TEE ($Z/2$, Y, $Z/2$) or a PI ($Y/2$, Z, $Y/2$). The number of lumps selected for a given problem depends on the required accuracy. As a general rule the accuracy improves as the number of lumps n_l increases. In some cases one lump is sufficient while in others we need considerably more. The magnitude of the impedances in each lump are not necessarily equal. That is, the impedances in lump 1 (Z_1 and Y_1) need not equal the impedances in lump 2 (Z_2 and Y_2) or in any other lump. Instead, the impedance values in each lump depend upon the method used in the approximation. Here we consider three distinct methods: (a) lumping by length, (b) Taylor series expansion, and (c) infinite products.

Each method is first described in a general way. Then we adapt each to apply to the calculation of the potential at the end of an open-circuited line. A comparison between the distributed theory [Eq. (3.16)] and the approximations provides valuable insight into the effects of lumping.

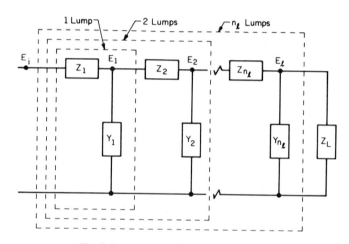

Fig. 3.4 Lumped transmission line model.

3.3.1 Lumping by Length

When lumping is based on length each lump is identical. Furthermore, the sum of the series impedances and the sum of the shunt admittances in the lumps are equal to Zl and Yl of the transmission line. Thus if there are two lumps, for example,

$$Z_1 = Z_2 = Z(l/2) \quad \text{and} \quad Y_1 = Y_2 = Y(l/2)$$

In general,

$$Z_1 + Z_2 + \cdots + Z_n = Zl \quad \text{and} \quad Y_1 + Y_2 + \cdots + Y_n = Yl$$

where $Z_1, Z_2 = Z(l/n_l)$ and $Y_1, Y_2 \cdots = Y(l/n_l)$ and n_l is the number of lumps. With these provisions the matrix representation of the first lump of the model shown in Fig. 3.4 is

$$\begin{bmatrix} E_i \\ I_i \end{bmatrix} = \begin{bmatrix} 1 + ZY(l/n_l)^2 & Z(l/n_l) \\ Y(l/n_l) & 1 \end{bmatrix} \begin{bmatrix} E_1 \\ I_1 \end{bmatrix} \qquad (3.17)$$

3.3 Lumped Parameter Approximations

We may write a similar matrix between the first and second lump, and then the second and third lump. As a result the lumped model selected may be represented after n_l lumps by

$$\begin{bmatrix} E_i \\ I_i \end{bmatrix} = \begin{bmatrix} 1 + ZY(l/n_l)^2 & Z(l/n_l) \\ Y(l/n_l) & 1 \end{bmatrix}^{n_l} \begin{bmatrix} E_l \\ I_l \end{bmatrix} \quad (3.18)$$

Now the two important properties of the transmission line are the propagation constant per unit length Γ and the characteristic impedance Z_c. The propagation constant as defined in Eq. (3.7) is retained for all the approximate methods. On the other hand the characteristic impedance previously calculated for the distributed line [Eq. (3.8)] changes for the lumped line. To calculate the characteristic impedance of the lumped line (Z_{cn_l}) we recall that it is the impedance for which the load and input impedances are equal. Manipulation of Eq. (3.17) for the condition that $E_i/I_i = E_1/I_1 = Z_{cn_l}$ then yields

$$Z_{cn_l} = \frac{Zl}{2n_l} \left\{ 1 + \left[1 + \frac{4n_l^2}{(\Gamma l)^2} \right]^{1/2} \right\} \quad (3.19)$$

Thus the ratio of the lumped to distributed characteristic impedance is

$$Z_{cn_l}/Z_c = (1/2n_l)\{\Gamma l + [(\Gamma l)^2 + 4n_l^2]^{1/2}\} \quad (3.20)$$

Since the propagation factor is a complex quantity the lumped and distributed characteristic impedances differ in both magnitude and phase. However, these differences become small when the number of lumps is large.

To observe the effect of lumping we compare the lumped approximation with the distributed result for an open-circuited line ($Z_L = \infty$). Equation (3.16) gives the potential at the end of the distributed line for this condition. The corresponding results for the lumped approximation may be obtained from Eq. (3.18). For lines represented by one, two, or three lumps the relations are

$$\left(\frac{E_l}{E_i}\right)_{1l} = [1 + (\Gamma l)^2]^{-1} \quad \text{(1 lump)} \quad (3.21a)$$

$$\left(\frac{E_l}{E_i}\right)_{2l} = [1 + (\tfrac{3}{4})(\Gamma l)^2 + (\tfrac{1}{16})(\Gamma l)^4]^{-1} \quad \text{(2 lumps)} \quad (3.21b)$$

$$\left(\frac{E_l}{E_i}\right)_{3l} = [1 + (\tfrac{2}{3})(\Gamma l)^2 + (\tfrac{5}{81})(\Gamma l)^4 + (\tfrac{1}{729})(\Gamma l)^6]^{-1} \quad \text{(3 lumps)} \quad (3.21c)$$

Figure 3.5 shows the comparison between the distributed line [Eq. (3.16)] and the line approximated by one and two lumps [Eqs. (3.21a) and (3.21b)]. To make the comparison we arbitrarily select the real part of the propagation factor αl as 0.1 while the imaginary part, the propagation phase βl, is allowed to vary. In the distributed case the first resonant peak occurs at $\beta l = \pi/2$, and thereafter other peaks occur at intervals of 2π. The one lump approximation has only one resonant peak at $\beta l = 1$. There are two resonant peaks for the two-lump approximation, the first at $\beta l = 1.25$ and the second at $\beta l = 3.25$. Clearly then the approximation improves as the number of lumps increases. However, a good approximation with a small number of lumps is desirable. The lumping by length method requires a large number of lumps to obtain an accurate representation of the line. For example, in the open-circuited case, the approximation is within 10% only when βl is restricted to less than 0.4 for one lump and less than 0.6 for two lumps. Since in a practical case of high-frequency transmission βl may equal 100, the lumped approximation is often not appropriate.

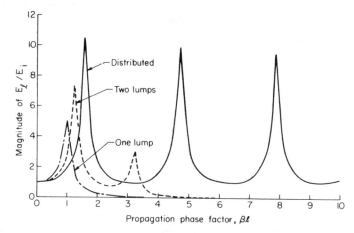

Fig. 3.5 The effect of lumping by length ($\alpha l = 0.1$, $Z_L = \infty$).

3.3.2 Taylor Series Expansion

The Taylor series expansion method is conceptually different from the lumping by length method. In length lumping we began with a circuit representation and determined the equivalent matrix. In the Taylor series method, on the other hand, the starting point is the distributed matrix given in Eq. (3.10). The method consists simply of approximating each matrix element by a Taylor series. As a result the Taylor series approximation to the transmission line is

$$\begin{bmatrix} E_i \\ I_i \end{bmatrix} = \begin{bmatrix} 1 + (\Gamma l)^2/2! + (\Gamma l)^4/4! + \cdots & Z_c[\Gamma l + (\Gamma l)^3/3! + \cdots] \\ Y_c[\Gamma l + (\Gamma l)^3/3! + \cdots] & 1 + (\Gamma l)^2/2! + (\Gamma l)^4/4! + \cdots \end{bmatrix} \begin{bmatrix} E_l \\ I_l \end{bmatrix} \quad (3.22)$$

Now the lumped circuit which represents the Taylor series approximation [Eq. (3.22)] is not easy to find. Consider, for example, the one-lump TEE section shown in Fig. 3.6. The matrix for this section is

$$\begin{bmatrix} E_i \\ I_i \end{bmatrix}_{T_1} + \begin{bmatrix} 1 + ZYl^2/2 & Zl + Z^2Yl^3/4 \\ Yl & 1 + ZYl^2/2 \end{bmatrix} \begin{bmatrix} E_l \\ I_l \end{bmatrix} \quad (3.23)$$

3.3 Lumped Parameter Approximations

The characteristic impedance for a single-lump TEE section is

$$Z_{cT} = [Z/Y + (Zl)^2/4]^{1/2} \tag{3.24}$$

If we apply Eq. (3.24), Eq. (3.23) may be written in terms of the propagation factor as

$$\begin{bmatrix} E_i \\ I_i \end{bmatrix} = \begin{bmatrix} 1 + (\Gamma l)^2/2 & Z_{cT}[(\Gamma l)^2 + (\Gamma l)^4/4] \\ Y_{cT}[(\Gamma l)^2 + (\Gamma l)^4/4] & 1 + (\Gamma l)^2/2 \end{bmatrix} \begin{bmatrix} E_l \\ I_l \end{bmatrix} \tag{3.25}$$

Equation (3.25) is close to a two-term approximation of the Taylor series matrix presented in Eq. (3.22). However, a comparison of Eqs. (3.25) and (3.22) shows that the one-lump TEE section does not represent the Taylor series matrix

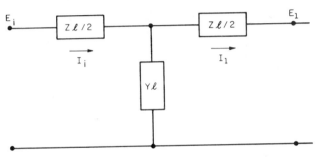

Fig. 3.6 Single-lump TEE section.

exactly. The matching of the Taylor expansion with a specific lumped circuit becomes increasingly difficult as more terms are included. Oldenburger and Goodson [17] have shown, in fact, that Taylor expansions to fifth degree or higher do not preserve the poles of the function and produce unstable systems. Obviously then, in this case, there is no passive circuit that is equivalent to the Taylor series expansion.

Although the precise lumped circuit is not known, we may still use the Taylor approximation with terms less than fifth degree. For the open-circuited line the two- and three-term Taylor expansions yield

$$\left(\frac{E_l}{E_i}\right)_{T_2} = [1 + (\Gamma l)^2/2]^{-1} \quad \text{(2-term)} \tag{3.26a}$$

$$\left(\frac{E_l}{E_i}\right)_{T_3} = [1 + (\Gamma l)^2/2 + (\Gamma l)^4/24]^{-1} \quad \text{(3-term)} \tag{3.26b}$$

Figure 3.7 compares the distributed line results [Eq. (3.16)] with the two- and three-term series approximations given in Eqs. (3.26a) and (3.26b). Once again αl is arbitrarily selected at 0.1. As in the length lumped method the resonant peaks of the series method do not necessarily occur at the same values of βl where there are peaks in the distributed case. However, the approximation of the magnitude of E_i/E_o is somewhat better in the series method. The agreement between distributed and series results is within 10% for βl less than 0.6 for two terms and βl less than 1.5 for three terms.

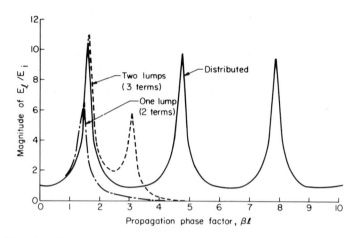

Fig. 3.7 The effect of lumping by Taylor series approximation ($\alpha l = 0.1$, $Z_L = \infty$).

3.3.3 Infinite Products

In the infinite-product method of Oldenburger and Goodson [17] the lumping is in effect, done by resonant peaks rather than by length. However, as in the series method the equivalent lumped circuit is difficult to obtain.

To illustrate the method let us suppose that we want to express $\cosh \Gamma l$ in terms of a series of infinite products. The roots of $\cosh \Gamma l$ occur when

$$\cosh \Gamma l = \cosh(\alpha l + j\beta l) = 0 \tag{3.27}$$

Equation (3.27) may be expanded to

$$\cosh \alpha l \cos \beta l + j \sinh \alpha l \sin \beta l = 0 \tag{3.28}$$

Two simultaneous equations in αl and βl are obtained from Eq. (3.28) by setting the real and imaginary parts equal to zero. This yields the roots of $\cosh \Gamma l$ as

$$\alpha l = 0, \qquad \beta l = \pm \frac{2m+1}{2} \pi \tag{3.29}$$

where m is an integer that takes on the values 0, 1, 2, ..., etc. Thus there are complex conjugate roots for each value of m and we may write $\cosh \Gamma l$ as

$$\cosh \Gamma l = \prod_{m=0}^{\infty} \left(1 + \frac{\Gamma l}{j((2m+1)/2)\pi}\right)\left(1 - \frac{\Gamma l}{j((2m+1)/2)\pi}\right) \tag{3.30a}$$

3.4 Propagation Models for Circular Fluid Lines

or

$$\cosh \Gamma l = \prod_{m=1}^{\infty} \left(1 + \frac{4(\Gamma l)^2}{\pi^2 (2m-1)^2}\right) \quad (3.30b)$$

In a similar way the infinite-product representation for $\sinh \Gamma l$ may be derived as

$$\sinh \Gamma l = \Gamma l \prod_{m=1}^{\infty} \left(1 + \frac{(\Gamma l)^2}{m^2 \pi^2}\right) \quad (3.31)$$

For the open-circuited line the one- and two-product approximations are therefore

$$\left(\frac{E_l}{E_i}\right)_{P_1} = [1 + 4(\Gamma l)^2/\pi^2]^{-1} \quad (3.32a)$$

$$\left(\frac{E_l}{E_i}\right)_{P_2} = [(1 + 4(\Gamma l)^2/\pi^2)(1 + 4(\Gamma l)^2/9\pi^2)]^{-1} \quad (3.32b)$$

Equations (3.32a), (3.32b), and (3.16) are shown on Fig. 3.8 for $\alpha l = 0.1$. We observe that the resonant peaks of the infinite-product approximation occur at the same values of βl as the peaks for the distributed line. In addition, the magnitude of E_l/E_i for the approximation is closer to the distributed result.

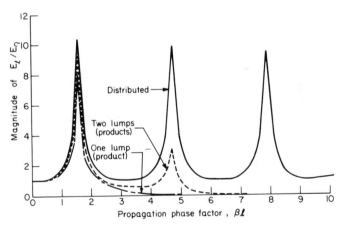

Fig. 3.8 The effect of lumping by infinite products ($\alpha l = 0.1$, $Z_L = \infty$).

3.4 PROPAGATION MODELS FOR CIRCULAR FLUID LINES

To determine propagation models for fluid lines, the fluid equations must be placed in the form given in Eq. (3.2). This requires the following five basic assumptions:

(a) *Laminar Flow.* The Reynolds number in the circular line is less than 2000.
(b) *Axial Symmetry.* There are no tangential velocity components.

(c) *Small Amplitude Signals.* The Mach number is low (negligible dc flow) and the density variations are small compared to the average density. The signals are, therefore, of acoustic type.

(d) *Small Viscous Forces from Compressibility.* The precise assumption is that $\omega v/a^2 \ll 1$ (where ω is the circular frequency, v is the kinematic viscosity, and a is the speed of sound). This assumption removes the axial gradients of velocity and temperature from the momentum and energy equations.

(e) *Radius Small Compared to Wavelength.* This is usually expressed as $\omega r_w/a \ll 1$ (where r_w is the radius of the circular line). The assumption implies that the pressure is uniform over the cross section and that the radial velocity is, therefore, zero [12].

The pertinent fluid equations are the axial momentum equation, the energy equation, the continuity equation, and the state equations. With the five basic assumptions the fluid equations are considerably simplified. The axial momentum equation becomes

$$\frac{\partial u}{\partial t} = -\frac{1}{\rho}\frac{\partial p}{\partial x} + v\left[\frac{1}{r}\frac{\partial}{\partial r}\left(r\frac{\partial u}{\partial r}\right)\right] \quad (3.33)$$

where u is the axial velocity, t the time, ρ the average density, p is static pressure, and r is the radial dimension. Since the mean velocity is very small, the static pressure is approximately equal to the total pressure. The energy equation reduces to

$$\rho c_p \frac{\partial T}{\partial t} = \frac{\partial p}{\partial t} + k\left[\frac{1}{r}\frac{\partial}{\partial r}\left(r\frac{\partial T}{\partial r}\right)\right] \quad (3.34)$$

where c_p is the specific heat at constant pressure, T is the absolute temperature, and k is the thermal conductivity of the fluid. The continuity equation is

$$\partial \rho/\partial t + \rho\, \partial u/\partial x = 0 \quad (3.35)$$

and the state equations are

$$p = \rho R_g T \quad \text{(gas)} \quad (3.36a)$$

$$dp = \beta_M\, d\rho/\rho \quad \text{(liquid)} \quad (3.36b)$$

where β_M is the bulk modulus of the liquid and R_g is the gas constant. Equations (3.33)–(3.36) are the basic fluid equations. From these equations we want to determine the propagation constant and the characteristic impedance of fluid transmission lines.

Goodson and Leonard [14] discuss three distinct line propagation models: (a) the lossless line, (b) the line with average friction and no heat transfer, and (c) the line with distributed friction and heat transfer. The first two models provide additional assumptions that reduce the complexity of the basic fluid equations In the following subsections we consider the three line models separately.

3.4 Propagation Models for Circular Fluid Lines

3.4.1 Lossless Line Model

The additional assumptions for the lossless line model are that the fluid is inviscid ($v = 0$) and that there is no heat transferred between the fluid and walls of the line. With these assumptions Eqs. (3.33)–(3.36) reduce to

$$\partial p/\partial x = -\rho\, \partial u/\partial t \qquad (3.37a)$$

$$\partial u/\partial x = -(1/\rho a^2)\, \partial p/\partial t \qquad (3.37b)$$

where a is the speed of sound in free space and is equal to $(\gamma p/\rho)^{1/2}$ for gases and $(\beta_M/\rho)^{1/2}$ for liquids (γ is the ratio of specific heats of the gas). For this model the velocity is uniform over the cross section and the volume flow $q = Au$ (where A is the cross-sectional area). Equation (3.37) may be written in terms of pressure and volume flow as

$$\partial p/\partial x = -(\rho/A)\, \partial q/\partial t \qquad (3.38a)$$

$$\partial q/\partial x = -(A/\rho a^2)\, \partial p/\partial t \qquad (3.38b)$$

The Laplace transform of Eq. (3.38) then yields

$$\partial P/\partial x = -(\rho s/A)Q \qquad (3.39a)$$

$$\partial Q/\partial x = -(As/\rho a^2)P \qquad (3.39b)$$

where P and Q are the transforms of p and q.

Equation (3.39) is now in exactly the same form as the general transmission line equation (3.2) with P representing the potential and Q representing the flow. By comparison with Eq. (3.2) we observe that the fluid impedance and admittance per unit length for the lossless line are

$$Z = \rho s/A \qquad (3.40a)$$

$$Y = As/\rho a^2 \qquad (3.40b)$$

In general the series impedance and shunt admittance per unit length will have the forms $Z = R + j\omega L$ and $Y = G + j\omega C$, where R and L are the series resistance and inertance per unit length and G and C are the shunt conductance and capacitance per unit length. If Eq. (3.40) is expressed in the frequency domain ($s = j\omega$) we recognize that $R = G = 0$, $L_a = \rho/A$, and $C_a = A/\rho a^2$ for the lossless line. The subscripts 'a' refer to the adiabatic values of inductance and capacitance that result from the neglect of heat transfer in the model. Now reference to equation (3.7) will show that the propagation constant for the lossless line (Γ_a) is:

$$\Gamma_a = s/a = j\omega/a \qquad (3.41a)$$

or

$$\Gamma_a = (L_a C_a)^{1/2} s = j\omega(L_a C_a)^{1/2} \qquad (3.41b)$$

Since there is no attenuation in the lossless case, $\alpha = 0$, and the propagation constant $\Gamma_a = j\beta$. From Eq. (3.41a) we observe that the phase constant per unit length β is equivalent to ω/a.

The characteristic impedance of the lossless line (Z_{ca}) from Eqs. (3.8) and (3.40) is

$$Z_{ca} = (\rho a/A) = (L_a/C_a)^{1/2} \tag{3.42}$$

Thus, oddly enough, the characteristic impedance of the lossless line is purely resistive [no reactive part in Eq. (3.42)].

The speed of wave propagation c_0 is defined as

$$c_0 \equiv \omega/\beta \tag{3.43}$$

and for the lossless line the speed of wave propagation is equal to the speed of sound in free space ($c_0 = a$).

3.4.2 Average Friction Model (Circular)

In the average friction model we again assume that no heat transfer takes place. This model, however, uses the average viscous effect [18] so that Eq. (3.33) becomes

$$\frac{\partial p}{\partial x} = -\rho \frac{\partial u}{\partial t} + \frac{2\pi\mu}{A} \int_0^{r_w} \frac{\partial}{\partial r}\left(r \frac{\partial u}{\partial r}\right) dr \tag{3.44}$$

where μ is the absolute viscosity. The integration of the friction term then reduces Eq. (3.44) to

$$\frac{\partial p}{\partial x} = -\rho \frac{\partial u}{\partial t} + \frac{2\pi\mu r_w}{A} \left(\frac{\partial u}{\partial r}\right)_{r=r_w} \tag{3.45}$$

To evaluate the velocity gradient at the wall of the fluid line ($r = r_w$) we assume the fully developed Hagen–Poiseuille velocity distribution $u(r) = 2\bar{u}(1 - r^2/r_w^2)$, where $\bar{u}$ is average velocity. The gradient at the wall $(\partial u/\partial r)_{r=r_w} = -4\bar{u}$, and Eq. (3.45) therefore becomes

$$\partial p/\partial x = -(\rho \, \partial u/\partial t + 8\pi\mu\bar{u}/A) \tag{3.46}$$

Now if we replace the average velocity with the volume flow ($q = \bar{u}A$) and take the Laplace transform, Eq. (3.46) changes to

$$\partial P/\partial x = -(\rho s/A + 8\pi\mu/A^2)Q \tag{3.47}$$

Equation (3.47) is the series impedance equation [Eq. (3.2a)] for the average friction model. The shunt impedance equation [Eq. (3.21)] is the same as for the lossless line [Eq. (3.39b)]. Thus the series impedance and shunt admittance for the average friction line model are

$$Z = \rho s/A + 8\pi\mu/A^2 \tag{3.48a}$$

$$Y = As/\rho a^2 \tag{3.48b}$$

3.4 Propagation Models for Circular Fluid Lines

Equation (3.48) for the average friction model only differs from Eq. (3.40) for the lossless model in the resistive portion of the series impedance, $R = 8\pi\mu/A^2$. The values of G, L, and C remain the same. Thus the propagation factor and the characteristic impedance for the average friction case are

$$\Gamma = (L_a C_a)^{1/2} s (R/L_a s + 1)^{1/2} \tag{3.49a}$$

$$Z_c = (L_a/C_a)^{1/2} (R/L_a s + 1)^{1/2} \tag{3.49b}$$

In the frequency domain the real and imaginary parts of the propagation constant are

$$\alpha = (\omega/a)[\tfrac{1}{2}(R^2/\omega^2 L_a^2 + 1)^{1/2} - \tfrac{1}{2}]^{1/2} \tag{3.50a}$$

$$\beta = (\omega/a)[\tfrac{1}{2}(R^2/\omega^2 L_a^2 + 1)^{1/2} + \tfrac{1}{2}]^{1/2} \tag{3.50b}$$

From Eq. (3.50b), then, the speed of wave propagation (the phase velocity) for this case is

$$c_0 = \frac{a}{\{\tfrac{1}{2}[(R^2/\omega^2 L_a^2) + 1]^{1/2} + \tfrac{1}{2}\}^{1/2}} \tag{3.51}$$

Equations (3.49)–(3.51) reduce to the lossless case when $R = 0$.

3.4.3 Distributed Friction Model (Circular)

The present distributed friction model developed slowly over a period going back to 1926. Crandall [4] considered the distributed friction in the momentum equation and obtained a series impedance for the transmission line. This impedance consisted of a frequency-dependent resistance and inertance. Daniels [5] derived the shunt admittance of cylindrical and other enclosures from the continuity and energy equations. This work included the effects of heat transfer to the isothermal walls of the enclosure. From these separate parts a complete model of the fluid transmission line was introduced by Iberall [6] and Nichols [8] in the frequency domain and Brown [10] in the Laplace domain. In this model the walls of the transmission line are assumed to be isothermal and there is heat transfer between the fluid and walls. Thus Eqs. (3.33) and (3.34) may be written directly in terms of the Laplace transform as

$$\frac{\partial^2 U}{\partial r^2} + \frac{1}{r}\frac{\partial U}{\partial r} = \frac{s}{\nu} U + \frac{1}{\mu}\frac{\partial P}{\partial x} \tag{3.52a}$$

$$\frac{\partial^2 \theta}{\partial r^2} + \frac{1}{r}\frac{\partial \theta}{\partial r} = \frac{N_p s}{\nu}\left(\theta - \frac{P}{\rho c_p}\right) \tag{3.52b}$$

with the boundary conditions that

$$r = 0, \quad \partial U/\partial r = 0, \quad \partial \theta/\partial r = 0$$

$$r = r_w, \quad U = 0, \quad \theta = 0$$

where U and θ are the Laplace transforms of u and T, N_p represents the Prandtl number, and θ refers to T above ambient. The solutions of Eq. (3.52) are

$$U = -\frac{1}{\rho s}\frac{\partial P}{\partial x}\left[1 - \frac{J_0[j(s/v)^{1/2}r]}{J_0[j(s/v)^{1/2}r_w]}\right] \quad (3.53a)$$

$$\theta = \frac{P}{\rho c_p}\left[1 - \frac{J_0[j(N_p s/v)^{1/2}r]}{J_0[j(N_p s/v)^{1/2}r_w]}\right] \quad (3.53b)$$

and J_0 is a Bessel function of the first kind of zeroth order. To arrange Eq. (3.53a) in the series impedance form of Eq. (3.2a), the volume flow Q is expressed as

$$Q = 2\pi \int_0^{r_w} Ur\, dr \quad (3.54)$$

The substitution of Eq. (3.53a) into Eq. (3.54) leads to the series impedance

$$Z = L_a s \Big/ \left[1 - \frac{2J_1(j(s/v)^{1/2}r_w)}{j(s/v)^{1/2}r_w J_0(j(s/v)^{1/2}r_w)}\right] \quad (3.55)$$

where J_1 is a Bessel function of the first kind of first order. Equation (3.55) in the frequency domain ($s = j\omega$) is the equation first presented by Crandall [4]. If Eq. (3.55) is separated into real and imaginary parts, we find that the resistance and inertance are both frequency dependent. At low frequencies the series impedance is predominantly resistive and the resistance is essentially the same as the value developed for the average friction model [see Eq. (3.48a)]. As

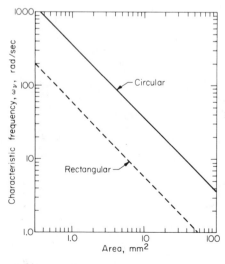

Fig. 3.9 Characteristic frequencies of circular and rectangular transmission lines.

3.4 Propagation Models for Circular Fluid Lines

frequency increases the resistance increases, although the ratio of resistance to inertive reactance decreases. The high-frequency inertance is equal to the adiabatic inertance of the average friction model. The inertance increases by a factor of one third for low frequencies. However, this increase is relatively unimportant because the series impedance is primarily resistive at these frequencies. Since the impedance depends on the frequency range where the transmission line will be used, Nichols [8] introduced a reference frequency, which is called the viscous characteristic frequency (ω_v) and is defined for circular lines as

$$\omega_v \equiv R/L_a = 8v\pi/A \tag{3.56}$$

The characteristic frequency is the frequency at which the resistive and inertive reactance portions of the series impedance are equal. Thus when the frequency is much smaller than the characteristic frequency the line is resistive and when it is much higher the line is inertive. Figure 3.9 shows the characteristic frequency as a function of cross-sectional area. For example a transmission line 1.0 mm in diameter ($A = 0.785 \times 10^{-6}$ m^2) has a characteristic frequency of 476 rad/sec (76 Hz). From Eqs. (3.55) and (3.56) the series impedance in the frequency domain is

$$\frac{Z}{\omega L_a} = j \bigg/ \left\{ 1 - \frac{2J_1[j^{3/2}(8\omega/\omega_v)^{1/2}]}{j^{3/2}(8\omega/\omega_v)^{1/2}J_0[j^{3/2}(8\omega/\omega_v)^{1/2}]} \right\} \tag{3.57}$$

and depends on the frequency ratio ω/ω_v. The Stokes number N_{sk} [19] equals $32\omega/\omega_v$.

To find the shunt admittance of the fluid transmission line when the medium is a gas, Eqs. (3.35) and (3.36a) are rewritten in terms of Laplace transforms and combined to yield

$$\partial Q/\partial x = -C_a sP[1 - R_g \bar{\theta}/c_p] \tag{3.58}$$

where $\bar{\theta}$ represents the transform of the average temperature in the line and is obtained by averaging Eq. (3.53b) over the line area. The substitution of this average into Eq. (5.38) results in an equation of the form described in the shunt admittance equation (3.2b) and therefore leads to

$$Y = C_a s \left[1 + \frac{2(\gamma - 1)J_1(j(N_p s/v)^{1/2} r_w)}{j(N_p s/v)^{1/2} r_w J_0(j(N_p s/v)^{1/2} r_w)} \right] \tag{3.59}$$

or in the frequency domain

$$\frac{Y}{\omega C_a} = j \left[1 + \frac{2(\gamma - 1)J_1(j^{3/2}(8N_p \omega/\omega_v)^{1/2})}{j^{3/2}(8N_p \omega/\omega_v)^{1/2} J_0(j^{3/2}(8N_p \omega/\omega_v)^{1/2})} \right] \tag{3.60}$$

For a liquid $\gamma \cong 1$ and the admittance is the same as in the other transmission line models.

The shunt conductance and capacitive susceptance [real and imaginary parts of Eq. (3.59)] are also frequency dependent. At frequencies well below the characteristic frequency the capacitance is isothermal. That is, the capacitance is 1.4 times the adiabatic value of the lossless and the average friction models. The shunt conductance in this region, however, is close to zero as obtained from the other models. When the frequency is considerably above the characteristic frequency the capacitance reduces to the adiabatic value and the conductance continually increases. However, the ratio of the conductance to the capacitive susceptance has a maximum near the characteristic frequency.

From Eqs. (3.57) and (3.60) the propagation factor and the characteristic impedance for the distributed friction case are

$$\frac{\Gamma}{\Gamma_a} = \left\{\left[1 + \frac{2(\gamma-1)J_1(j^{3/2}\sqrt{N_p}\,F)}{j^{3/2}\sqrt{N_p}\,FJ_0(j^{3/2}\sqrt{N_p}\,F)}\right] \bigg/ \left[1 - \frac{2J_1(j^{3/2}F)}{j^{3/2}FJ_0(j^{3/2}F)}\right]\right\}^{1/2} \quad (3.61\text{a})$$

$$\frac{Z_c}{Z_{ca}} = \left\{\left[1 - \frac{2J_1(j^{3/2}F)}{j^{3/2}FJ_0(j^{3/2}F)}\right]\left[1 + \frac{2(\gamma-1)J_1(j^{3/2}\sqrt{N_p}\,F)}{j^{3/2}\sqrt{N_p}\,FJ_0(j^{3/2}\sqrt{N_p}\,F)}\right]\right\}^{-1/2} \quad (3.61\text{b})$$

where $F = [8\omega/\omega_v]^{1/2}$.

Although these equations are derived for zero dc flow, the experimental results of Krishnaiyer and Lechner [20] show that they hold even with some dc flow.

3.4.4 Approximation to the Distributed Model

The distributed model equations [Eqs. (3.61a) and (3.61b)] are difficult to apply directly. For this reason there have been numerous attempts to find reasonable approximations. When $(\omega/\omega_v) \ll 1$ (low frequency), Eqs. (3.57), (3.60), and (3.61) (for air) are approximately equal to

$$\frac{Z}{\omega L_a} = \frac{1}{\omega/\omega_v} + j\frac{4}{3} \quad (3.62\text{a})$$

$$\frac{Y}{\omega C_a} = j1.4 \quad (3.62\text{b})$$

$$\left(\frac{\Gamma}{\Gamma_a}\right) = \left(j\frac{1.4}{\omega/\omega_v} - \frac{4}{3}\right)^{1/2} \quad (3.62\text{c})$$

$$\left(\frac{Z_c}{Z_{ca}}\right) = \left(\frac{1}{j1.4\omega/\omega_v} + 0.95\right)^{1/2} \quad (3.62\text{d})$$

In this regime the transmission line has negligible inertance. It consists of resistance and capacitive reactance.

3.4 Propagation Models for Circular Fluid Lines

When $\omega/\omega_v \geq 1$, Brown [10, 11], uses a good approximation of the Bessel functions to obtain the Taylor series expansions

$$\frac{Z}{\omega L_a} = j\left(1 + \frac{2}{j^{1/2}F} + \frac{3}{jF^2} + \frac{\frac{15}{4}}{j^{3/2}F^3}\right) \quad (3.63a)$$

$$\frac{Y}{\omega C_a} = j\left[1 + (\gamma - 1)\left(\frac{2}{(jN_p)^{1/2}F} - \frac{1}{jN_pF^2} - \frac{\frac{1}{4}}{(jN_p)^{3/2}F^3}\right)\right] \quad (3.63b)$$

$$\frac{\Gamma}{\Gamma_a} = 1 + \frac{A_1}{j^{1/2}F} + \frac{A_2}{jF^2} + \frac{A_3}{j^{3/2}F^3} \quad (3.63c)$$

$$\frac{Z_c}{Z_{ca}} = \frac{1}{1 + A_4/j^{1/2}F + A_5/jF^2 + A_6/j^{3/2}F^3} \quad (3.63d)$$

where $F = (8\omega/\omega_v)^{1/2}$ and $A_1, \ldots, A_6$ are functions of the specific heat ratio and the Prandtl number N_p. The constants $A_1, \ldots, A_6$ are given in Table 3.1 for air and liquid.

Table 3.1
Constants in Brown's Approximation

	Air	Liquid
A_1	1.478	1.000
A_2	1.078	1.000
A_3	1.058	0.875
A_4	-0.520	-1.000
A_5	-0.880	0.000
A_6	0.640	0.130

Brown's approximation is quite good over a broad range (whenever the frequency is greater than the characteristic frequency).

Karam and Franke [9] suggest a high-frequency approximation ($\omega/\omega_v \gg 1$) based on Nichols' [8] circuit model. This approximation leads to

$$\frac{Z}{\omega L_a} = 1 + \frac{1}{j2(\omega/\omega_v)^{1/2}} \quad (3.64a)$$

$$\frac{Y}{\omega C_a} = 1 + \frac{\gamma - 1}{j2(N_p\omega/\omega_v)^{1/2}} \quad (3.64b)$$

$$\frac{\Gamma}{\Gamma_a} = \left[\left(1 + \frac{1}{2j(\omega/\omega_v)^{1/2}}\right)\left(1 + \frac{(\gamma - 1)}{2j(N_p\omega/\omega_v)^{1/2}}\right)\right]^{1/2} \quad (3.64c)$$

$$\frac{Z_c}{Z_{ca}} = \left[\frac{1 + 1/2j(\omega/\omega_v)^{1/2}}{1 + (\gamma - 1)/2j(N_p\omega/\omega_v)^{1/2}}\right]^{1/2} \quad (3.64d)$$

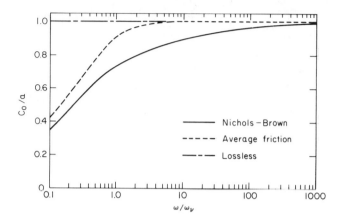

Fig. 3.10. Speed of wave propagation in circular lines (air).

3.4.5 Comparison of Models

The previous section presented the lossless, average friction, and distributed friction (Nichols–Brown) models for the propagation factor (Γ) and characteristic impedance (Z_c) of circular transmission lines. In this section we compare Γ and Z_c for the various models.

Figures 3.10 and 3.11 shows how each model affects the propagation factor. In Fig. 3.10 the normalized speed of wave propagation (inverse of $\beta a/\omega$) is plotted against the normalized frequency ω/ω_v. For the lossless model, the propagation speed always equals the speed of sound in free space. However, friction causes a significant reduction in propagation speed at low frequencies. The average friction model is fairly accurate at very low frequencies but tends to

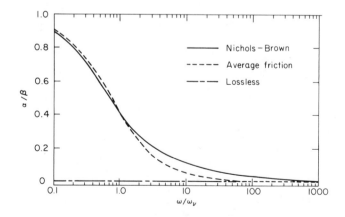

Fig. 3.11 Attenuation–phase ratio in circular lines (air).

3.4 Propagation Models for Circular Fluid Lines

underestimate the friction as the frequency increases. For example the average friction model propagation speed reaches $0.98a$ at $\omega/\omega_v = 3.0$, whereas the Nichols–Brown model does not reach $0.98a$ until $\omega/\omega_v = 300$. Since this latter model is in good agreement with experimental results, lines transmitting frequency information must operate at high frequencies to prevent dispersion, which is the distortion of the transmitted signal that results from the dependence of propagation speed on frequency at the low and intermediate frequencies.

Figure 3.11 shows the ratio of the propagation attenuation factor to the propagation phase factor as a function of frequency. Quite often this ratio is given in decibels [8], which is the multiplication of the ordinate by 54.575. At low frequencies the ratio approaches unity (54.575 dB) and at high frequency the ratio approaches zero. The Nichols–Brown model has more attenuation at the higher frequencies ($\omega/\omega_v > 2$) than the other models. The average friction model almost coincides with the lossless model when $\omega/\omega_v = 50$. Actually this is very misleading—because both the attenuation factor α and the phase factor β increase monotonically with frequency. (See Appendix A.) Since the phase factor has a larger rate of increase, the ratio decreases.

Figure 3.12 and 3.13 show the magnitude and phase of the normalized characteristic admittance (Y_c/Y_{ca}) as a function of frequency. For the lossless case the line is a pure conductance [see Eq. (3.42)]. The models with friction have both conductance and susceptance. The susceptance increases as the frequency decreases and is always more capacitive than inertive.

Although the average friction and the Nichols–Brown model have relatively small percentage differences in propagation factor (Figs. 3.10 and 3.11), the hyperbolic operators [Eq. (3.10)] accentuate these differences. For example, Fig. 3.14 shows the magnitude response of a blocked pneumatic line with length l, fixed at a/ω_v. The first two resonant peaks of the lossless line occur at $\omega/\omega_v = \pi/2$ and $3\pi/2$, and their magnitudes increase without bound. The

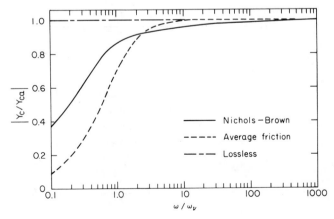

Fig. 3.12 Magnitude of characteristic admittance for circular lines (air).

86   3 Distributed Fluid Passive Components

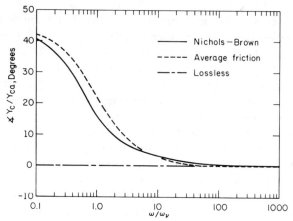

Fig. 3.13 Phase of characteristic admittance for circular lines (air).

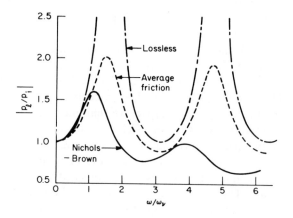

Fig. 3.14 Magnitude response of blocked circular lines (air).

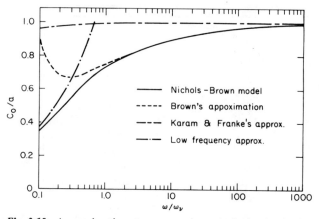

Fig. 3.15 Approximations to propagation velocity in circular lines (air).

average friction model produces finite-amplitude peaks at slightly lower frequencies. The exact (Nichols–Brown) model on the other hand has considerably more attenuation and the peaks occur at significantly lower frequencies.

Figure 3.15 shows the propagation speed of the exact model and the various approximations. Brown's approximation (dashed line) is very close for all frequencies above the characteristic frequency. Karam and Franke's approximation is valid when $\omega/\omega_v = 1000$ and the low-frequency approximation holds best when $\omega/\omega_v < 0.1$.

3.5 PROPAGATION MODELS FOR RECTANGULAR FLUID LINES

Figure 3.16 is a schematic drawing of a rectangular transmission line. The line has a depth h, a width b, and a length l. The x coordinate represents the axial direction and the transverse coordinates are y and z. To derive propagation models for rectangular lines we assume (a) laminar flow, (b) small-amplitude signals, (c) small viscous forces from compressibility, and (d) the transverse dimensions (b and h) are small compared to a/ω. Thus the pressure is uniform over each cross section and the transverse velocities are negligible.

For this case the axial momentum and energy equations reduce to

$$\frac{\partial u}{\partial t} = -\frac{1}{\rho}\frac{\partial p}{\partial x} + v\left(\frac{\partial^2 u}{\partial y^2} + \frac{\partial^2 u}{\partial z^2}\right) \tag{3.65a}$$

$$\frac{\partial T}{\partial t} = \frac{1}{\rho c_p}\frac{\partial p}{\partial t} + \frac{k}{\rho c_p}\left(\frac{\partial^2 T}{\partial y^2} + \frac{\partial^2 T}{\partial z^2}\right) \tag{3.65b}$$

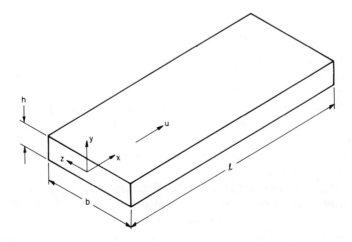

Fig. 3.16 Schematic of rectangular transmission line.

The continuity and state equations are the same as in the circular line [Eqs. (3.35) and (3.36)].

The lossless model (inviscid and without heat transfer) produces the same results previously obtained for the circular line [Eqs. (3.38)–(3.43)]. Therefore we consider only the average and distributed friction models for the rectangular line in this section.

3.5.1 Average Friction Model (Rectangular)

The average friction models for rectangular and circular lines are very similar. They differ only in the resistive portion of the series impedance. The resistance of the circular line ($R = 8\pi\mu/A^2$) is given in Eq. (3.48a) (second term). For the rectangular line with fully developed flow, Cornish [21] derives the resistance as

$$R = \frac{12\mu\sigma/A^2}{1 - (192/\pi^5)(1/\sigma)[\tanh \pi\sigma/2 + (1/3^5)\tanh 3\pi\sigma/2 + \cdots]} \quad (3.66)$$

where σ is the aspect ratio of the line ($\sigma = b/h$) and A, the cross-sectional area, equals bh. Although the series given in Eq. (3.66) is rapidly convergent, the expression is somewhat unwieldy. We may obtain an approximate form for the resistance of the rectangular line by relating the viscous terms in Eq. (3.65a) and reducing the equation by one dimension. In particular the assumption that $z = \sigma y$ yields a steady state form of Eq. (3.65a) as

$$\partial p/\partial x = \mu[1 + 1/\sigma^2][\partial^2 u/\partial y^2] \quad (3.67)$$

Equation (3.67) has the same form as the flow between parallel plates with the correction factor $(1 + 1/\sigma^2)$. The solution of Eq. (3.67) for the velocity profile and the subsequent integration of that profile leads to the approximate resistance R_a, where

$$R_a = (12\mu\sigma/A^2)[1 + 1/\sigma^2] \quad (3.68)$$

Figure 3.17 shows the exact and approximate rectangular line resistances that are given in Eqs. (3.66) and (3.68). In each case the resistance of a constant-area cross section increases as the aspect ratio increases. The exact solution always produces higher resistance than the approximate solution. This is due to the neglect of the corners in the approximate form. The largest percentage error between the solutions is 15.3% and occurs when the aspect ratio is unity. As the aspect ratio increases the corner effects become less significant and the two solutions approach each other.

The propagation factor and the characteristic impedance for the rectangular line with average friction can also be obtained from Eq. (3.49). In this case the rectangular line resistance [Eq. (3.66) or (3.68)] is used instead of the circular line resistance. The inductance and capacitance have the same form for both circular and rectangular sections.

3.5 Propagation Models for Rectangular Fluid Lines

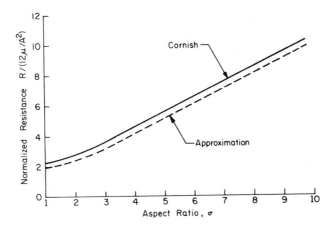

Fig. 3.17 Fully developed laminar resistance of rectangular line.

3.5.2 Distributed Friction Model (Rectangular)

Schaedel [15] obtained an exact series solution of Eqs. (3.65a) and (3.65b). This solution gives the series impedance and shunt admittance of the rectangular line with distributed friction and heat transfer to the walls as

$$\frac{Z}{\omega L_a} = \left[2\sigma^2(\omega/\omega_v) \sum_{i=1}^{\infty} \frac{(\psi_1/\sigma) - \tanh(\psi_1/\sigma)}{\alpha_i^2 \psi_1^3} \right]^{-1} \quad (3.69a)$$

$$\frac{Y}{\omega C_a} = j\left[\gamma - 2j\sigma^2 \frac{N_p \omega}{\omega_v}(\gamma - 1) \sum_{i=1}^{\infty} \frac{(\psi_2/\sigma) - \tanh(\psi_2/\sigma)}{\alpha_i^2 \psi_2^3} \right] \quad (3.69b)$$

where

$$\psi_1 = (\alpha_i^2 + j\sigma\omega/\omega_v)^{1/2}$$
$$\psi_2 = (\alpha_i^2 + j\sigma N_p \omega/\omega_v)^{1/2}$$
$$\alpha_i = [(2i-1)/2]\pi$$
$$\omega_v = 4v/A \quad \text{(for rectangular sections)}$$

The propagation factor and characteristic impedance for this case can be found by substitution of Eqs. (3.69a) and (3.69b) into Eqs. (3.7) and (3.8). The resulting formulation is extremely cumbersome. For this reason we consider two alternate approaches for the rectangular line with distributed friction and heat transfer.

The first approach is due to Healey and Carlson [16]. They consider the rectangular line as a circular line with the same hydraulic mean diameter. The hydraulic radius r_h for a rectangle is

$$r_h = bh/(b + h) \quad (3.70)$$

If r_h is used in place of r_w, the propagation factor and characteristic impedance of the rectangular line is approximately equal to Eqs. (3.61a) and (3.61b) for the circular line with F replaced by F' and

$$F' = \left[\frac{4\omega}{\omega_v}\frac{\sigma}{(\sigma+1)^2}\right]^{1/2} \tag{3.71}$$

where $\omega_v = 4v/A$ as in Schaedel's solution. Equation (3.71) can also be applied to the approximations for the circular line expressed in Eqs. (3.62)–(3.64).

Another approach to the rectangular line is to use the approximation $z = \sigma y$ (as in the average friction case) and thereby eliminate one dimension from the momentum and energy equations [Eqs. (3.65a) and (3.65b)]. This approach yields

$$\frac{Z}{\omega L_a} = \left[1 - \frac{\tanh(j^{1/2}F'')}{j^{1/2}F''}\right]^{-1} \tag{3.72a}$$

$$\frac{Y}{\omega C_a} = 1 + \frac{(\gamma-1)\tanh(j^{1/2}N_p^{1/2}F'')}{j^{1/2}N^{1/2}F''} \tag{3.72b}$$

$$\frac{\Gamma}{\Gamma_a} = \left[\frac{1 + [(\gamma-1)\tanh(j^{1/2}N_p^{1/2}F'')]/j^{1/2}N_p^{1/2}F''}{1 - \tanh(j^{1/2}F'')/j^{1/2}F''}\right]^{1/2} \tag{3.72c}$$

$$\frac{Z_c}{Z_{ca}} = \left[\left(1 + \frac{(\gamma-1)\tanh(j^{1/2}F''\sqrt{N_p})}{j^{1/2}F''\sqrt{N_p}}\right)\left(1 - \frac{\tanh(j^{1/2}F'')}{j^{1/2}F''}\right)\right]^{-1/2} \tag{3.72d}$$

where

$$F'' = \left[\frac{\omega}{\omega_v}\left(\frac{1}{\sigma + 1/\sigma}\right)\right]^{1/2}, \quad \omega_v = \frac{4v}{A}$$

When the aspect ratio becomes very large the modified two-dimensional result [Eqs. (3.72a) and (3.72b)] and Schaedel's exact solution reduce to the same equations.

Figure 3.18 shows the speed of wave propagation for three rectangular distributed friction models. When the aspect ratio is unity, Schaedels' model and the Nichols–Healey–Carlson (hydraulic diameter) model fall along the same line. The modified two-dimensional model gives a larger propagation speed at all frequencies. This is a consequence of the underestimation of the line resistance. When $\sigma = 10$ the modified two-dimensional approach and Schaedel's model are only slightly different, whereas the hydraulic diameter model predicts higher speed until $\omega/\omega_v = 20$ and thereafter has lower speed.

The ratio of the attenuation factor to the phase factor is shown in Fig. 3.19. When the aspect ratio is unity the modified two-dimensional model and the exact model coincide until ω/ω_v reaches about 3. The hydraulic diameter model predicts a higher attenuation ratio at the lower frequencies. At an aspect ratio of 10 the situation is reversed. Now the hydraulic diameter model shows a

3.5 Propagation Models for Rectangular Fluid Lines

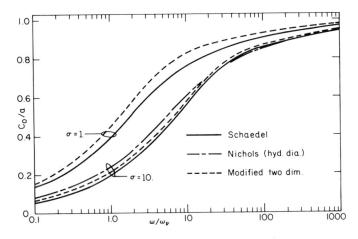

Fig. 3.18 Speed of wave propagation in rectangular lines (air).

smaller attenuation ratio. In this case the modified two-dimensional model coincides with Schaedel's model.

Figures 3.20 and 3.21 show the magnitude and phase of the characteristic admittance of the rectangular transmission line. When the aspect ratio is unity the modified two-dimensional approach has too high an admittance. Again this reflects the neglect of the corners in the formulation. When $\sigma = 10$, Schaedel's solution and the two-dimensional approximation are almost identical. Here the hydraulic diameter model shows considerable divergence from the exact theory.

Discrepancies between the exact theory and the approximations do not appear to be large on Figs. 3.18–3.21. However, the hyperbolic functions in the transfer matrix [Eq. (3.10)] may amplify the differences as previously mentioned

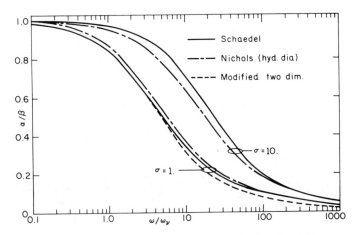

Fig. 3.19 Attenuation–phase ratio in rectangular lines (air).

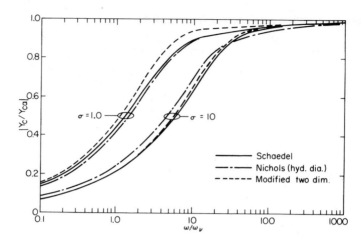

Fig. 3.20 Magnitude of characteristic admittance for rectangular lines (air).

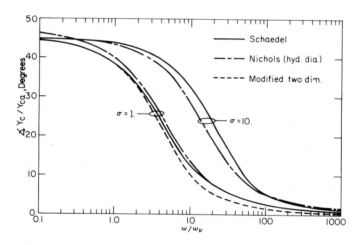

Fig. 3.21 Phase of characteristic admittance for rectangular lines (air).

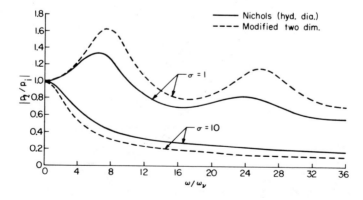

Fig. 3.22 Magnitude response of blocked rectangular lines (air).

3.6 Lumped Modeling of Fluidic Circuits

in regard to the circular models (Fig. 3.14). Figure 3.22 shows the frequency response of a blocked rectangular line of length, $l = 2\pi a/\omega_v$, At $\sigma = 1$ the hydraulic diameter model has more attenuation then the two-dimensional model and the reverse is true when $\sigma = 10$. Schaedel's exact solution lies between the two when $\sigma = 1$ and coincides with the two-dimensional approach when $\sigma = 10$.

3.6 LUMPED MODELING OF FLUIDIC CIRCUITS

We have previously discussed the lumping of transmission lines in general (Section 3.3). However, at that time we had not yet developed the particular fluid line models presented in Sections 3.4 and 3.5. Let us now, therefore, determine appropriate lumped models for some practical fluidic circuits.

The appropriate lumped model for a fluid transmission line depends on the phase shift βl between two points at the extreme ends of the line. When we want to model the line with a single lump, βl must be much less than unity. For the average friction case discussed in Section 3.4.2, Eq. (3.50b) gives βl as

$$\beta l = \frac{\omega l}{a} \left[\frac{1}{2} \left(\frac{R^2}{\omega^2 L_a^2} + 1 \right)^{1/2} + \frac{1}{2} \right]^{1/2} \quad (3.73a)$$

or, in terms of the characteristic frequency,

$$\beta l = \frac{\omega l}{a} \left[\frac{1}{2} \left(\frac{1}{[\omega/\omega_v]^2} + 1 \right)^{1/2} + \frac{1}{2} \right]^{1/2} \quad (3.73b)$$

When ω/ω_v is large (lossless line) Eq. (3.73) reduces to

$$\beta l = \omega l/a \quad (3.74)$$

Now if we apply the criterion for a single lump ($\beta l \ll 1$) we obtain the condition that

$$l \ll a/\omega = \lambda/2\pi \quad (3.75)$$

where λ is the wavelength. Thus, to be able to model with a single lump means that the line length must be short compared to the wavelength. If, for example, the desired operating frequency is 330 rad/sec and the fluid is air ($a = 330$ m/sec), the line length must be much less than one meter for a single lump. In this case any line of length 10 cm or less could be represented by a single-lump model.

Suppose that instead of being large ω/ω_v is small (i.e., $\omega/\omega_v = 0.1$). Then the criterion for a single lump is

$$l \ll 0.425 a/\omega \quad (3.76)$$

and the permissible length for a single lump is restricted to 42.5% of the lossless case.

3.6.1 Lumped Input Impedance of a Transmission Line Terminated by an Orifice

In fluidics we often need an impedance model for the loads or vents associated with an amplifier (an impedance to ground). A common geometric configuration for this impedance is a transmission line terminated by an orifice. Figure 3.23 shows a line of length l and diameter d_1 that is terminated by an orifice of

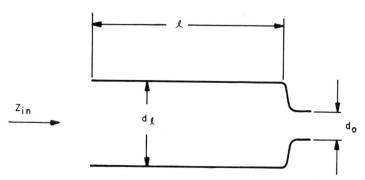

Fig. 3.23 Line terminated with orifice.

diameter d_o. We want to derive a single-lump equivalent impedance model for this component.

We assume at the outset that the transmission line is essentially lossless. This means, in effect, that the transmission line resistance is so much smaller than the orifice resistance that it can be neglected. Now if the pressure–flow relation across the orifice is of the form $p = K_o q^2$, then the linearized orifice resistance R_o is

$$R_o = 2K_o q \tag{3.77a}$$

where, for a well-rounded orifice ($K_L = 0$), Eq. (2.13) gives approximately

$$K_o = (\rho/2A_o^2)[1 - (A_o/A_1)^2] \tag{3.77b}$$

From equation (3.11) the input impedance of a lossless line terminated by the resistance R_o is

$$Z_{in} = \frac{R_o \cosh j\omega l(LC)^{1/2} + (L/C)^{1/2} \sinh j\omega l(LC)^{1/2}}{R_o (C/L)^{1/2} \sinh j\omega l(LC)^{1/2} + \cosh j\omega l(LC)^{1/2}} \tag{3.78}$$

where L and C are the values per unit length. Equation (3.78) may be written in terms of the circular functions as

$$Z_{in} = \frac{R_o[1 + \tan^2 \omega l(LC)^{1/2}] + j[(L/C)^{1/2} - R_o^2(C/L)] \tan \omega l(LC)^{1/2}}{1 + (R_o^2 C/L) \tan^2 \omega l(LC)^{1/2}} \tag{3.79}$$

To obtain a single-lumped equivalent model for the impedance represented by Eq. (3.79) we express the input impedance as

$$Z_{in} = R_{eq} + j\omega l L_{eq} + 1/j\omega l C_{eq} \tag{3.80}$$

3.6 Lumped Modeling of Fluidic Circuits

where R_{eq}, lL_{eq}, and lC_{eq} are single-lumped values. Now if we approximate, we find for small values of $\omega l(LC)^{1/2}$ that

$$\tan \omega l(LC)^{1/2} = \omega l(LC)^{1/2} + (\omega l)^3 (LC)^{3/2}/3 \tag{3.81}$$

and from Eqs. (3.79) and (3.80) after considerable algebraic maneuvers we obtain

$$\frac{R_{eq}}{(1/\omega Cl)} = \frac{(\omega R_o Cl)}{1 + (\omega R_o Cl)^2} \tag{3.82a}$$

$$\frac{C}{C_{eq}} = \frac{(\omega R_o Cl)^2}{1 + (\omega R_o Cl)^2} \tag{3.82b}$$

$$\frac{L_{eq}}{L} = \frac{1 + \tfrac{2}{3}\{(\omega R_o Cl)^4/[1 + (\omega R_o Cl)^2]\} - (\omega R_o Cl)^2/3}{1 + (\omega R_o Cl)^2} \tag{3.82c}$$

Figure 3.24 shows the equivalent lumped component values given in Eq. (3.82). The equivalent resistance of the line terminated by an orifice exhibits a maximum at $\omega R_o Cl = 1$. The equivalent resistance R_{eq} is zero when $R_o = 0$ (the open line has an orifice of line diameter). Strangely enough the equivalent resistance (for $\omega \neq 0$) is also zero for a blocked line. In the open-line case the reactance is completely inertive and $L_{eq} = L$. In the blocked-line case $C_{eq} = C$ and $L_{eq} = L/3$. The intermediate values are given in Eq. (3.82) and in Fig. 3.24. For example, if the operating frequency and the line and orifice dimensions produce $\omega R_o Cl = 2$, the lumped input impedance of the line would be

$$Z_{in} = (0.4/\omega lC) + j(0.36\omega lL) + [1/j(1.25\omega lC)] \tag{3.83}$$

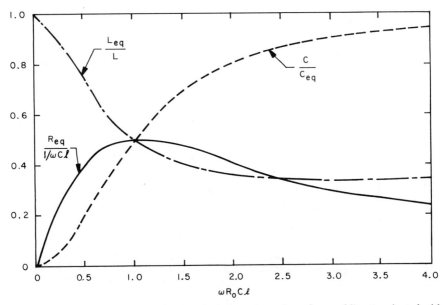

Fig. 3.24 Equivalent R, L, and C for single-lump input impedance of line terminated with orifice.

3.6.2 Errors Due to Lumping with TEE Sections

In many fluidic applications we need a lumped circuit representation for a transmission line rather than an equivalent input impedance. We have observed previously in Section 3.3.2 that a TEE-section lump (Fig. 3.6) provides a good representation of the Taylor series expansion for hyperbolic functions. Now we

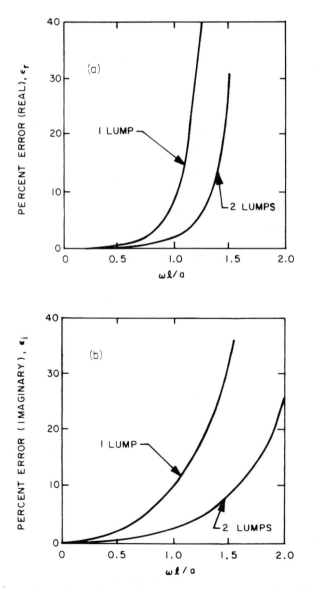

Fig. 3.25 TEE-section lumping errors: (a) real, (b) imaginary.

will show in a more quantitative way, the errors involved due to lumping with TEE sections.

For a lossless line terminated with the linear resistance R_L, Eq. (3.14) may be rewritten in terms of the frequency domain pressures as

$$(P_i/P_1) = \cos \omega l(LC)^{1/2} + j(1/R_L)(L/C)^{1/2} \sin \omega l(LC)^{1/2} \quad (3.84)$$

From Eq. (3.74) and the series expansion for the circular functions, Eq. (3.84) becomes

$$\frac{P_i}{P_1} = \left[1 - \frac{(\beta l)^2}{2!} + \frac{(\beta l)^4}{4!} - \cdots\right] + \frac{j}{R_L}\left[\omega L l\left(1 - \frac{(\beta l)^2}{3!} + \frac{(\beta l)^4}{5!} - \cdots\right)\right] \quad (3.85)$$

We may now model the lossless line with TEE-section lumps (Fig. 3.6). Each $Z/2$ is equivalent to $j\omega Ll/2n_1$ and Y is equivalent to $j\omega Cl/n_1$ (where n_1 is the number of identical lumps). The relation between P_i and P_1 for a one-lump TEE section is from Eq. (3.23):

$$\left[\frac{P_i}{P_1}\right]_T = \left[1 - \frac{(\beta l)^2}{2}\right] + \frac{j}{R_L}\left[\omega L l\left(1 - \frac{(\beta l)^2}{4}\right)\right] \quad (3.86)$$

Now we define the lumping error ε_1 as

$$\varepsilon_1 = \frac{(P_i/P_1) - (P_i/P_1)_T}{(P_i/P_1)_T} \quad (3.87)$$

It is convenient to compare the real and imaginary terms in Eqs. (3.85) and (3.86) separately. Thus, there are two lumping errors, one for the real part of P_i/P_1 and one for the imaginary part. Figure 3.25 shows the lumping errors of TEE-section models with one and two lumps. We may observe that a TEE-section model with one lump is in error by only 10% at $\omega l/a$ equal to unity. If two lumps are used the error at $\omega l/a = 1$ is reduced to 2.5%. Thus, the criterion given in Eq. (3.75) is very conservative. In most fluidic applications a single-lump approximation will be entirely adequate. The following sections consider cases where we work directly with the distributed equations.

3.7 CIRCUIT THEORY

Fluidic integrated or breadboard circuits contain interconnecting lines that often have branch points and sudden changes in cross section. To predict the signal variables throughout these circuits we need some circuit rules in addition to the propagation models and the transmission line theory. Franke *et al.* [22]

demonstrate that Kirchhoff's circuit laws are valid for small-amplitude fluid signals with small throughflows. They present a concise formulation for the signal variables at any point in a fluidic line network. We cover the same material here in somewhat greater detail.

As a starting point let us consider a simple branch circuit (Fig. 3.26). The input signal is applied at the entrance to line section 1. Thus, we may imagine that line section 1 enters the branch point and line sections 2 and 3 leave the

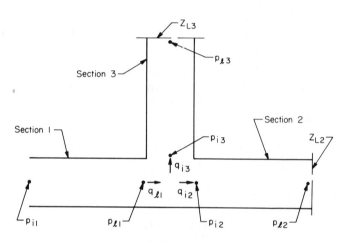

Fig. 3.26 Branch circuit.

branch point. The entrance variables to each line section are designated with the subscript i and the exit variables with the subscript l. We assume that at the branch point the pressure is uniform and the volume flow is conserved. Thus

$$P_{l1} = P_{i2} = P_{i3} \tag{3.88a}$$

$$Q_{l1} = Q_{i2} + Q_{i3} \tag{3.88b}$$

As a consequence the load impedance on line section 1 is

$$Z_{L1} = Z_{i2} Z_{i3}/(Z_{i2} + Z_{i3}) \tag{3.89}$$

where Z_{i2} and Z_{i3} are the input impedance of transmission lines with various arbitrary terminations [Eq. (3.11)]. From this information the pressure transfer function is available for line sections 2 and 3 through the application of Eq. (3.14) between the entrance and exit of these line sections. The result is

$$\frac{P_l}{P_i} = \frac{1}{(Z_c/Z_L) \sinh \Gamma l + \cosh \Gamma l} \tag{3.90}$$

The transfer functions for combined line sections is the product of the individual transfer functions. For example, suppose we wish to determine the pressure transfer function between sections l_2 or l_3 and section i_1. Then because of the assumption of uniform pressure at the branch point [Eq. (3.88a)] we may write

$$\frac{P_{12}}{P_{i1}} = \left(\frac{P_{12}}{P_{i2}}\right)\left(\frac{P_{11}}{P_{i1}}\right) \tag{3.91a}$$

or

$$\frac{P_{13}}{P_{i1}} = \left(\frac{P_{13}}{P_{i3}}\right)\left(\frac{P_{11}}{P_{i1}}\right) \tag{3.91b}$$

In a similar way if the individual line sections chosen in Fig. 3.26 also had branch points the overall transfer functions could be obtained from

$$\frac{P_{12}}{P_{i1}} = \prod_{n=1}^{\infty} \frac{P_{ln}}{P_{in}} \tag{3.92a}$$

or

$$\frac{P_{13}}{P_{i1}} = \prod_{n=1}^{\infty} \frac{P_{ln}}{P_{in}} \tag{3.92b}$$

where the products are only taken across the line segments that lie in the path between the terminals. The procedure is to start at the terminations of the circuits (i.e., cross sections l2 and l3) and work backward toward the input. Of course, we must remember to calculate the correct load impedance [parallel combination of input impedances, Eq. (3.89)] for line segments on the input side of a branch point. If the pressure transfer functions are desired for points that are not at terminations or junctions, we may apply the same procedure as indicated in Eqs. (3.91) and (3.92), except that some of the individual line transfer functions will come from Eq. (3.14) instead of Eq. (3.90). The examples given in the following sections will clarify the procedure.

3.8 FREQUENCY RESPONSE OF FLUID LINE CIRCUITS

The propagation models for circular lines [Eq. (3.61)] and rectangular lines [Eq. (3.69)] are not readily applied to practical applications. In addition, the distributed transmission matrix [Eq. (3.10)] increases the complexity. Healey and Carlson [16] present graphs of the transmission matrix elements for circular transmission lines. However, the accuracy of the graphical approach is quite limited, and especially so if interpolation is required. As a consequence we present the propagation models in tabular form. To conserve space the quantities are given in a normalized form. The normalization of the propagation factor is

$$\Gamma l = [\omega_v l/a]\phi \tag{3.93}$$

where

$$\phi = [\omega/\omega_v][(Z/\omega L_a)(Y/\omega C_a)]^{1/2}$$

Appendixes A and B give the normalized characteristic impedance (Z_c/Z_{ca}), the normalized characteristic admittance (Y_c/Y_{ca}), and the normalized propagation factor (ϕ) for circular and rectangular lines as functions of ω/ω_v respectively. The following examples will demonstrate the use of these tables.

Example 1. Change of Section A 5.0-mm diameter circular line 2.0-m long (Fig. 3.27) is connected to another circular line of 1.0-mm diameter and 1.0-m

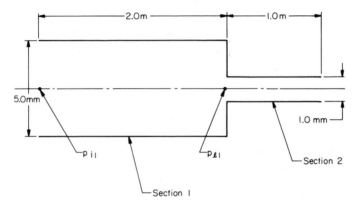

Fig. 3.27 Sudden contraction, Example 1.

length. A low-amplitude sinusoidal frequency of 1240 rad/sec is applied at the entrance to the larger line (section 1). The fluid is air. Find:

(a) the magnitude of the pressure ratio between the two ends of the 5.0-mm diameter line $(|P_1/P_i|)$ when the downstream end of the smaller line is open

(b) the magnitude of this ratio when the downstream end of the smaller line is closed;

(c) the lowest frequency at which the magnitude P_1/P_i has a resonant peak. (Note: For this configuration the peaks are only slightly dependent on whether the small tube is blocked or open.)

Solution

From Eq. (3.90) we may write that

$$P_{i1} = P_{11}\left[\cosh \Gamma l_1 + \frac{Z_{ca1}(Z_{c1}/Z_{ca1})\sinh \Gamma l_1}{Z_{L1}}\right]$$

3.8 Frequency Response of Fluid Line Circuits

(a) *Open Tube*

$$Z_{L1} = Z_{i2} = Z_{ca2}(Z_{c2}/Z_{ca2}) \tanh \Gamma l_2 \quad \text{[Eq. (3.12)]}$$

$v(\text{air}) = 14.85 \times 10^{-6} \text{ m}^2/\text{sec}$

$a(\text{air}) = 330 \text{ m/sec}$

$\omega_{v1} = 8\pi v/A_1 = 19.0 \text{ rad/sec}, \qquad \omega/\omega_{v1} = 1240/19.0 = 65.3$

$\omega_{v2} = 8\pi v/A_2 = 475.0 \text{ rad/sec}, \qquad \omega/\omega_{v2} = 1240/475 = 2.6$

$\dfrac{\omega_{v1} l_1}{a} = \dfrac{(19.)(2.)}{330} = 0.115, \qquad \dfrac{\omega_{v2} l_2}{a} = \dfrac{(475.)(1.)}{330} = 1.439$

From Appendix A at

$$\omega/\omega_{v1} = 65.3 \rightarrow Z_{c1}/Z_{ca1} = 1.016 - j0.018$$
$$\phi_1 = 3.124 + j68.244$$

at

$$\omega/\omega_{v2} = 2.6 \rightarrow Z_{c2}/Z_{ca2} = 1.075 - j0.143$$
$$\phi_2 = 0.755 + j3.179$$

Thus

$$\Gamma l_1 = \alpha_1 l_1 + j\beta_1 l_1 = (\omega_{v1} l_1/a)\phi_1$$
$$= 0.360 + j7.862$$

$$\Gamma l_2 = \alpha_2 l_2 + j\beta_2 l_2 = (\omega_{v2} l_2/a)\phi_2$$
$$= 1.087 + j4.576$$

$$\frac{P_{i1}}{P_{11}} = \cosh \Gamma l_1 + \left(\frac{Z_{ca1}}{Z_{ca2}}\right) \frac{(Z_{c1}/Z_{ca1}) \sinh \Gamma l_1}{(Z_{c2}/Z_{ca2}) \tanh \Gamma l_2}$$

$\cosh \Gamma l_1 = \cosh \alpha_1 l_1 \cos \beta_1 l_1 + j \sinh \alpha_1 l_1 \sin \beta_1 l_1$
$\qquad = 0.000 + j(0.368)$

$\sinh \Gamma l_1 = \sinh \alpha_1 l_1 \cos \beta_1 l_1 + j \cosh \alpha_1 l_1 \sin \beta_1 l_1$
$\qquad = 0 + j1.066$

$\tanh \Gamma l_2 = (\tanh \alpha_2 l_2 + j \tan \beta_2 l_2)/(1 + j \tanh \alpha_2 l_2 \tan \beta_2 l_2)$
$\qquad = 1.244 + j0.077$

$\dfrac{Z_{c1}/Z_{ca1}}{Z_{c2}/Z_{ca2}} = \dfrac{1.016 - j0.018}{1.075 - j0.143} = 0.931 + j0.107$

$Z_{ca1}/Z_{ca2} = A_2/A_1 = 1/25$

$$\frac{P_{i1}}{P_{11}} = \left[j(0.368) + \frac{(0.932 + j0.107)(j1.066)}{25(1.244 + j0.077)} \right]$$

$P_{i1}/P_{11} = -0.002 + j0.400$

$|P_{11}/P_{i1}| = 2.500$

(b) *Closed Tube*

$$Z_{L1} = Z_{i2} = Z_{ca2}(Z_{c2}/Z_{ca2}) \coth \Gamma l_2 \quad \text{[Eq. (3.13)]}$$

$$\coth \Gamma l_2 = 1/\tanh \Gamma l_2$$
$$= 0.801 - j0.050$$

$$\frac{P_{i1}}{P_{11}} = \left[j0.368 + \frac{(0.932 + j0.107)(j1.066)}{25(0.801 - j0.050)} \right]$$

$$P_{i1}/P_{11} = -0.009 + j0.417$$
$$|P_{11}/P_{i1}| = 2.396$$

Thus the contribution of the small tube does not change the result significantly.

(c) *First Resonant Peak.* The first peak will occur when $\beta_1 l_1 = 1.57$. Therefore

$$\text{Im } \phi_1 = 1.57/(\omega_{v1} l_1/a) = 13.635$$

From Appendix A,

$$\text{Im } \phi_1 = 13.635 \rightarrow \omega/\omega_{v1} = 12.347$$
$$\omega = (12.347)(19.) = 234.6 \quad \text{rad/sec}$$

Example 2. Branch Circuit The fluidic branch circuit shown in Fig. 3.28 consists of line segments 1, 2, and 3 with rectangular cross sections. The segments are all 1.0 mm deep. The in-line segments (1 and 2) are 3.0 mm wide and are 0.15 and 0.05 m long, respectively. The branch segment is 2.0 mm wide and

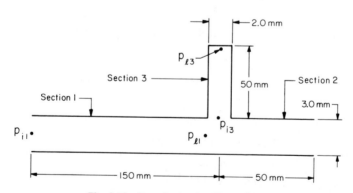

Fig. 3.28 Branch circuit—Example 2.

0.05 m long. A small-amplitude sinusoidal fluid air signal at a frequency of 800 rad/sec is applied at the entrance to line segment 1. If the branch outlet section is blocked find the magnitude of the pressure ratio (P_{13}/P_{i1}) when (a) line segment 2 is open to atmosphere; (b) line segment 2 is loaded with its characteristic impedance.

3.8 Frequency Response of Fluid Line Circuits

Solution

The input impedance of line segment 2 is

$$Z_{i2}/Z_{ca2} = (Z_{c2}/Z_{ca2})(W)$$

where

$$W = \frac{(Z_{12}/Z_{ca2}) \cosh \Gamma l_2 + (Z_{c2}/Z_{ca2}) \sinh \Gamma l_2}{(Z_{12}/Z_{ca2}) \sinh \Gamma l_2 + (Z_{c2}/Z_{ca2}) \cosh \Gamma l_2}$$

The input impedance of line segment 3 is

$$Z_{i3}/Z_{ca3} = (Z_{c3}/Z_{ca3}) \coth \Gamma l_3$$

The load impedance on line segment 1 is the parallel combination of the input impedances of segments 2 and 3. Thus

$$\frac{Z_{l1}}{Z_{ca1}} = \frac{(Z_{c2}/Z_{ca2})(W)(Z_{c3}/Z_{ca3})(Z_{ca3}/Z_{ca2}) \coth \Gamma l_3}{(Z_{c2}/Z_{ca2})W + (Z_{c3}/Z_{ca3})(Z_{ca3}/Z_{ca2}) \coth \Gamma l_3}$$

$$P_{i1} = P_{11}\left[\cosh \Gamma l_1 + \frac{(Z_{c1}/Z_{ca1}) \sinh \Gamma l_1}{Z_{l1}/Z_{ca1}}\right]$$

$$P_{i3} = P_{13} \cosh \Gamma l_3, \qquad P_{13}/P_{i1} = (P_{13}/P_{i3})(P_{11}/P_{i1})$$

$$\omega_{v1} = \omega_{v2} = 4v/A_1 = 19.8 \quad \text{rad/sec}$$

$$\omega_{v3} = 4v/A_3 = 29.7 \quad \text{rad/sec}$$

$$\omega/\omega_{v1} = 800/19.8 = 40.4, \qquad \omega/\omega_{v3} = 800/29.7 = 26.9$$

$$\frac{\omega_{v1} l_1}{a} = \frac{(19.8)(0.15)}{330} = 0.009, \qquad \frac{\omega_{v2} l_2}{a} = \frac{(19.8)(0.05)}{330} = 0.003$$

$$\frac{\omega_{v3} l_3}{a} = \frac{(29.7)(0.05)}{330} = 0.0045, \qquad \frac{Z_{ca3}}{Z_{ca2}} = \frac{A_2}{A_3} = 1.5$$

(a) *Line Segment 2 Open*

$$W = \tanh \Gamma l_2$$

from Appendix B, at $\sigma = 3.0$,

$$\omega/\omega_{v1} = 40.4 \quad \rightarrow \quad \phi_1 = 9.401 + j47.689 = \phi_2$$

$$Z_{c1}/Z_{ca1} = Z_{c2}/Z_{ca2} = 1.062 - j0.107$$

at $\sigma = 2.0$,

$$\omega/\omega_{v3} = 26.9 \quad \rightarrow \quad \phi_3 = 7.030 + j32.496$$

$$Z_{c3}/Z_{ca3} = 1.071 - j0.123$$

Thus

$$\Gamma l_1 = \phi_1(\omega_{v1} l_1/a) = 0.085 + j0.429$$
$$\Gamma l_2 = \phi_2(\omega_{v2} l_2/a) = 0.028 + j0.143$$
$$\Gamma l_3 = \phi_3(\omega_{v3} l_3/a) = 0.032 + j0.146$$
$$\cosh \Gamma l_1 = 0.913 + j0.035$$
$$\sinh \Gamma l_1 = 0.077 + j0.418$$
$$\tanh \Gamma l_2 = 0.029 + j0.144$$
$$\coth \Gamma l_3 = 1.350 - j6.500$$

$$Z_{i2}/Z_{ca2} = (Z_{c2}/Z_{ca2})(W)$$
$$= (1.062 - j0.107)(0.029 + j0.144) = 0.046 + j0.150$$

$$\frac{Z_{11}}{Z_{ca}} = \frac{(0.046 + j0.150)(1.5)(1.071 - j0.123)(1.350 - j6.500)}{(0.046 + j0.150) + (1.5)(1.071 + j0.123)(1.350 - j6.500)}$$
$$= 0.047 + j0.152$$

$$\frac{P_{i1}}{P_{11}} = \left[0.913 + j0.035 + \frac{(1.062 - j0.107)(0.077 + j0.418)}{0.047 + j0.152} \right]$$
$$= 3.770 + j0.091$$

$$P_{i3}/P_{13} = \cosh \Gamma l_3 = 0.989 + j0.004$$

$$\frac{P_{13}}{P_{i1}} = \left(\frac{P_{13}}{P_{i3}}\right)\left(\frac{P_{11}}{P_{i1}}\right) = \frac{1}{(3.770 + j0.091)(0.989 + j0.004)}$$
$$= 0.268 - j0.008$$

$$|P_{13}/P_{i1}| = 0.268$$

(b) *Line Segment 2—Matched ($W = 1$)*

$$\frac{Z_{11}}{Z_{ca1}} = \frac{(1.5)(1.0623 - j0.107)(1.071 - j0.123)(1.350 - j6.500)}{(1.0623 - j0.107) + 1.5(1.071 - j0.123)(1.350 - j6.500)}$$
$$= 1.023 - j0.023$$

$$\frac{P_{i1}}{P_{11}} = \left[0.913 + j0.035 + \frac{(1.062 - j0.107)(0.077 + j0.418)}{1.023 - j0.203} \right]$$
$$= 0.951 + j0.469$$

$$\frac{P_{13}}{P_{i1}} = \left(\frac{P_{13}}{P_{i3}}\right)\left(\frac{P_{11}}{P_{i1}}\right) = \frac{1}{(0.989 + j0.004)(0.951 + j0.469)}$$
$$= 0.853 - j0.426$$

$$|P_{13}/P_{i1}| = 0.953$$

3.9 IMPULSE AND STEP RESPONSE OF FLUID LINES

The formulation of the propagation factor and transmission matrix [combination of Eqs. (3.7), (3.10), (3.55), and (3.59)] for the Laplace domain is not readily inverted into the time domain. At present, results are only available for the matched (semi-infinite) transmission line, and even in this case approximations were required to simplify the propagation model.

As indicated in Eq. (3.15c), the transfer function for the matched line is

$$P_1/P_i = e^{-\Gamma l} \tag{3.94}$$

For the propagation factor [exponent in Eq. (3.94)] Brown [10, 11] applies the circular line approximation [Eq. (3.63c)] in the form

$$\Gamma l = \frac{sl}{a}\left[1 + A_1\left(\frac{v}{sr_w^2}\right)^{1/2} + A_2\left(\frac{v}{sr_w^2}\right) + A_3\left(\frac{v}{sr_w^2}\right)^{3/2}\right] \tag{3.95}$$

where the A's come from Table 3.1.

When the transient times of interest are short (high frequencies), the contribution of the A_3 term in Eq. (3.95) is small and can be neglected. Thus the substitution of Eq. (3.95) into Eq. (3.94) yields

$$P_1/P_i = \exp(-st_f)\exp[-A_1 t_f(vs/r_w^2)^{1/2}]\exp(-A_2 t_f v/r_w^2) \tag{3.96}$$

where t_f is the nominal transmission time l/a. The first exponential term in Eq. (3.96) represents a pure time delay and the last exponential term is the signal attenuation. The shape of the transient response, therefore, is determined solely by the second exponential term. When the input is a unit impulse ($P_i = 1$) the inverse transform of Eq. (3.96) is

$$p_1 = \frac{A_1 t_f (v/r_w^2)^{1/2}}{[\pi(t-t_f)^3]^{1/2}} \exp\left[-\frac{A_1^2 t_f^2 v}{4r_w^2(t-t_f)} - \frac{A_2 t_f v}{r_w^2}\right] H(t-t_f) \tag{3.97}$$

where $H(t-t_f)$ is the unit step function delayed by t_f. For the unit step input ($P_i = 1/s$) Eq. (3.96) inverts to

$$p_1 = \exp\left(-\frac{A_2 t_f v}{r_w^2}\right)\left[\text{erfc}\,\frac{A_1 t_f(v/r_w^2)^{1/2}}{2(t-t_f)^{1/2}}\right] H(t-t_f) \tag{3.98}$$

where erfc is the complimentary error function.

Karam [13] provides a semiempirical formulation similar to Eq. (3.98) that is based on an asymptotic approximation between the long-time and short-time step responses. This form is somewhat more convenient to use and is in good agreement with experimental results. On the other hand, the results obtained by Kantola [12] are presented in infinite series and are more difficult to apply.

3.10 THE METHOD OF CHARACTERISTICS

3.10.1 Introduction

Dynamic analysis of fluidic lines and line networks is often an important part of fluidic system design. Fluid transmission line theory provides an effective analytical method for determining the frequency response of lines that have small-amplitude (acoustic-type) signals. Unless the lines are matched, however, transient (step and pulse) response is difficult to obtain by this method. In addition, sizable errors can result from the application of transmission line theory to fluidic circuits that operate with finite-amplitude signals. Thus we need another method of analysis to find the transient response with or without finite amplitudes and the frequency response with finite amplitudes. In the remainder of this chapter we consider dynamic analysis by the method of characteristics.

Steady and unsteady compressible flow problems have been solved by the method of characteristics for many years. The method is based on a particular condition that satisfies second-order partial differential equations of the hyperbolic type. Courant and Friedrichs [23] give a complete mathematical description of the theory of characteristics. Shapiro [24], Kantrowitz [25], and Owczarek [26] discuss the method in considerable detail and also give particular examples from the field of gas dynamics. Some of the typical examples are the sudden discharge of gas from a tube and the flow of gas in a shock tube. The method of characteristics that is presented in [24–26] and also by Rudinger [27] emphasizes the graphical approach. This gives an excellent understanding of the characteristics but limits their application to rather simple problems. The graphical constructions for complex problems involving friction and line networks becomes extremely tedious and quite impractical. Now, however, the advent and accessibility of high-speed digital computers have led to renewed interest in characteristics. Streeter and Wylie [28] present a comprehensive treatment of characteristics for hydraulic transients with special emphasis on computer calculations. Computational procedures have also been reported by Benson *et al.* [29] and Manning [30] for compressible flow problems. All the foregoing treatments neglect frequency-dependent effects. Recently Zielke [31] has proposed a method for taking this into account for frictional properties and Brown [32] has developed a quasi-method of characteristics which considers both frequency-dependent friction and heat transfer effects.

The purpose of this portion is to indicate the procedure for finding the dynamic response of fluidic lines and networks by the computerized method of characteristics.

(1) Section 3.11 deals with the basic characteristic equations for small-amplitude signals. The computerized method of characteristics uses a finite

grid size. As a consequence the method produces a filtering effect on the fluidic line under consideration. The limitation in response due to grid size is considered and some typical results for step and pulse input signals are given.

(2) Section 3.12 treats the quasi method of characteristics developed by Brown [32] and Zielke [31]. The consideration of time dependency requires a computer with large storage capacity. However, the results are in closer agreement with experiments. The quasi method and the method of characteristics are compared to show the advantages of the quasi method.

(3) Section 3.13 considers the effects of finite-amplitude signals. In the past, problems with large signals were handled by applying the method of characteristics from a graphical viewpoint. The computerized method introduces some small errors but extends the method to complex problems that would not be feasible graphically.

(4) Section 3.14 contains a derivation of the "matching" impedance obtained from the method of characteristics. Here the connection between transmission line theory and the method of characteristics may be observed. The matching of a finite-amplitude step with an orifice termination is also described.

(5) Section 3.15 demonstrates briefly the extension of the method of characteristics to branch networks and to lines with lumped terminations.

3.11 THE METHOD OF CHARACTERISTICS FOR SMALL-AMPLITUDE SIGNALS

The method of characteristics is based on the assumption of one-dimensional flow. The momentum equation may therefore be written as

$$\frac{\rho}{A}\frac{\partial q}{\partial t} + \frac{\rho u}{A}\frac{\partial q}{\partial x} + \frac{\partial p}{\partial x} = \left(\overline{\frac{\partial \tau}{\partial y}}\right)_w \qquad (3.99)$$

where p is the static pressure, ρ is the fluid density, u is the velocity, A is the area of the cross section, q is the volume flow ($q = uA$), x is the linear distance, t is the time, and $\overline{(\partial \tau/\partial y)}_w$ is the average frictional force per unit volume. The continuity equation with convective velocity term ($u\, \partial \rho/\partial x$) is

$$\partial \rho/\partial t + (\rho/A)\, \partial q/\partial x + u\, \partial \rho/\partial x = 0 \qquad (3.100)$$

In the method of characteristics a linear combination of the momentum and continuity equations [Eqs. (3.99) and (3.100)] changes the partial derivatives to total derivatives under certain conditions. To accomplish this, however, there must be a known relation between the static pressure and the density. One possible form for this relation (see Kirshner [18, p. 22]) is

$$\frac{d\rho}{\rho} = \frac{1}{\gamma}\frac{dp}{p} - \frac{dS}{\gamma c_v} \qquad (3.101)$$

where S is the entropy of the fluid, γ is the ratio of specific heats, and c_v is the specific heat at constant volume. The substitution of Eq. (3.101) into Eq. (3.100) yields

$$\frac{1}{\gamma p}\left(\frac{\partial p}{\partial t}+u\frac{\partial p}{\partial x}\right)+\frac{1}{A}\frac{\partial q}{\partial x}-\frac{1}{\gamma c_v}\left(\frac{\partial S}{\partial t}+u\frac{\partial S}{\partial x}\right)=0 \qquad (3.102)$$

If Eq. (3.102) is multiplied by an arbitrary constant K_1, and the result added to Eq. (3.99), we obtain

$$\frac{K_1}{\gamma p}\left[\frac{\partial p}{\partial t}+\left(\frac{\gamma p}{K_1}+u\right)\frac{\partial p}{\partial x}\right]+\frac{\rho}{A}\left[\frac{\partial q}{\partial t}+\left(\frac{K_1}{\rho}+u\right)\frac{\partial q}{\partial x}\right]=\left(\overline{\frac{\partial \tau}{\partial y}}\right)_w+\frac{K_1}{\gamma c_v}\left[\frac{\partial S}{\partial t}+u\frac{\partial S}{\partial x}\right] \qquad (3.103)$$

The pressure, flow, and entropy variables are functions of both time and distance. The total derivatives of these variables therefore have the form

$$\frac{d}{dt}=\frac{\partial}{\partial t}+\frac{dx}{dt}\frac{\partial}{\partial x} \qquad (3.104)$$

To transform Eq. (3.103) into an expression containing total derivatives requires that the following conditions be satisfied in accordance with Eq. (3.104):

$$\left[\frac{dx}{dt}\right]_{p,q}=\gamma p/K_1+u=K_1/\rho+u \qquad (3.105a)$$

$$\left[\frac{dx}{dt}\right]_S=u \qquad (3.105b)$$

Thus the selection of the arbitrary constant K_1 as $\pm(\gamma p\rho)^{1/2}$ changes Eqs. (3.103) and (3.105) to

$$\frac{\rho}{A}\frac{dq}{dt}+\frac{1}{a}\frac{dp}{dt}=\left(\overline{\frac{\partial \tau}{\partial y}}\right)_w+\frac{\rho a}{\gamma c_v}\frac{dS}{dt} \qquad (3.106a)$$

when

$$(dx/dt)_{p,q}=u+a,\ (dx/dt)_S=u,$$

$$\frac{\rho}{A}\frac{dq}{dt}-\frac{1}{a}\frac{dp}{dt}=\left(\overline{\frac{\partial \tau}{\partial y}}\right)_w-\frac{\rho a}{\gamma c_v}\frac{dS}{dt} \qquad (3.106b)$$

when

$$(dx/dt)_{p,q}=u-a,\ (dx/dt)_S=u,$$

where a is the sonic velocity $(\gamma p/\rho)^{1/2}$. The lines of slope dx/dt on the xt plane are the position characteristics, and the differential relations between pressure, flow, and entropy are the compatibility conditions. These differential relations should not be confused with differential equations. They are valid only when the variables change along the characteristic directions.

3.11 The Method of Characteristics for Small-Amplitude Signals

Let us consider Eq. (3.106) in terms of circuit components. We recall that the adiabatic inertance per unit length L_a equals ρ/A, the adiabatic capacitance per unit length C_a equals $A/\gamma p$, and the average laminar frictional resistance per unit length R equals $-\overline{(\partial \tau/\partial y)_w}/q$. In normalized form, Eq. (3.106) then becomes

$$N_k \, dq'/dt' + dp'/dt' + q' = a'^2 \, dS'/dt'$$

when (3.107a)

$$(dx'/dt')_{p', q'} = (u/a) + 1, \quad (dx'/dt')_{S'} = u/a,$$

$$N_k \, dq'/dt' - dp'/dt' + q' = -a'^2 \, dS'/dt'$$

when (3.107b)

$$(dx'/dt')_{p', q'} = (u/a) - 1, \quad (dx'/dt')_{S'} = u/a,$$

where

$$x' = x/l, \qquad p_f = Rlq_f$$
$$t' = ta/l, \qquad a_f = (\gamma p_f/\rho)^{1/2}$$
$$p' = p/p_f, \qquad N_k = (L_a/C_a)^{1/2}/Rl = Z_{ca}/Rl$$
$$q' = q/q_f,$$
$$S' = S/c_v,$$
$$a' = a/a_f,$$

l is the length of the line, Z_{ca} is the adiabatic characteristic impedance, N_k is the characteristic number, and the subscript f refers to reference conditions.

The direct application of Eq. (3.107) to practical problems has some shortcomings and limitations. The main difficulty lies in a proper formulation for the entropy term. Many times this term is neglected in which case the results may be quite inaccurate. Another difficulty is that the friction term should be time dependent, whereas friction is independent of time in Eq. (3.107). Nevertheless, the basic method presented in this section is valuable because it will demonstrate the calculation procedure and the general trends to the dynamic response. In Section 3.12 the use of the quasi-method of characteristics improves the results.

3.11.1 The Effect of Mesh Size

The method of characteristics operates in the time domain. Figure 3.29 shows the characteristic lines on the $x't'$ plane for small-amplitude signals without throughflow ($u/a = 0$) or heat transfer ($dS'/dt' = 0$). There are two families of lines; the positive slope lines ($dx'/dt' = 1$) and the negative slope lines ($dx'/dt' = -1$). The signal variables propagate along these lines; that is, they change discontinuously across the direction lines and can be calculated only at the nodal points of the mesh. In the interior of any mesh diamond (e.g., *BEIF* of

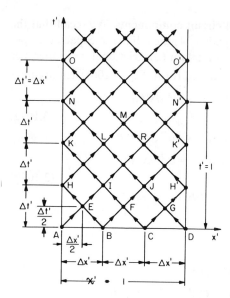

Fig. 3.29 Characteristic grid for small-amplitude signals.

Fig. 3.29) the signal variables have a constant value. The number of cells into which the signal path is divided is the mesh size n. For purpose of illustration, on Fig. 3.29 the mesh size is 3. Now, although signal discontinuities do propagate along the characteristic lines, the method of characteristics acts to filter the signal variables. The situation is analogous to that shown in Fig. 3.30 where a

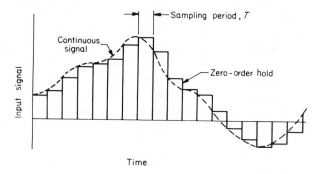

Fig. 3.30 Sampling followed by zero-order hold.

continuous signal is sampled periodically and each sample is passed into a zero-order or boxcar hold. As in the method of characteristics the signal variables do not change between sampling periods. Thus the frequency spectrum of the zero-order hold and the method of characteristics are the same. In the method of characteristics the sampling period ΔT is related to the mesh size n by

$$\Delta T = l/na \qquad (3.108)$$

3.11 The Method of Characteristics for Small-Amplitude Signals

From Kuo [33] the magnitude of the frequency response G_H of a zero-order hold is

$$G_H = (\Delta T) \left| \frac{\sin(\omega \cdot \Delta T/2)}{\omega \cdot \Delta T/2} \right| \tag{3.109}$$

where ω is the angular frequency of the applied signals. Equation (3.109) is also the frequency response imposed by the method of characteristics. Figure 3.31 shows the magnitude response of the zero-order hold plotted against the

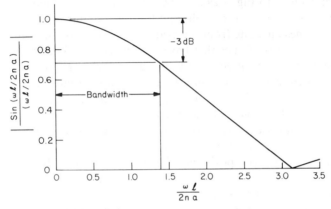

Fig. 3.31 Frequency response of zero-order hold.

dimensionless frequency $\omega l/(2na)$. The response is down 3 dB when the dimensionless frequency equals 1.39. Thus the bandwidth ω_b, due to the discontinuities of the method of characteristics, is

$$\omega_b = 2.78 na/l \tag{3.110}$$

This bandwidth must be appreciably greater than that of the system being examined.

As an example, when a 10-mesh characteristic grid is used to solve the dynamic response of a 1-m-long line with air as the working fluid, the bandwidth limitation due to the mesh is 9180 rad/sec. Since the bandwidth is proportional to to the mesh size, doubling the mesh doubles the bandwidth, etc. It is possible, therefore, to treat lines with high-frequency signal components by using a very fine mesh.

3.11.2 Small-Amplitude Response without Throughflow or Heat Transfer

When there is no throughflow or heat transfer, Eq. (3.107) becomes

$$N_k \, dq'/dt' + dp'/dt' + q' = 0, \quad dx'/dt' = 1 \tag{3.111a}$$

$$N_k \, dq'/dt' - dp'/dt' + q' = 0, \quad dx'/dt' = -1 \tag{3.111b}$$

The values of the signal variables p' and q' in Eq. (3.111) propagate only along the characteristic lines, $dx'/dt' = \pm 1$. These lines form a grid in the $x't'$ plane. There are two basic ways of selecting the grid: (a) a grid of characteristics or (b) a grid of specified time intervals. Streeter and Wylie [28] and others have shown that computer calculations are more readily performed with fixed time intervals. Therefore, in this treatment we always use this method. However, for the case under consideration now (small amplitude without throughflow) the two methods are identical anyway.

Let us return now to Fig. 3.29 and the mesh with $n = 3$. Along the line where time equals zero there are three cells and four grid points, A, B, C, and D. The characteristic lines emanate from the grid points. The interior points (B and C) are the starting points of lines that have both positive ($B\Gamma$) and negative slopes (BE). A single line of positive slope begins at entrance grid point A and one of negative slope from exit grid point D. Each line continues until it reaches a left or right boundary. When this happens the slope reverses. The points of intersection are called *nodal points*. These are the only points in the $x't'$ plane at which the signal variables (p' and q') change and where they may be calculated. Notice that the time period between intersections is $\Delta t'/2$ and that the intersections are staggered in position. This follows directly from the position characteristic equations ($dx'/dt' = \pm 1$). From the initial conditions of the problem the signal variables are both known at points B, C, and D along the line. We must select the boundary conditions according to the problem we wish to solve. At the entrance points A, H, K, N, O, etc., and the exit points H', K', N', O', etc., we must supply either a relation between the signal variables p' and q' or specify the value of either variable. At the entrance we may assign a signal variable to apply any desired input signal to the line, For example, if we want the pressure step response of a line, we would fix the entrance point pressures at a constant value (i.e., $p_A' = p_H' = p_K' = p_N' = \cdots = 1.0$). At the exit a relation of the type $p = R_L q$ loads the line with a linear resistance. For a blocked line we merely specify that $q_D' = q_{H'}' = q_{K'}' = \cdots = 0$.

After the initial and boundary conditions are specified we can calculate the signal variables at $t' = \Delta t'/2$ for points E, F, and G. Following this we proceed to $t' = \Delta t'$ to calculate points H, I, J, and H' and continue on in this manner at time increments of $\Delta t'/2$. The way in which the calculation is performed depends on the location of the point. For any arbitrary interior point M (Fig. 3.29) a second-order finite difference approximation to Eq. (3.111) yields

$$N_k[q_M' - q_L'] + [p_M' - p_L'] + \left[\frac{q_M' + q_L'}{2}\right]\frac{\Delta t'}{2} = 0 \qquad (3.112a)$$

$$N_k[q_M' - q_R'] - [p_M' - p_R'] + \left[\frac{q_M' + q_R'}{2}\right]\frac{\Delta t'}{2} = 0 \qquad (3.112b)$$

where the subscripts L and R refer to location of the points (left and right) at the previous time increment and $\Delta t' = 1/n$.

3.11 The Method of Characteristics for Small-Amplitude Signals

The solution of the simultaneous equations (3.112a) and (3.112b) gives the values of signal variables at interior point M as

$$p_M' = \frac{p_L' + p_R'}{2} + \left(N_k - \frac{1}{2}\frac{\Delta t'}{2}\right)\left(\frac{q_L' - q_R'}{2}\right) \quad (3.113a)$$

$$q_M' = \frac{(p_L' - p_R') + \left(N_k - \frac{1}{2}\frac{\Delta t'}{2}\right)[q_L' + q_R']}{2\left(N_k + \frac{1}{2}\frac{\Delta t'}{2}\right)} \quad (3.113b)$$

Thus to calculate the signal variables at any interior point we must know the variables to the left and right from the previous time step. The calculations are made across all the points at a given time step before proceeding to the next time step.

At a left boundary a single compatibility relation of the type given in Eq. (3.112b) is available. Thus, for example, at point H at the left boundary we obtain information only from point E so that

$$N_k[q_H' - q_E'] - [p_H' - p_E'] + \left(\frac{q_H' + q_E'}{2}\right)\left(\frac{\Delta t'}{2}\right) = 0 \quad (3.114)$$

To calculate the variables at this boundary we need another relation between them. Often only the value p_H' is specified and we assume that the signal source has zero impedance. However, if we wish, we may supply a relation between p_H' and q_H' that takes the actual source impedance into account. In addition, by selecting appropriate values of p' at each successive point on the left boundary we may apply any form of input signal (i.e., step, pulse, ramp, sine wave, etc.).

At a right boundary, the compatability relation given in Eq. (3.112a) applies. For point H', on the right boundary the equation becomes

$$N_k[q_{H'}' - q_G'] + [p_{H'}' - p_G'] + \left(\frac{q_{H'}' + q_G'}{2}\right)\left(\frac{\Delta t'}{2}\right) = 0 \quad (3.115)$$

Point H' receives information on the signal variables only from point G. To determine the signal variables at H', an additional relation between $p_{H'}'$ and q_H' is required and this is usually of the form $p_{H'}' = Z_L q_H'/Rl$. The load impedance Z_L is usually purely resistive. Loads with reactive parts add complexity to the solution but they may be included. The cases of a line terminated by a distributed line of another size or a lumped volume are considered in Section 3.15.

Figure 3.32 shows the pressure step response of a blocked line ($Z_L = \infty$) calculated with a mesh of $n = 25$ for various values of the characteristic number N_k. The computer programs for this case and for the other cases considered in subsequent sections are given in Appendix E in the back of the book. When N_k is small (i.e., 0.2) the resistance of the line is large relative to the inertance and the blocked end pressure monotonically approaches the steady state value.

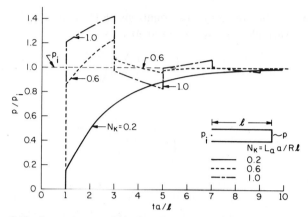

Fig. 3.32 Step response of blocked line by the method of characteristics.

For $N_k = 0.6$ the inertance portion of the line impedance increases and the response contains both overshoots and undershoots. The maximum values of the overshoots and undershoots occur when $ta/l = 1, 3, 5, \ldots$. These times correspond to the transport time of the wavefront propagating at the sonic velocity. The largest overshoot occurs just before the second wave reaches the end of the line ($ta/l = 3$) and has the magnitude $p/p_i = 1.22$ for $N_k = 1.0$.

Figure 3.33 shows the method of characteristics solutions (mesh = 25) for the blocked-line response to a square pressure pulse. The pulse duration $ta/l = 2$. The response is identical to the step response until $ta/l = 3$. At this time information transmitted through the characteristic lines reaches the closed end and indicates that the input signal has been removed. Thereafter the signal approaches zero. The approach is monotonic when $N_k = 0.2$, but the more reactive lines ($N_k = 0.6$ and 1.0) oscillate about zero as they proceed toward the steady state.

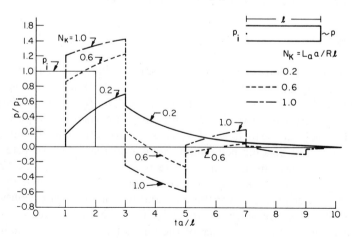

Fig. 3.33 Rectangular pulse response of blocked line by the method of characteristics.

3.11 The Method of Characteristics for Small-Amplitude Signals

The responses shown in Figs. 3.32 and 3.33 have discontinuous changes in the signal variables that are considerably more abrupt than experimental responses. The discrepancy results from the neglect of heat transfer and the use of time-independent friction. It is interesting to note, however, that the results obtained (Figs. 3.32 and 3.33) apply also to electrical lines. In this case, N_k is computed from the known properties of the electrical line.

3.11.3 Small-Amplitude Response with Throughflow

From Eq. (3.107) the appropriate equations for the case of throughflow without heat transfer are

$$N_k\, dq'/dt' + dp'/dt' + q' = 0, \qquad dx'/dt' = (u/a) + 1 \qquad (3.116a)$$

$$N_k\, dq'/dt' - dp'/dt' + q' = 0, \qquad dx'/dt' = (u/a) - 1 \qquad (3.116b)$$

The amount of the throughflow must be known in advance. The p' and q' variables in Eq. (3.116) are the small-amplitude dynamic signals that are superimposed on the steady throughflow. The variables bear the same relation to each other as they do without throughflow [Eq. (3.111)]. The difference lies in the altered direction of the position lines. Thus we may expect that the effect of throughflow is to distort the response.

Figure 3.34 shows the grid that is used to apply the method of characteristics

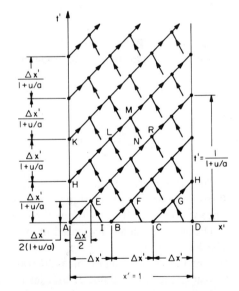

Fig. 3.34 Characteristic grid for small-amplitude signals with throughflow.

with throughflow. The nodal points are staggered at intervals of $\Delta x'/2$ and placed at intervals $\Delta t'/2$. In this case, however, $\Delta t' = \Delta x'/(1 + u/a)$. The grid has been adjusted so that the rightward traveling position lines emanate from nodal points and pass through nodal points. The leftward traveling lines, on the other hand, begin at intermediate points between the nodes of the previous time step (i.e., point I between points A and B). The location of the intermediate point depends on the amount of throughflow. The magnitudes of the intermediate point signal variables are determined by a linear interpolation from the nodal points. For example, the signal variables at point N between nodal points L and R are

$$p_N' = \frac{p_R' + (u/a)p_L'}{1 + u/a} \tag{3.117a}$$

$$q_N' = \frac{q_R' + (u/a)q_L'}{1 + u/a} \tag{3.117b}$$

In the absence of throughflow ($u/a = 0$), point N coincides with nodal point R and $p_N' = p_R'$, $q_N' = q_R'$. Now to calculate the signal variables at interior point M we use the second-order finite-difference approximation of Eq. (3.116) to obtain

$$N_k[q_M' - q_L'] + [p_M' - p_L'] + \left(\frac{q_M' + q_L'}{2}\right)\left(\frac{\Delta t'}{2}\right) = 0 \tag{3.118a}$$

$$N_k[q_M' - q_N'] - [p_M' - p_N'] + \left(\frac{q_M' + q_N'}{2}\right)\left(\frac{\Delta t'}{2}\right) = 0 \tag{3.118b}$$

where we recall that $\Delta t' = 1/[n(1 + u/a)]$. From Eqs. (3.117) and (3.118), the signal variables at an interior node with throughflow are

$$p_M' = \frac{[(1 + 2u/a)p_L' + p_R'] + [N_k - \tfrac{1}{2}\Delta t'/2][q_L' - q_R']}{2(1 + u/a)} \tag{3.119a}$$

$$q_M' = \frac{[p_L' - p_R'] + [N_k - \tfrac{1}{2}\Delta t'/2][(1 + 2u/a)q_L' + q_R']}{2(1 + u/a)(N_k + \tfrac{1}{2}\Delta t'/2)} \tag{3.119b}$$

Thus the interior node variables depend again only on the variables at L and R from the previous time step. Equation (3.119) is more general than Eq. (3.113) since it applies with or without throughflow.

The right boundary is treated in exactly the same way as given in Eq. (3.115). The left boundary, however, requires an interpolation between known values. Figure 3.35 presents a graphical diagram of the interpolation at the left boundary. Consider the nodal point H and let us suppose that p_H' is known and that we must find q_H'. Now there is no leftward traveling characteristic line from a previous node that passes through the node H on the left boundary. The closest characteristic line, the one through E, intersects the left boundary at Z. To compute q_H' we must first compute q_Z'. This is accomplished by first finding p_Z'

3.11 The Method of Characteristics for Small-Amplitude Signals

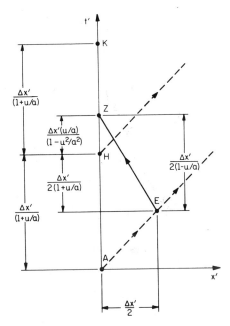

Fig. 3.35 Left boundary interpolation for small-amplitude signals with throughflow.

from a linear interpolation between p_H' and p_K' and then using an equation of the type given in Eq. (3.118b). The results are

$$p_z' = \frac{[1 - 2u/a]p_H' + 2(u/a)p_K'}{1 - (u/a)^2} \quad (3.120a)$$

$$N_k[q_z' - q_E'] - [p_z' - p_E'] + \left(\frac{q_z' + q_E'}{2}\right)\frac{\Delta x'}{2(1 - u/a)} = 0 \quad (3.120b)$$

From Eq. (3.120a) we obtain p_z'. Then q_z' is determined from Eq. (3.120b) and finally a linear interpolation between q_z' and q_A' yields

$$q_H' = \frac{q_z' - (u/a)q_A'}{1 - u/a} \quad (3.121)$$

Figure 3.36 shows the effect of throughflow on the step response of a line terminated by a pure resistance R_L that equals $3Rl$. The signals with throughflow propagate rightward at higher velocity. Thus the throughflow appears to reduce friction at the leading edge of the step. In effect the retardation caused by friction is partially balanced by the increase in signal propagation. After reflections from the termination, however, the signal information travels slower than previously and the friction seems larger. As a result, the overshoot at the leading edge increases as throughflow increases but the remainder of the response is attenuated.

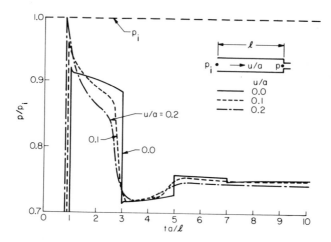

Fig. 3.36 The effect of throughflow on the step response of a line terminated with an orifice ($N_k = 1.0$).

3.12 THE QUASI METHOD OF CHARACTERISTICS FOR SMALL-AMPLITUDE SIGNALS

Zielke [31] extended the method of characteristics to include the effects of time-dependent friction. Brown [32] formulated the "quasi method" of characteristics which takes into account both time-dependent friction and heat transfer. We present, here, a modified development of Brown's approach.

Let us begin with the transmission line equations

$$\partial P/\partial x = -ZQ \qquad (3.122a)$$

$$\partial Q/\partial x = -YP \qquad (3.122b)$$

where the signal variables P and Q are functions of the Laplace operator s, and Z and Y are the series impedance and shunt admittance per unit length. The impedance and admittance are expressable in terms of their average real part, their adiabatic imaginary part, and a time-dependent part:

$$Z = R + L_a s + Z_1(s) \qquad (3.123a)$$

$$Y = C_a s + Y_1(s) \qquad (3.123b)$$

where L_a is the inertance per unit length, C_a is the adiabatic capacitance per unit length, and $Z_1(s)$ and $Y_1(s)$ are time-dependent quantities with both real and imaginary parts. The substitution of Eq. (3.123) into Eq. (3.122) yields

$$\partial P/\partial x = -[R + L_a s + Z_1(s)]Q \qquad (3.124a)$$

$$\partial Q/\partial x = -[C_a s + Y_1(s)]P \qquad (3.124b)$$

3.12 The Quasi Method of Characteristics for Small-Amplitude Signals

The inverse transform of Eq. (3.124) is

$$L_a \partial q/\partial t + \partial p/\partial x = -Rq - \mathcal{L}^{-1}\{Z_1(s)Q\} \tag{3.125a}$$

$$\partial q/\partial x + C_a \partial p/\partial t = -\mathcal{L}^{-1}\{Y_1(s)P\} \tag{3.125b}$$

where the symbol $\mathcal{L}^{-1}$ is the inverse Laplace transform. Equation (3.125a) is analogous to the momentum equation and Eq. (3.125b) is analogous to the continuity equation. If we multiply Eq. (3.125b) by the arbitrary constant K_2 and add the result to Eq. (3.125a),

$$L_a\left(\frac{\partial q}{\partial t} + \frac{K_2}{L_a}\frac{\partial q}{\partial x}\right) + K_2 C_a\left(\frac{\partial p}{\partial t} + \frac{1}{K_2 C_a}\frac{\partial p}{\partial x}\right) = -Rq - \mathcal{L}^{-1}\{Z_1(s)Q\}$$
$$- K_2 \mathcal{L}^{-1}\{Y_1(s)P\} \tag{3.126}$$

To place Eq. (3.126) in the form of a total differential relationship, Eq. (3.104) indicates that

$$dx/dt = K_2/L_a = 1/K_2 C_a \tag{3.127}$$

From Eq. (3.127), $K_2 = \pm(L_a/C_a)^{1/2}$ and thus Eq. (3.126) becomes

$$L_a\, dq/dt + (L_a C_a)^{1/2}\, dp/dt + Rq = -\mathcal{L}^{-1}\{Z_1(s)Q\}$$
$$-(L_a/C_a)^{1/2}\mathcal{L}^{-1}\{Y_1(s)P\}, \quad dx/dt = a \tag{3.128a}$$

$$L_a\, dq/dt - (L_a C_a)^{1/2}\, dp/dt + Rq = -\mathcal{L}^{-1}\{Z_1(s)Q\}$$
$$+(L_a/C_a)^{1/2}\mathcal{L}^{-1}\{Y_1(s)P\}, \quad dx/dt = -a \tag{3.128b}$$

since $a = 1/(L_a C_a)^{1/2}$. With the aid of the relation for the characteristic impedance $Z_{ca} = aL_a$, Eq. (3.128) takes the form

$$Z_{ca}\, dq/dt + dp/dt + Raq = -Z_{ca}\mathcal{L}^{-1}\{W_z(s)sQ\} - \mathcal{L}^{-1}\{W_y(s)sP\}, \quad dx/dt = a \tag{3.129a}$$

$$Z_{ca}\, dq/dt - dp/dt + Raq = -Z_{ca}\mathcal{L}^{-1}\{W_z(s)sQ\} + \mathcal{L}^{-1}\{W_y(s)sP\}, \quad dx/dt = -a \tag{3.129b}$$

where the weighting factors $W_z(s) \equiv Z_1(s)/sL_a$ and $W_y(s) \equiv Y_1(s)/sC_a$. Since multiplication in the frequency domain is equivalent to convolution in the time domain, Eq. (3.129) transforms to

$$Z_{ca}\, dq/dt \pm dp/dt + Raq = -Z_{ca}\int_0^t W_z(\tau)\frac{dq}{d\tau}(t-\tau)\, d\tau$$
$$\mp \int_0^t W_y(\tau)\frac{dp}{d\tau}(t-\tau)\, d\tau, \quad \frac{dx}{dt} = \pm a \tag{3.130}$$

where τ is a dummy variable, the derivatives

$$\frac{dq}{d\tau}(t-\tau) \quad \text{and} \quad \frac{dp}{d\tau}(t-\tau)$$

indicate functions of $(t-\tau)$ and

$$\frac{dq}{d\tau}(t-\tau) \equiv \frac{dq}{dt}(t-\tau)\bigg|_{t=\tau}, \quad \frac{dp}{d\tau}(t-\tau) \equiv \frac{dp}{dt}(t-\tau)\bigg|_{t=\tau}$$

The weighting factors $W_z(t)$ and $W_y(t)$ depend on the shape of the cross section, the heat transfer, and the velocity profile in the passage. Expressions for the weighting factors are derived in Appendixes C and D, where it is also shown that the factors in terms of the normalized parameters are

$$W_z(t') = (R/K_s L_a)\phi_z(t'/N_k K_s) \tag{3.131a}$$

$$W_y(t') = \frac{(\gamma-1)R}{N_p K_s L_a}\phi_y\left(\frac{t'}{N_p K_s N_k}\right) \tag{3.131b}$$

where N_p is the Prandtl number, K_s is a shape factor (which equals 8 for a circular section and 12 for a rectangular section), and ϕ_z and ϕ_y are the normalized weighting factors. Figure 3.37a and 3.37b show the normalized weighting

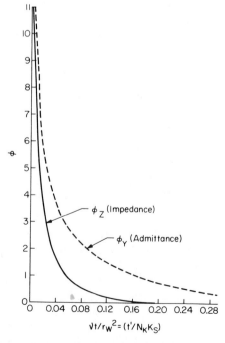

Fig. 3.37a Normalized weighting factors for circular lines (air).

3.12 The Quasi Method of Characteristics for Small-Amplitude Signals

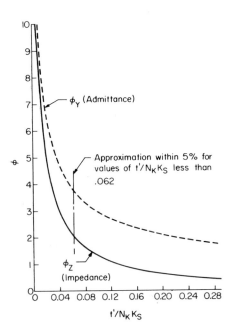

Fig. 3.37b Normalized weighting factors for rectangular lines (air).

factors for circular and rectangular lines when air is the fluid. These factors both decrease monotonically with an increase in time. Thus in the integral terms of Eq. (3.130), the weighting functions give greatest influence to the most recent changes in the signal variables. In normalized form, Eq. (3.130) is

$$N_k \frac{dq'}{dt'} \pm \frac{dp'}{dt'} + q' = -\frac{1}{K_s} \int_0^{t'} \phi_z \frac{dq'}{d\tau'}(t' - \tau') \, d\tau'$$

$$\mp \int_0^{t'} \phi_y \frac{dp'}{d\tau'}(t' - \tau') \, d\tau', \qquad \frac{dx'}{dt'} = \pm 1 \qquad (3.132)$$

A comparison of quasi-characteristic equation (3.132) with characteristic equation (3.111) reveals that the left-hand sides of the two equations are identical. However, the right-hand side of Eq. (3.132) has two nonzero terms. The first of these terms represents the effects of the time-dependent friction. The second term is an integral formulation of the heat transfer effects. The effects of these terms will become evident when we compare the results obtained by the quasi method with those from the ordinary method. First, however, let us indicate how the integral terms in Eq. (3.132) are evaluated.

Figure 3.38 shows the grid of characteristics lines for the quasi method of characteristics without throughflow. The dashed lines are the position lines already discussed in connection with the time-independent characteristics (Fig. 3.29). The left-hand side of Eq. (3.132) is evaluated along these position lines exactly as before. The integrals, on the other hand, are evaluated along lines of fixed position $(dx'/dt' = 0)$. For example, suppose we want to determine

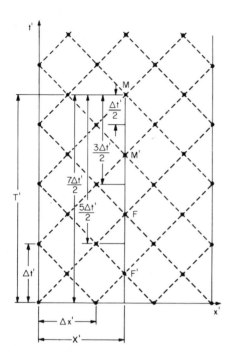

Fig. 3.38 Characteristic grid for small-amplitude signals with heat transfer.

the signal variables at point M with coordinates X', T'. The contribution of the integral terms depends on the values of the signal variables at the nodes along the line $MM'FF'$. The approximate value of each integral is

$$\int_0^{t'} \phi_z \frac{dq'}{d\tau'} d\tau' = \sum_{i=1}^{m} \phi_z \left(\frac{[2i-1]\Delta t'/2}{N_k K_s} \right)$$

$$\times [q'(X', T' - (i-1)\Delta t') - q'(X', T' - i\Delta t')] \quad (3.133a)$$

$$\int_0^{t'} \phi_y \frac{dp'}{d\tau'} d\tau' = \sum_{i=1}^{m} \phi_y \left(\frac{[2i-1]\Delta t'/2}{N_k K_s N_p} \right)$$

$$\times [p'(X', T' - (i-1)\Delta t') - p'(X', T' - i\Delta t')] \quad (3.133b)$$

where i are the integer values between 1 and the integer m, and m must be large enough to traverse all the nodes back to the initial time. The arguments of the weighting functions depend on the difference in time from the point in question

3.12 The Quasi Method of Characteristics for Small-Amplitude Signals

(*M*) to the time midway between the previous nodal points whose contribution is being estimated. Thus at point *M* the first three terms in the integral approximation are

$$\int_0^{t'} \phi_z \frac{dq'}{d\tau'} d\tau' = \phi_z\left(\frac{\Delta t'/2}{N_k K_s}\right)[q_M' - q_{M'}'] + \phi_z\left(\frac{3\,\Delta t'/2}{N_k K_s}\right)[q_{M'}' - q_F']$$

$$+ \phi_z\left(\frac{5\,\Delta t'/2}{N_k K_s}\right)[q_F' - q_{F'}'] + \cdots \quad (3.134a)$$

$$\int_0^{t'} \phi_y \frac{dp'}{d\tau'} d\tau' = \phi_y\left(\frac{\Delta t'/2}{N_k N_p K_s}\right)[p_M' - p_{M'}'] + \phi_y\left(\frac{3\,\Delta t'/2}{N_k N_p K_s}\right)[p_{M'}' - p_F']$$

$$+ \phi_y\left(\frac{5\,\Delta t'/2}{N_k N_p K_s}\right)[p_F' - p_{F'}'] + \cdots \quad (3.134b)$$

The number of terms required to obtain a close approximation to the integrals in Eq. (3.132) depends on the signal waveform. Although the weighting function decreases with time, the derivatives of the signal variables (approximated by the difference) may have large values at particular points in time. Thus, for example, a step or pulse signal may contribute significantly to the integrals for the entire duration of the response. As a result, computers with large storage capacities are needed to apply the quasi method of characteristics.

The computational procedure for the quasi method is the same as for the ordinary method except for the inclusion of the integral terms. Figures 3.39 and 3.40 compare the two methods for the response of a blocked line to step and pulse inputs. The increased friction and heat transfer effects of the quasi method lead to increased attenuation and time shift in the response.

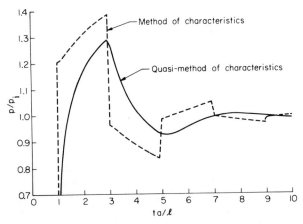

Fig. 3.39 Comparison of blocked-line step response by the method and quasi method of characteristics ($N_k = 1.0$).

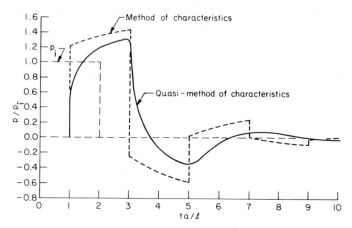

Fig. 3.40 Comparison of blocked-line pulse response by the method and quasi method of characteristics ($N_k = 1.0$).

Figure 3.41 shows the quasi method results for the magnitude of the first two overshoots and the first undershoot due to a step input when the line is blocked. The independent variable is the characteristic number N_k. When N_k is less than 0.45 there are no overshoots and the signal approaches the steady state value monotonically. For N_k between 0.45 and 0.55 there is a small first overshoot but practically no first undershoot. As N_k increases the magnitudes of the overshoots and undershoots continually increase. From Fig. 3.41 and the transport times ($ta/l = 1, 3, 5, \ldots$) for full wave reflections, it is possible to get a quick estimate of the shape of the step response.

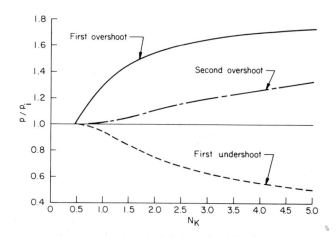

Fig. 3.41 Results of step inputs to blocked lines by the quasi method of characteristics.

3.13 THE METHOD OF CHARACTERISTICS FOR LARGE-AMPLITUDE SIGNALS

When the flow is compressible, the p and q variables do not completely determine the state of the fluid. For this reason characteristic equations (3.111) and (3.132) are difficult to apply directly to lines with large-amplitude signals. We may avoid the additional difficulties by using the dependent signal variables u and a for compressible flow problems.

The characteristic equations for the u and a variables are developed by first relating the local speed of sound a to the static pressure. The combination of Eq. (3.101) and the relation $a = [\gamma p/\rho]^{1/2}$ yields

$$\frac{dp}{p} = \left(\frac{2\gamma}{\gamma - 1}\right)\frac{da}{a} - \frac{dS}{(\gamma - 1)c_v} \tag{3.135}$$

Equation (3.135) can now be used to eliminate the pressure variable from Eqs. (3.99) and (3.102) with the result

$$\rho\frac{\partial u}{\partial t} + \rho u \frac{\partial u}{\partial x} + \left(\frac{2}{\gamma - 1}\right)\rho a \frac{\partial a}{\partial x} - \frac{\rho a^2}{\gamma(\gamma - 1)c_v}\frac{\partial S}{\partial x} = \left(\overline{\frac{\partial \tau}{\partial y}}\right)_w \tag{3.136a}$$

$$\left(\frac{2}{\gamma - 1}\right)\left(\frac{1}{a}\frac{\partial a}{\partial t} + \frac{u}{a}\frac{\partial a}{\partial x}\right) + \frac{\partial u}{\partial x} - \frac{1}{c_v(\gamma - 1)}\left(\frac{\partial S}{\partial t} + u\frac{\partial S}{\partial x}\right) = 0 \tag{3.136b}$$

The multiplication of Eq. (3.136b) by K_3 and subsequent addition to Eq. (3.136a) leads to

$$\frac{2K_3}{(\gamma - 1)a}\left[\frac{\partial a}{\partial t} + \left(u + \frac{a^2}{K_3}\right)\frac{\partial a}{\partial x}\right] + \left[\frac{\partial u}{\partial t} + (u + K_3)\frac{\partial u}{\partial x}\right]$$

$$= \frac{1}{\rho}\left(\overline{\frac{\partial \tau}{\partial y}}\right)_w + \frac{K_3}{R_g}\left(\frac{\partial S}{\partial t} + u\frac{\partial S}{\partial x}\right) + \frac{a^2}{\gamma R_g}\frac{\partial S}{\partial x} \tag{3.137}$$

where R_g is the universal gas constant. When $K_3 = \pm a$, Eq. (3.137) transforms to the characteristic equation sets

$$\frac{du}{dt} + \left(\frac{2}{\gamma - 1}\right)\frac{da}{dt} = \frac{1}{\rho}\left(\overline{\frac{\partial \tau}{\partial y}}\right)_w + \frac{a}{R_g}\frac{dS}{dt} + \frac{a^2}{\gamma R_g}\frac{\partial S}{\partial x}$$

$$(dx/dt)_{u,a} = u + a, \qquad (dx/dt)_S = u \tag{3.138a}$$

$$\frac{du}{dt} - \left(\frac{2}{\gamma - 1}\right)\frac{da}{dt} = \frac{1}{\rho}\left(\overline{\frac{\partial \tau}{\partial y}}\right)_w - \frac{a}{R_g}\frac{dS}{dt} + \frac{a^2}{\gamma R_g}\frac{\partial S}{\partial x}$$

$$(dx/dt)_{u,a} = u - a, \qquad (dx/dt)_S = u \tag{3.138b}$$

A square law frictional relation is usually more compatible with large-amplitude signals. Thus we assume that $(\partial \tau/\partial y)_w = -f\rho u^2/2d$, where f is the friction factor and d the line diameter. Now we may normalize equation sets (3.138a) and (3.138b) with $x' = x/l$, $a' = a/a_f$, $u' = u/a_f$, and $t' = ta_f/l$ (where a_f is a reference sonic velocity) to obtain

$$\frac{du'}{dt'} + \left(\frac{2}{\gamma-1}\right)\frac{da'}{dt'} = -K_4 u'^2 + \frac{a'}{R_g}\frac{dS}{dt'} + \frac{a'^2}{\gamma R_g}\frac{\partial S}{\partial x'}$$

$$(dx'/dt')_{u',a'} = u' + a', \qquad (dx'/dt')_S = u' \qquad (3.139a)$$

$$\frac{du'}{dt'} - \left(\frac{2}{\gamma-1}\right)\frac{da'}{dt'} = -K_4 u'^2 - \frac{a'}{R_g}\frac{dS}{dt'} + \frac{a'^2}{\gamma R_g}\frac{\partial S}{\partial x'}$$

$$(dx'/dt')_{u',a'} = u' - a', \qquad (dx'/dt')_S = u' \qquad (3.139b)$$

where $K_4 = fl/(2d)$.

Rudinger [27] demonstrates the application of the heat transfer effects. In this treatment, however, we neglect heat transfer so that the characteristic equation sets (3.139) reduce to

$$\frac{du'}{dt'} + \left(\frac{2}{\gamma-1}\right)\frac{da'}{dt'} = -K_4 u'^2, \qquad \frac{dx'}{dt'} = u' + a' \qquad (3.140a)$$

$$\frac{du'}{dt'} - \left(\frac{2}{\gamma-1}\right)\frac{da'}{dt'} = -K_4 u'^2, \qquad \frac{dx'}{dt'} = u' - a' \qquad (3.140b)$$

Figure 3.42 shows the grid (with fixed time intervals and positions) that is

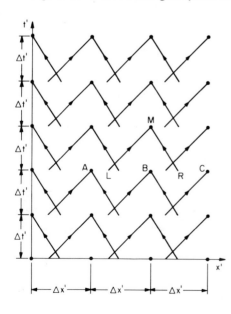

Fig. 3.42 Characteristic grid for large-amplitude signals.

3.13 The Method of Characteristics for Large-Amplitude Signals

used to calculate the variables in equation sets (3.140). The nodal points are represented by solid dots and the variables are only computed for these points. The characteristic lines that pass through the nodal points do not emanate from the nodal points of a previous time step. This is unlike the small-amplitude case and is a direct consequence of the appreciable magnitude of the velocity u'. As a result the signal variables corresponding to the characteristic lines must be obtained by interpolation between the previous time nodal points. For example, to calculate the variables at point M requires the variables at points L and R. These, in turn, are obtained by interpolation between A and B, and B and C, respectively. A first-order finite difference approximation to equation sets (3.140) gives

$$u_M' + 5a_M' = u_L' + 5a_L' - K_4 u'^2(\Delta t') \tag{3.141a}$$

$$u_M' - 5a_M' = u_R' - 5a_R' - K_4 u'^2(\Delta t') \tag{3.141b}$$

where the gas has a specific heat ratio (γ) of 1.4. Linear interpolation yields the signal variables at points L and R as

$$u_{L,R}' = (u_B' + \theta_n Y_{L,R})/(1 - \theta_n V_{L,R}) \tag{3.142a}$$

$$a_{L,R}' = (a_B' - \theta_n Y_{L,R})/(1 - \theta_n V_{L,R}) \tag{3.142b}$$

where

$$\theta_n = \Delta t'/\Delta x'$$
$$Y_L = u_A' a_B' - u_B' a_A'$$
$$Y_R = u_C' a_B' - u_B' a_C'$$
$$V_L = u_A' + a_A' - u_B' - a_B'$$
$$V_R = u_C' - a_C' - u_B' + a_B'$$

The signal variables at interior points are calculated by substituting the interpolated values from Eqs. (3.142) into Eqs. (3.141) and then solving the resulting simultaneous equations. To determine the boundary node signal variables we use Eq. (3.141a) for a right boundary and Eq. (3.141b) for a left boundary. After the appropriate interpolations from Eq. (3.142) we then obtain one equation in the two unknown signal variables at each boundary. Thus, as in the small-amplitude case, we require an additional condition at each boundary. We often use constant energy as the left boundary condition and therefore

$$a_{LB}'^2 + 0.2 u_{LB}'^2 = a_E'^2 \tag{3.143}$$

where a_E' is a preselected constant which depends on the energy level and the subscripts refer to the left boundary. When the line is terminated by an orifice of diameter d_o, the continuity equation with the assumption of isentropic flow through the orifice leads to

$$\frac{u_{RB}' a_{RB}'^5}{[5(a_{RB}'^2 - 1) + u_{RB}'^2]^{1/2}} = \frac{d_o^2}{d^2} \tag{3.144}$$

where the subscripts refer to the right boundary. The simultaneous solution of Eqs. (3.141a) and (3.144) provides the right boundary signal variables. However, due to the order of Eq. (3.144), the solution is obtained numerically as indicated by the ROOT subprogram in Appendix E.

Figure 3.43 shows the normalized pressure ($p_l' = a_l'^7$) at the end of a blocked line that has an input step of $a_E' = 1.03 [p_E' = 1.23]$. The solid line is the response calculated from a graphical representation of the characteristic lines. The dashed line is the computer solution with fixed time intervals. In this case the mesh size is 40 and $\theta = 0.75$. Note that θ must always be chosen less than the smallest value of $|1/(u' + a')|$ or else the interpolation will fall outside the range of the closest previous points and the solution will become unstable. Benson et al. [29] present a figure similar to Fig. 3.43 in which they show that a mesh method with variable time intervals produces even closer agreement to the graphical (nonmesh) method.

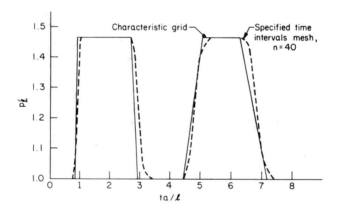

Fig. 3.43 Comparison of mesh and nonmesh method of characteristics for large-amplitude signals ($\theta = 0.75$) on blocked frictionless line.

Figure 3.44 shows the method of characteristic solutions for the step response of a frictionless line terminated by an orifice ($d_o/d = 0.2$). The curves represent three different entrance stagnation pressure levels ($p_E' = 1.035, 1.110,$ and 1.188). The ordinate is the normalized pressure in the line immediately upstream of the orifice and the abscissa is the normalized time. For $p_E' = 1.188$ the response has overshoots and undershoots. When $p_E' = 1.110$, the step response is essentially also a step function. The presentation given in Section 3.14.2 shows that this latter curve is a "matched" condition and can be calculated analytically. The small disturbances on the response curves (particularly at $p_E' = 1.110$) are the results of the interpolation procedure. The disturbances occur near the time that the wave reflection from the entrance reaches the exit (i.e., $ta_0/l = 3, 5, 7, \ldots$). During this time period interpolation causes information to propagate ahead of the graphical characteristic line. The disturbances can be reduced by

3.13 The Method of Characteristics for Large-Amplitude Signals

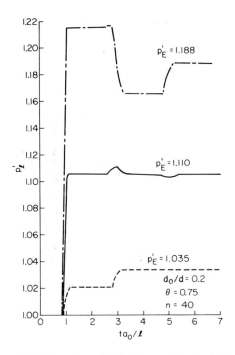

Fig. 3.44 Large-amplitude signals on frictionless line terminated by an orifice ($d_o/d = 0.2$).

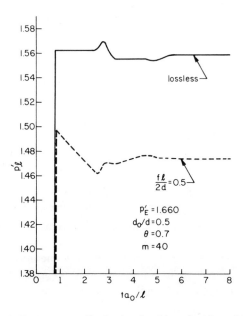

Fig. 3.45 Large-amplitude signals with and without friction.

increasing the mesh ratio θ, but care must be exercised to prevent the solution from becoming unstable.

Figure 3.45 shows the step response of a terminated line ($d_o/d = 0.5$) with and without friction. The normalized input stagnation pressure is 1.660. This pressure level is close to the "matched" condition for a frictionless line. The dashed and solid lines represent the response with and without friction, respectively. Friction acts to reduce the average pressure level and also alters the shape of the response. The initial plateau is no longer horizontal but decays monotonically. Lines with small orifice terminations ($d_o/d < 0.2$) do not show significant frictional effects since the line velocities are small. These lines can be treated as frictionless with respect to "matching" (see Section 3.14.2). On the other hand, in the line shown in Fig. 3.45 friction has a considerable influence.

3.14 MATCHING

In transmission line theory matching occurs when the line is terminated by its characteristic impedance. At the matched condition there are no signal reflections back into the line. Thus the load or terminating impedance that causes no reflections or a cancellation of reflections is the characteristic impedance. Since matching is equivalent to the condition of no reflections, the method of characteristics can be used to determine the characteristic (matched) impedance of a line. In this section we derive the characteristic impedance from the method of characteristics for a few simple cases. In the case of small-amplitude signals, the matching impedance approaches the characteristic impedance obtained analytically from transmission line theory.

3.14.1 Small-Amplitude Matching with Average Friction

The characteristic impedance $Z_c(s)$, given by transmission line theory, for a line with average friction is from Eq. (3.49b)

$$Z_c(s) = Z_{ca}[1 + R/L_a s]^{1/2} \tag{3.145}$$

where s is the Laplace operator.

Let us now determine the characteristic impedance by the method of characteristics and compare the result with Eq. (3.145). Since the method of characteristics operates in the time domain, however, we will have to take the Laplace transform of the result.

To derive the "matched" impedance from the method of characteristics, refer to the mesh arrangement shown in Fig. 3.46. We assume that the fluid line is perturbed by a small step change in flow at $x' = 0$. This means that $q_A' = q_H' = q_k' \neq 0$. To calculate the characteristic impedance, recall that it is also the input impedance of an infinite line (i.e., no reflections back to the input). Thus, it is

3.14 Matching

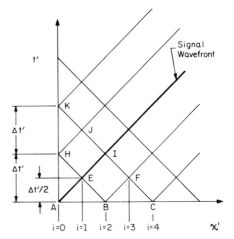

Fig. 3.46 Matching from the characteristic grid for small-amplitude signals.

only necessary to find p_A', p_H', and p_k' and to divide by the magnitude of the flow variable. For simplicity, consider the first-order finite-difference approximation to Eqs. (3.111a) and (3.111b):

$$N_k q' + p' = M q_L' + p_L' \tag{3.146a}$$

$$N_k q' - p' = M q_R' - p_R' \tag{3.146b}$$

where $M = N_k - \Delta t'/2$ and the subscripts L and R refer to points from the previous time step ($\Delta t'/2$) to the left and right of the point to be calculated.

The matched condition fixes the values of all the signal variables to the right of the signal wavefront line (Fig. 3.46) equal to zero (i.e., $q_B' = q_F' = q_C' = p_B' = p_F' = p_C' = 0$). This makes the line appear infinitely long from the standpoint of the signal variables at the line entrance (points A, H, K, etc.). If Eqs. (3.146) are applied consecutively to points along the line AEI, etc., the pressure and flow become

$$p' = \left(\frac{M + N_k}{2}\right)^i N_k^{1-i} q_A' \tag{3.147a}$$

$$q' = \left(\frac{M + N_k}{2}\right)^i N_k^{-i} q_A' \tag{3.147b}$$

where i takes on integer values 0, 1, 2, ... at each time step. For example, at point A, $i = 0$, at E, $i = 1$, and at I, $i = 2$. The signal variables along the wavefront are therefore

$$p_A' = N_k q_A', \qquad\qquad q_A' = q_A' \tag{3.148a}$$

$$p_E' = [(M + N_k)/2] q_A', \qquad q_E' = [(M + N_k)/2N_k] q_A' \tag{3.148b}$$

$$p_I' = \left(\frac{M + N_k}{2}\right)^2 \frac{1}{N_k} q_A', \qquad q_I' = \left(\frac{M + N_k}{2N_k}\right)^2 q_A' \tag{3.148c}$$

The impedance (p'/q') of all the points on the wavefront line is the same and is equal to

$$p'/q' = N_k \tag{3.149}$$

or, in terms of the dimensional signal variables, is

$$p/q = Z_{ca} \tag{3.150}$$

Equation (3.150) indicates that the characteristic impedance along the wavefront is the same as the characteristic impedance of a lossless line. This result was obtained because, in effect, there are no frictional losses in a motionless fluid. Repeated application of Eq. (3.146) yields the signal variables at the nodes behind the wavefront (i.e., H, J, K). The results for the impedance at points H and K are

$$Z_H = Z_{ca}[\tfrac{3}{2} - \tfrac{1}{2}(1 - Rt/2L_a)^2] \tag{3.151a}$$

$$Z_k = Z_{ca}[\tfrac{15}{8} - \tfrac{3}{4}(1 - Rt/4L_a)^2 - \tfrac{1}{8}(1 - Rt/4L_a)^4] \tag{3.151b}$$

or, in expanded form,

$$Z_H = Z_{ca}\left[1 + \frac{Rt}{2L_a} - \frac{1}{8}\left(\frac{Rt}{L_a}\right)^2\right] \tag{3.152a}$$

$$Z_k = Z_{ca}\left[1 + \frac{1}{2}\left(\frac{Rt}{L_a}\right) - \frac{3}{32}\left(\frac{Rt}{L_a}\right)^2 + \frac{1}{128}\left(\frac{Rt}{L_a}\right)^3 - \frac{1}{2048}\left(\frac{Rt}{L_a}\right)^4\right] \tag{3.152b}$$

The Laplace transforms of Eqs. (3.152) are

$$Z_H(s) = Z_{ca}\left(\frac{1}{s} + \frac{1}{2}\frac{R}{L_a s^2} - \frac{1}{4}\frac{R^2}{L_a^2 s^3}\right) \tag{3.153a}$$

$$Z_k(s) = Z_{ca}\left(\frac{1}{s} + \frac{1}{2}\frac{R}{L_a s^2} - \frac{3}{16}\frac{R^2}{L_a^2 s^3} + \frac{3}{64}\frac{R^3}{L_a^3 s^4} - \frac{3}{256}\frac{R^4}{L_a^4 s^5}\right) \tag{3.153b}$$

The impedances given in Eqs. (3.153) were calculated from the characteristic equations with a step input. The transmission line theory result [Eq. (3.145)], on the other hand, is the impedance for an impulse. It is not practical to apply an impulse in the time domain where the characteristic method operates. Since an impulse is the derivative of a step function, we can bring the results into correspondence by differentiating Eq. (3.153). This is equivalent to multiplication by s, so that Eq. (3.153) becomes

$$Z_{cH}(s) = Z_{ca}[1 + \tfrac{1}{2}(R/L_a s) - \tfrac{1}{4}(R/L_a s)^2] \tag{3.154a}$$

$$Z_{ck}(s) = Z_{ca}\left[1 + \frac{1}{2}\left(\frac{R}{L_a s}\right) - \frac{3}{16}\left(\frac{R}{L_a s}\right)^2 + \frac{3}{64}\left(\frac{R}{L_a s}\right)^3 - \frac{3}{256}\left(\frac{R}{L_a s}\right)^4\right] \tag{3.154b}$$

3.14 Matching

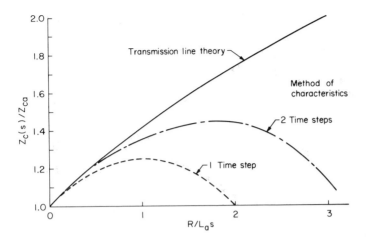

Fig. 3.47 Comparison of the characteristic impedance derived from the method of characteristics and from transmission line theory.

Figure 3.47 shows the ratio of the characteristic impedance to the adiabatic characteristic impedance as a function of $R/(L_a s)$. The curves designated as "1 time step" and "2 time steps" represent the impedances at points H and K, respectively [Eqs. (3.154a) and (3.154b)]. The impedance given by transmission line theory [Eq. (3.145)] is also shown. As more time steps are considered the result from the method of characteristics approaches the transmission line result. When the line is frictionless, $R/(L_a s) = 0$ and the characteristic impedance equals the adiabatic characteristic impedance. Under this condition the method of characteristics with any number of time steps is equivalent to the analytical result. Thus we may emphasize again that a frictionless line has a purely resistive characteristic impedance and can be matched with a resistive termination. This also holds true for finite-amplitude signals, as we shall see in the following section. However, lines with friction cannot be matched with a resistive termination except at a particular frequency.

3.14.2 Finite-Amplitude Matching of a Frictionless Line with a Step Input

Figure 3.48 shows a duct terminated by an orifice. At the entrance, a shutter separates the duct from a constant energy source. The sudden removal of the shutter starts a wave down the duct. This wave is represented by the rightward-going solid line in the $x't'$ plane. At the end of the line a portion of the wave strikes the end wall and reflects as a compressive wave. A portion of the wave surrounding the axis encounters the orifice and reflects as an expansive wave. Although the compressive wave reflection travels slightly faster than the expansive wave reflection it is still possible to get an approximate cancellation of the waves if the orifice area is of the correct size (matched). When the aperture size is not the matched area, the net result is a reflective wave that is represented by

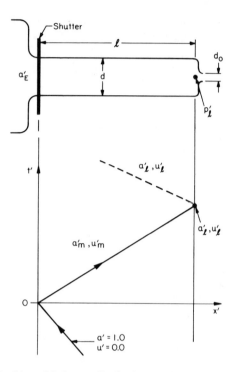

Fig. 3.48 Matching of finite-amplitude signals with an orifice termination.

the leftward-going dashed line. Keto [34] has shown that, for each energy level at the input, there is an orifice termination that causes approximate cancellation of the compression and expansion portions of the reflected wave. As a consequence, step inputs may appear to propagate without reflections if the duct is properly terminated. Let us use the method of characteristics to derive the orifice size that matches a frictionless line.

Consider the $x't'$ plane illustrated in Fig. 3.48. The first step in deriving the "matched" orifice is to find expressions for the signal variables throughout the plane. The signal variables a_M' and u_M' to the left of the rightward-traveling solid line are determined from the conditions:

(1) constant energy at the left boundary [Eq. (3.143)]

$$a_M'^2 + 0.2 u_M'^2 = a_E'^2 \tag{3.155}$$

(2) the initial conditions of the signal variables in the duct, $u' = 0$, $a' = 1$, which produce a leftward wave [Eq. (3.141b)] such that

$$u_M' - 5 a_M' = -5 \tag{3.156}$$

The simultaneous solution of Eqs. (3.155) and (3.156) yields

$$u_M' = \tfrac{5}{6}[(6 a_E'^2 - 5)^{1/2} - 1] \tag{3.157a}$$

$$a_M' = \tfrac{1}{6}[(6 a_E'^2 - 5)^{1/2} + 5] \tag{3.157b}$$

3.14 Matching

Equation (3.157) represents the signal variables in the region of the $x't'$ plane bounded by the rightward solid line and the leftward dashed line. Since, from Eq. (3.141a), $u_M' + 5a_M' = u_1' + 5a_1'$, the signal variables after reflection from the termination may be expressed as

$$u_1' = \tfrac{5}{6}[(2 - V)(6a_E'^2 - 5)^{1/2} + (4 - 5V)] \tag{3.158a}$$

$$a_1' = (V/6)[(6a_E'^2 - 5)^{1/2} + 5] \tag{3.158b}$$

where $V = a_1'/a_M'$. The necessary condition for a match is that $V = 1$ (i.e., $a_M' = a_1'$, $u_M' = u_1'$). That is, the wavefront must behave as if it were moving into stagnant fluid ($u' = 0$, $a' = 1$). For values of V other than unity the overshoot or undershoot of the first plateau in the response is set. Thus, in effect, Eq. (3.158) supplies the signal variables after the wavefront reaches the termination, whether the orifice matches or not. From the continuity equation the signal variables at the right boundary are related to the orifice size [Eq. (3.144)] by

$$\left(\frac{d_o}{d}\right)^2 = \frac{u_1' a_1'^5}{[5(a_1'^2 - 1) + u_1'^2]^{1/2}} \tag{3.159}$$

Equations (3.158) and (3.159) relate the orifice size to the energy level. Figure 3.49 shows the area ratio $(d_o/d)^2$ for a perfect match ($V = 1$) as a function of a_E'. When a_E' approaches unity the match requires a small orifice and eventually approaches the blocked condition. The maximum value of a_E' is 1.095. This is the energy level at which the flow through the matched orifice $[(d_o^2/d^2)_M = 0.591]$ reaches sonic velocity. When the terminating area ratio is larger than the matched area ratio the duct reaches steady state in staircase fashion (recall Fig. 3.44). Orifices smaller than the matched orifice produce alternate overshoots and undershoots in the response.

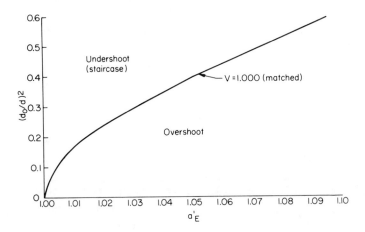

Fig. 3.49 The matched orifice for a step input to a lossless line.

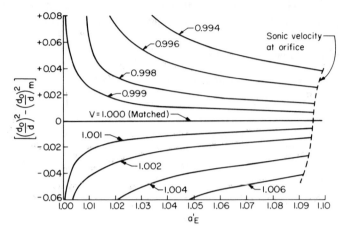

Fig. 3.50 Overshoot and undershoot caused by a mismatched orifice.

Figure 3.50 shows the magnitude of the first plateau in the response (V) when the orifice is not matched. The ordinate is the difference between the actual and matched areas of the terminating orifice and represents the degree of mismatch. The abscissa is the source energy level. At low energy levels the response is less sensitive to mismatching than at higher levels. For example, a duct terminated with $d_o^2/d^2 = 0.120$ and subject to a step change in energy equivalent to $a_E' = 1.01$ should produce an overshoot. From Eqs. (3.158) and (3.159) we find that, for a perfect match at this energy level, $d_o^2/d^2 = 0.162$. The degree of mismatch is the difference in area ratios $-(d_o^2/d^2)_M + d_o^2/d^2$ or -0.042. From Fig. 3.50, the ratio V for these conditions equals 1.002 and this means that the first plateau of the response is larger than for the matched condition. Note that if this same degree of mismatch (-0.042) occurred when the energy level was 1.091 the value of V would be 1.006 and the signal variables would have considerably larger first overshoot.

3.15 THE METHOD OF CHARACTERISTICS FOR FLUIDIC LINE CIRCUITS

The previous sections pertain to the dynamics of flow in a single line. In this section the small-amplitude method of characteristics is extended to include line networks. The extension consists essentially of supplying a new boundary condition. Three network elements are considered: (1) lines with a sudden change in section, (2) branches, and (3) lines terminated with a lumped volume. These three networks alone and in combination cover many of the line circuits that occur in fluidic systems.

3.15.1 Sudden Changes in Section

Figure 3.51 shows a network formed by two lines of different dimensions in series. This circuit is usually called a sudden enlargement or sudden contraction. Since the characteristic equations are valid only when the flow is one dimensional, the magnitude of the sudden change is limited. This is particularly important in the case of sudden enlargements where large separation regions may exist if the magnitude of the enlargement is too great. The problem is more serious with finite-amplitude signals.

The configuration shown in Fig. 3.51, with the aforementioned limitation, can be treated as two individual line segments. All that need be specified, in addition, is the equation for the variables at the junction section. For any single line, in a line network, Eq. (3.111) may be written as

$$(l_i/l_r)N_{ki}\, dq'/dt' \pm (R_r/R_i)\, dp'/dt' + q' = 0, \qquad dx'/dt' = \pm 1 \qquad (3.160)$$

where R is line resistance and the subscript r refers to the reference line and the subscript i the other line. The procedure is to designate one of the lines as the reference line and then to select a mesh size for this line. Suppose, for example, the reference line is line 1 on Fig. 3.51 and the mesh size is n_1. Then, since the specified time increment must be the same in each line. $\Delta x_1'$ equals $\Delta x_2'$. Thus

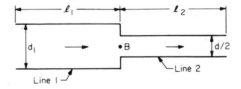

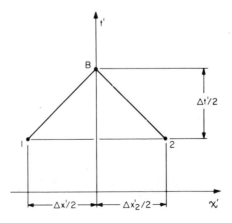

Fig. 3.51 Sudden change in section.

the condition that $l_1/n_1 = l_2/n_2$ must be satisfied. If there is no integer number n_2 that satisfies this condition, l_2 is adjusted. Thus it is not always possible to solve networks with the exact lengths desired. However, the choice of a large n_1 always permits a close approximation to the desired circuit dimensions.

With line 1 as the reference line the first-order finite-difference approximations to Eq. (3.160) yield

$$N_{k1}[q_B' - q_1'] + [p_B' - p_1'] + q_1'(\Delta t'/2) = 0 \quad (3.161a)$$

$$(l_2/l_1)N_{k2}[q_B' - q_2'] - (R_1/R_2)[p_B' - p_2'] + q_2'(\Delta t'/2) = 0 \quad (3.161b)$$

and the simultaneous solution of Eqs. (3.161a) and (3.161b) provide the junction variables as

$$p_B' = \frac{(1/N_{k1})p_1' + (R_1 l_1/R_2 l_2 N_{k2})p_2' + (1/N_{k1})(N_{k1} - \Delta t'/2)q_1' - (l_1/N_{k2} l_2)[(l_2/l_1)N_{k2} - \Delta t'/2]q_2'}{(1/N_{k1} + R_1 l_1/R_2 l_2 N_{k2})} \quad (3.162a)$$

$$q_B' = \frac{(N_{k1} - \Delta t'/2)q_1' + (R_1/R_2)[(l_2/l_1)N_{k2} - \Delta t'/2]q_2' + p_1' - p_2'}{N_{k2} + R_2 l_2 N_{k2}/R_1 l_1} \quad (3.162b)$$

3.15.2 Branches

The sudden change in section is a special case of the more general branch configuration. Figure 3.52 shows the signal variables involved at a branch point.

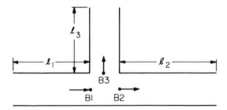

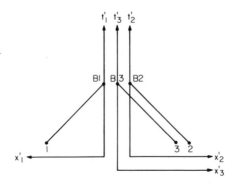

Fig. 3.52 Branch point.

3.15 The Method of Characteristics for Fluidic Line Circuits

The line designations are the same as in the sudden change of section except that a third line (the branch line, number 3) is added. If line 1 is the reference line, the application of Eq. (3.160) to the branch point results in

$$N_{k1}[q'_{B1} - q_1'] + [p'_{B1} - p_1'] + q_1' \Delta t'/2 = 0 \quad (3.163a)$$

$$(l_2/l_1)N_{k2}[q'_{B2} - q_2'] - (R_1/R_2)[p'_{B2} - p_2'] + q_2' \Delta t'/2 = 0 \quad (3.163b)$$

$$(l_3/l_1)N_{k3}[q'_{B3} - q_3'] - (R_1/R_3)[p'_{B3} - p_3'] + q_3' \Delta t'/2 = 0 \quad (3.163c)$$

The usual Kirchhoff law equations $p'_{B1} = p'_{B2} = p'_{B3}$ and $q'_{B1} = q'_{B2} + q'_{B3}$ provide enough information to determine the branch point signal variables. The branch point pressure is

$$p_B' = \frac{[(N_{k1} - \Delta t'/2)q_1' + p_1']/N_{k1} - \sum_{i=2}^{3}\{[(l_i/l_1)N_{ki} - \Delta t'/2]q_i' - (R_1/R_i)p_i'\}/(l_i N_{ki}/l_1)}{\sum_{i=1}^{3} R_1 l_1/(R_i l_i N_{ki})}$$

(3.164)

and the branch flows can now be determined.

Equation (3.164) reduces to Eq. (3.162a) for the special case of the sudden enlargement by setting the upper limit of the summation as 2 rather than 3.

3.15.3 Line Terminated by a Lumped Volume

Figure 3.53 shows a distributed line terminated by a lumped volume (i.e., all pressures are equal within the volume). The boundary signal variables are determined from one characteristic equation and one boundary condition. The characteristic equation for a particular time B on the boundary is

$$N_k q_B' + p_B' = (N_k - \Delta t'/2)q_L' \quad (3.165)$$

The boundary condition is the relation between pressure and flow in a lumped capacitance. In normalized form this is

$$p' = \left(\frac{l/a}{RlC}\right) \int_0^{t'} q' \, dt' \quad (3.166)$$

To obtain a relation between p_B' and q_B' from Eq. (3.166) requires the integration of all flows up to the time in question. A trapezoidal approximation to the integral yields

$$p_B' = \left(\frac{l/a}{RlC}\right)\frac{q_B'}{2}\Delta t' + \frac{\Delta t' l/a}{RlC}\sum_{i=0}^{B-1} q_i' \quad (3.167)$$

The boundary signal variables are then calculated from Eqs. (3.165) and (3.167).

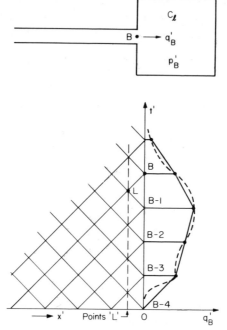

Fig. 3.53 Line terminated by lumped volume.

PROBLEMS

3.1 A transmission line has a series impedance $Zl = 1 + j0.01\omega$ and a shunt admittance $Yl = j0.2\omega$:

(a) What is the characteristic impedance of the line?

(b) For what frequency does the real portion of the characteristic impedance equal the imaginary portion?

(c) What happens to the characteristic impedance at high frequencies?

3.2 If a transmission line has a series impedance and a shunt admittance that are given by $Zl = j9\omega$ and $Yl = j\omega$:

(a) Find the characteristic impedance of the line.

(b) Find the input impedance for the blocked line.

(c) Find the input impedance for the open line.

(d) Characterize the type of impedance from parts a, b, and c (resistive, inertive, capacitive).

3.3 If the transmission line in Problem 3.2 is blocked:

(a) Find the frequency of the first resonant peak.

(b) Show that the relation between the pressures at the end and at the entrance for low frequencies is approximately

$$P_1/P_i = 1 + 4.5\omega^2$$

3.4 A transmission line has a diameter of 6.892 mm. The fluid medium is air for which $v = 14.85 \times 10^{-6}$ m^2/sec and $N_p = 0.7$. The operating frequency is 125 rad/sec:

(a) Show that the series impedance and shunt admittance for the line with average friction is
$$Z/\omega L = j + 0.080$$
$$Y/\omega C = j$$

(b) Show that the series impedance and shunt admittance for the line with distributed friction is approximately
$$Z/\omega L = 1.141j + 0.171$$
$$Y/\omega C = 1.067j + 0.063$$

(c) Find the series impedance and shunt admittance for the average and distributed friction cases when the frequency is raised to 12,500 rad/sec.

3.5 A lossless transmission line 3 mm in diameter and 1-m long is terminated by an orifice that has a linearized resistance of 2×10^6 kN-sec/m^5. If the line–orifice combination is put into a single lump, for what frequency will the real part of the lumped input impedance be a maximum? At this frequency what is the magnitude of the resistive portion of the lumped input impedance?

3.6 A 4.0-mm diameter line 2.0-m long is connected to a line of 2.0-mm diameter, 2.0-m long. A low-amplitude sinusoidal frequency of 1000 rad/sec is applied at the entrance to the larger line. The fluid is air. Find:

(a) The magnitude of the pressure ratio between the two ends of a 4.0-mm diameter line when the 2.0-mm diameter line is open.
(b) The magnitude of this ratio when the 2.0-mm diameter line is blocked.
(c) The frequencies of the first resonant peak for the smaller line blocked and open.

3.7

(a) Find the propagation factor for a circular line $d = 2.18$ mm and $l = 33.0$ m operating at 2000 rad/sec with air.
(b) Find the propagation factor for a square line (2.0 mm on a side) and $l = 33.0$ m operating at 1800 rad/sec.
(c) Find the propagation factor for a rectangular line (4.0 by 1.0 mm) and $l = 33.0$ m operating at 1800 rad/sec.

3.8 Show that the pressure P_m at the midpoint of a transmission line is:

(a) For an open line
$$P_m/P_i = \tfrac{1}{2} \text{sech } \Gamma l/2$$

(b) For a blocked line
$$P_m/P_i = \cosh \Gamma l/2 - \tanh \Gamma l \sinh \Gamma l/2$$

3.9 Find the magnitude and phase of the pressure at the midpoint of an open rectangular line (2.0 × 1.0 mm × 3.3 m) when $|P_i| = 0.1$ kN/m² and $\omega = 100$ rad/sec, assuming the line must be treated as a transmission line.

3.10 Find the magnitude and phase of the pressure in the line of Problem 3.9 from a single-lumped TEE-section model.

3.11 A matched transmission line with $d = 2.0$ mm and $l = 1.0$ m is subjected to a unit step input of small amplitude. Compare the pressure response at the end of the line calculated by transmission line theory and by the method of characteristics. The fluid medium is air ($v = 14.85 \times 10^{-6}$ m²/sec, $a = 330$ m/sec).

3.12 A finite-amplitude step in pressure $p_E' = 1.072$ is applied at the entrance to a frictionless air line of diameter $d = 5.0$ mm. What is the diameter of the orifice that must be attached to the end of the line so that there will be no reflections?

NOMENCLATURE

a	Speed of sound, m/sec
a_f	Reference speed of sound, m/sec
a'	Normalized speed of sound, a/a_f
a_E'	Normalized stagnation speed of sound
A	Cross-sectional area, m²
b	Width of rectangular line, m
c_0	Speed of wave propagation in line, m/sec
c_p	Specific heat at constant pressure, J/kg-°K
c_v	Specific heat at constant volume, J/kg-°K
C	Capacitance per unit length, m⁴/kN
C_a	Adiabatic capacitance per unit length, m⁴/kN
d	Diameter of line, m
d_o	Diameter of orifice, m
E	General across variable in frequency or Laplace domain
E_i	Input across variable in frequency or Laplace domain
f	Friction factor
G	Shunt conductance per unit length, m⁴/kN-sec
h	Depth of rectangular line, m
$H(t)$	Unit step function
i	Integer number
I	General through variable in frequency or Laplace domain
I_1	Input through variable in frequency or Laplace domain
j	$(-1)^{1/2}$, complex operator
J_0	Bessel function of first kind zeroth order
J_1	Bessel function of first kind first order
k	Thermal conductivity, J/m-sec-°K
K_1	Arbitrary proportionality constant, kN-sec/m³
K_2	Arbitrary proportionality constant, kN-sec/m⁵
K_3	Arbitrary proportionality constant, m/sec
K_4	Friction number, $fl/(2d)$
K_5	Shape factor

Nomenclature

l	Line length, m
L	Inertance per unit length, kN-sec²/m⁶
L_a	Adiabatic inertance per unit length, kN-sec²/m⁶
m	Integer constant
M	$[N_k - \Delta t'/2]$
n	Number of mesh divisions
n_l	Number of lumps
N_k	Characteristic number ($= Z_{ca}/Rl$)
N_p	Prandtl number ($= \mu c_p/k$)
N_{sk}	Stokes number ($= 32\omega/\omega_v$)
p	Static pressure, kN/m²
p_f	Reference pressure, kN/m²
p'	Normalized pressure, p/p_f
p_i	Static pressure at entrance to line section, kN/m²
p_l	Static pressure at end of line section, kN/m²
p_E'	Normalized absolute stagnation pressure
P	Static pressure in frequency or Laplace domain, (kN-sec)/m²
P_i	Static pressure at line entrance in frequency or Laplace domain, (kN-sec)/m²
P_l	Static pressure at line exit in frequency or Laplace domain, (kN-sec)/m²
q	Volume flow, m³/sec
q_f	Reference volume flow, m³/sec
q'	Normalized volume flow, q/q_f
Q	Volume flow in frequency or Laplace domain, m³
r	Radial dimension, m
r_w	Radius of line, m
r_h	Hydraulic radius, m
R	Resistance per unit length, kN-sec/m⁶
R_g	Gas constant, m²/sec²-°K
R_a	Approximate resistance per unit length, kN-sec/m⁶
R_o	Linearized orifice resistance, kN-sec/m⁵
R_L	Load resistance, kN-sec/m⁵
s	Laplace operator, sec⁻¹
S	Entropy, J/kg-°K
S'	Normalized entropy, S/c_v
t	Time, sec
t_f	Nominal transmission time ($= l/a_f$), sec
t'	Normalized time, t/t_f
T	Temperature, °K
ΔT	Sampling period, sec
T'	Normalized time at a specific point
u	Axial velocity, m/sec
$\bar{u}$	Average axial velocity, m/sec
u'	Normalized axial velocity, u/a
U	Axial velocity in frequency or Laplace domain, m
V	Overshoot or undershoot ratio
$W_z(s)$	Impedance weighting factor, sec
$W_Y(s)$	Admittance weighting factor, sec
x	Axial dimension, m
x'	Normalized axial dimension, x/l
y	Transverse dimension, m
Y	Shunt admittance per unit length, m⁴/kN-sec
Y_c	Characteristic admittance, m⁵/kN-sec
Y_{ca}	Adiabatic characteristic admittance, m⁵/kN-sec

z	Transverse dimension, m
Z	Series impedance per unit length, kN-sec/m^6
Z_c	Characteristic impedance, kN-sec/m^5
Z_{ca}	Adiabatic characteristic impedance, kN-sec/m^5
Z_{in}	Input impedance of a terminated line, kN-sec/m^5
Z_L	Load impedance, kN-sec/m^5
α	Attenuation factor per unit length, Nρ/m
β	Phase factor per unit length, rad/m
β_M	Bulk modulus, kN/m^2
γ	Ratio of specific heats
Γ	Complex propagation factor per unit length ($\alpha + j\beta$)
Γ_a	Adiabatic complex propagation factor per unit length
θ	Temperature in frequency or Laplace domain, °K-sec
θ_n	Mesh ratio
μ	Absolute viscosity, kg/m-sec
ν	Kinematic viscosity, m^2/sec
ρ	Density, kg/m^3
σ	Aspect ratio (b/h)
τ	Shear stress, kN/m^2
ϕ_Z	Normalized impedance weighting factor
ϕ_Y	Normalized admittance weighting factor
ω	Angular frequency, rad/sec
ω_v	Characteristic viscous frequency, rad/sec

REFERENCES

1. L. A. Ware and H. R. Reed, "Communications Circuits," Chapters 5 and 6, pp. 71–120. Wiley, New York, 1949.
2. R. W. P. King, "Transmission Line Theory." McGraw-Hill, New York, 1955.
3. J. W. S. Rayleigh, "The Theory of Sound," Vol. II, p. 327. Dover, New York, 1945.
4. J. B. Crandall, "Theory of Vibrating Systems and Sound." Van Nostrand-Reinhold, Princeton, New Jersey, 1926.
5. F. B. Daniels, Acoustical impedance of enclosures. *J. Acoust. Soc. Amer.* **19**, No. 4, 569 (1947).
6. A. S. Iberall, Attenuation of oscillatory pressures in instrument lines. *J. Res. Nat. Bur. Stand.* **45**, Res. Paper 2115 (1950).
7. C. P. Rohmann and E. C. Grogan, On the dynamics of pneumatic transmission lines. *Trans. ASME* **79**, 853–874 (1957).
8. N. B. Nichols, The linear properties of pneumatic transmission lines. *Trans. Instrum. Soc. Amer.* **1**, 5–14 (1962).
9. J. T. Karam and M. E. Franke, The frequency response of pneumatic lines. *J. Basic Eng. Trans. ASME Ser. D* **90**, No. 2, 371–378 (1967).
10. F. T. Brown, The transient response of fluid lines, *J. Basic Eng. Trans. ASME Ser. D*, **84**, No. 4, 547–553 (1962).
11. F. T. Brown, "Pneumatic Pulse Transmission with Bistable-Jet-Relay Reception and Amplification." Sc.D. Thesis, MIT, May 1962.
12. R. Kantola, "Transient Response of Fluid Lines Including Frequency Modulated Inputs." ASME Paper 70-WA/Flcs-1, November 1970. Published in *J. Basic Eng. Trans. ASME Ser. D* **93**, No. 2, 274 (1971).
13. J. T. Karam, Jr., "A Simple but Complete Solution for the Step Response of a Semi-Infinite Circular Fluid Transmission Line." ASME paper No. 71-WA/FE-10, December 1971, published in *J. Basic Eng. ASME Ser. D* **94**, No. 2, 455 (1972).

References

14. R. E. Goodson and R. G. Leonard, "A Survey of Modeling Techniques for Fluid Line Transients." ASME Paper 71-WA/FE-9; *J. Basic Eng. Ser. D* **94**, No. 2, 474 (1972).
15. H. Schaedel, A theoretical investigation of fluidic transmission lines with rectangular cross section. Paper K3, *Cranfield Conf.*, 3rd, Paper K3 (1968).
16. A. J. Healey and R. J. Carlson, "Frequency Response of Rectangular Pneumatic Transmission Lines." ASME Paper 69-WA/Flcs-5, Winter Meeting, November 1969.
17. R. Oldenburger and R. E. Goodson, Simplification of Hydraulic Line Dynamics by Use of Infinite Products. *J. Basic Eng. Trans. ASME Ser. D* **86**, 1–10 (1964).
18. J. M. Kirshner, "Fluid Amplifiers," Chapter 11, pp. 152–175. McGraw-Hill, New York, 1966.
19. F. R. Goldschmied, "An Experimental Study of Pulsating Flow of Incompressible Viscous Fluids in Rigid Pipes in the Intermediate Damping Range." NASA TMX-53719, 25 March 1968.
20. R. Krishnaiyer and T. J. Lechner, An experimental evaluation of fluidic transmission line theory. *Advan. Fluidics ASME Publ.* (May 1967).
21. R. J. Cornish, Flow in a pipe of rectangular cross-section. *Proc. Roy. Soc. A* **Cxx** (1928).
22. M. E. Franke, A. J. Malanowski, and P. S. Martin, "Effects of Temperature, End-Conditions, Flow, and Branching on the Frequency Response on Pneumatic Lines." ASME Paper 71-WA/AUT-5(Flcs), November 1971; *J. Dynam. Syst. Measurement Contr. Ser. G* **94**, No. 1, 15 (1972).
23. R. Courant and K. O. Friedrichs, "Supersonic Flow and Shock Waves." Wiley (Interscience), New York, 1948.
24. A. H. Shapiro, "The Dynamics and Thermodynamics of Compressible Fluid Flow," Vol. II, Chapters 23–24. Ronald Press, New York, 1954.
25. A. Kantrowitz, One-dimensional treatment of unsteady gas dynamics, Section C–350–412, "Fundamentals of Gas Dynamics," Vol. III, High Speed Aerodynamics and Jet Propulsion. Princeton Univ. Press, Princeton, New Jersey, 1958.
26. J. Owczarek, "Fundamentals of Gasdynamics." International Book Co., 1964.
27. G. Rudinger, "Wave Diagrams for Non-Steady Flow in Ducts." Van Nostrand-Reinhold, Princeton, New Jersey, 1955.
28. V. L. Streeter and E. B. Wylie, "Hydraulic Transients." McGraw-Hill, New York, 1967.
29. R. S. Benson, R. D. Garg, and D. Woollatt, A numerical solution of unsteady flow problems. *Int. J. Mech. Sci.* **6**, No. 1, 117–144 (1964).
30. J. R. Manning, Computerized method of characteristic calculations for unsteady pneumatic line flows. *J. Basic Eng.* 231–240 (June 1968).
31. W. Zielke, Frequency dependent friction in transient pipe flow. *J. Basic Eng. Trans. ASME Ser. D* **90**, No. 1, 109–115 (1968).
32. F. T. Brown, A quasi method of characteristics with application to fluid lines with frequency dependent wall shear and heat transfer. *J. Basic Eng.* 217–227 (June 1969).
33. B. C. Kuo, "Analysis and Synthesis of Sampled-Data Control Systems." Prentice-Hall, Englewood Cliffs, New Jersey, 1963.
34. J. R. Keto, "Transient Response of Some Flueric Components to Step Waves and Application to Matching," Fluerics 29. HDL-TR-1531, September 1970.

Chapter 4

JET FLOWS

4.1 INTRODUCTION

There are a number of aspects of jet flow that are important to fluidics. These include the jet velocity distribution for laminar and turbulent jets, and the effects of jet instability or transition. We shall confine our attention throughout this chapter to incompressible flow.

Most of the derivations which have been shown experimentally to be reasonably accurate are based on the boundary layer equations. In this connection there are two types of boundaries or boundary nomenclatures: fixed boundaries and free boundaries. Fixed boundaries in general are walls of some type along which the fluid moves. Free boundaries are boundaries between two streams at different velocities. The latter type of boundary, the free boundary (or shear layer), is significant in the case of jet flow. For some of the cases in which we are interested the jet itself is a stream moving at some velocity, whereas the second stream is at zero velocity and is the stagnant fluid in which the jet moves. In other cases, a second stream (and possibly even a third) may coexist at nonzero velocities. The theoretical derivations for most of the basic jet equations have been obtained using the concept of similarity.

Similarity is used in two slightly differing senses. Two flows are similar if we can normalize them so that in the normalized form the expressions are the same for both. On the other hand, similarity in the case of a single type of flow, such as that within a jet, is a condition in which at any two points within that flow normalization will allow the same expression to be used at both points.

This condition of self-similarity is also called self-preservation. In general the similarity conditions for jet flows hold only under the assumption that the jets issue from infinitesimal apertures (holes of infinitesimal radius or slits of infinitesimal width).

In all the cases we shall discuss, the jet stream issues into the same fluid medium. The resulting flow is often referred to as "submerged" jet flow.

4.2 LAMINAR-FREE JETS FROM INFINITESIMAL APERTURES

4.2.1 The Two-Dimensional Laminar Jet

In a paper written in 1933, Schlichting [1] obtained similarity solutions for the incompressible flow cases for both the two-dimensional and the axisymmetric laminar jet using the boundary layer equations. Bickley [2] has also solved the two-dimensional laminar jet. Because this technique is used for many cases of jet flow, we shall consider these cases in detail.

The technique of obtaining similarity solutions for nonlinear partial differential equations is in essence the same as the method of separation of variables used for linear partial differential equations. The basic concept is to break the partial differential equation up into ordinary differential equations by assuming that the dependent variable can be written as a product of functions of the independent variable.

The primary difference between linear and nonlinear equations appears after the ordinary differential equations have been obtained. In the case of linear equations, superposition allows us to choose as many product terms as necessary to meet the boundary conditions. For nonlinear equations, the boundary conditions must be met with a single product. Since this is not always (as a matter of fact not usually) possible, the technique often fails.

Our single product solution has the form

$$\psi = G(\alpha) F(\eta)$$

A glance at this shows why it is called a similarity solution. If we choose any two values of η, we obtain the same form of curve for ψ, namely some constant times $G(\alpha)$ so that the curves for different values of η will be similar. The same thing is, of course, true with respect to α.

We shall now illustrate the above procedure for the case of the two-dimensional incompressible laminar jet issuing from a narrow slit. The geometry is shown in Fig. 4.1.

If we assume that the pressure throughout the jet is ambient, the boundary layer momentum equation is

$$u\, \partial u/\partial x + v\, \partial u/\partial y = v\, \partial^2 u/\partial y^2 \tag{4.1}$$

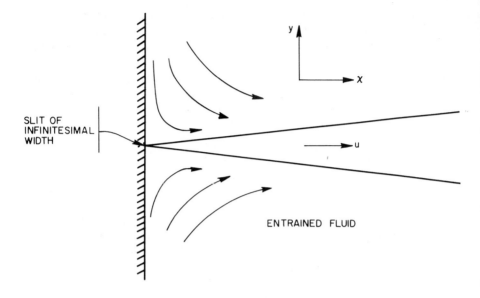

Fig. 4.1 Jet from an infinitesimal slit.

where $u = u(x, y)$ is the component of flow in the x direction, $v = v(x, y)$ is the component of flow in the y direction, and v is the kinematic viscosity. The continuity equation for steady flow is

$$\partial u/\partial x + \partial v/\partial y = 0 \tag{4.2}$$

The boundary conditions are

$$\text{at} \quad y = 0, \quad v = 0, \quad \partial u/\partial y = 0 \quad \text{as} \quad y \to \infty, \quad u \to 0 \tag{4.3}$$

In addition, since momentum is conserved

$$J = \rho \int_{-\infty}^{\infty} u^2 \, dy = \text{const} \tag{4.4}$$

where J is the momentum flux per unit depth. We can now satisfy Eq. (4.2) by using the substitutions

$$u = \partial \psi/\partial y \tag{4.5a}$$

$$v = -\partial \psi/\partial x \tag{4.5b}$$

At this point we attempt to separate into ordinary differential equations the partial differential equation that is obtained by inserting Eqs. (4.5) into Eq. (4.1). The process begins by looking for an appropriate coordinate system or coordinate systems. We thus assume

$$\psi = G(x) \, F(\eta) \tag{4.6a}$$

$$\eta = g(x) \cdot y \tag{4.6b}$$

4.2 Laminar-Free Jets from Infinitesimal Apertures

Although these are not the most general assumptions that can be made, they will prove to be adequate for all cases we shall discuss here.

Note that

$$\frac{\partial}{\partial y} = \frac{d\eta}{dy}\frac{\partial}{\partial \eta} = g(x)\frac{\partial}{\partial \eta} \qquad (4.7a)$$

$$\left(\frac{\partial}{\partial x}\right)_y = \left(\frac{\partial}{\partial x}\right)_\eta + \frac{d\eta}{dx}\frac{\partial}{\partial \eta}$$

$$\left(\frac{\partial}{\partial x}\right)_y = \left(\frac{\partial}{\partial x}\right)_\eta + y\,g'(x)\frac{\partial}{\partial \eta} \qquad (4.7b)$$

As is customary, the prime symbol indicates differentiation with respect to the argument.

Using Eqs. (4.5) and (4.7), we find that

$$u = \partial\psi/\partial y = gGF' \qquad (4.8a)$$

$$v = -(\partial\psi/\partial x)_y = -G'F - yg'GF' \qquad (4.8b)$$

Continuing in this way, the required partial derivatives with respect to the new variables are found. Inserting them into Eq. (4.1), we obtain

$$(gG' + g'G)F'^2 - gG'FF'' = vg^2 F''' \qquad (4.9)$$

If the coefficients of F and its derivatives are chosen to be proportional to each other, the equation will become ordinary; therefore, we let

$$gG' + g'G = a_1 gG' \qquad (4.10a)$$

$$gG' = a_2 vg^2 \qquad (4.10b)$$

From Eq. (4.10a) we can easily find

$$g = AG^{a_1 - 1} \qquad (4.11)$$

where A is a constant of integration, and, from (4.10b),

$$G' = a_2 vg \qquad (4.12)$$

so that

$$G^{2-a_1} = Aa_2(2 - a_1)vx \qquad (4.13)$$

Now if Eqs. (4.10) hold, Eq. (4.9) becomes

$$a_1 a_2 F'^2 - a_2 FF'' = F''' \qquad (4.14)$$

Let us now impose the condition of Eq. (4.4) that

$$\rho \int_{-\infty}^{\infty} u^2 \, dy = J = \text{const} \qquad (4.15)$$

Then, from (4.8a),
$$u = gGF'$$
and, from (4.6b),
$$\eta = y\, g(x)$$
or
$$d\eta = g(x)\, dy$$
so that (4.15) becomes
$$g(x)\, G^2(x)\, \rho \int_{-\infty}^{\infty} F'^2(\eta)\, d\eta = J = \text{const}$$
or
$$gG^2 = \text{const} = J \bigg/ \rho \int_{-\infty}^{\infty} F'^2(\eta)\, d\eta \tag{4.16}$$

Comparing this with Eq. (4.11), we see that
$$a_1 = -1 \tag{4.17}$$
whence
$$g = A/G^2, \qquad A = J \bigg/ \rho \int_{-\infty}^{\infty} F'^2(\eta)\, d\eta \tag{4.18}$$
and (4.13) becomes
$$G^3 = 3Aa_2\, vx, \qquad G = (3Aa_2\, vx)^{1/3} \tag{4.19}$$
Then, from (4.18) and (4.19),
$$g = A/(3Aa_2\, vx)^{2/3} \tag{4.20}$$
Consequently, using (4.6a) and (4.19), we find
$$\psi = (3Aa_2\, vx)^{1/3} F(\eta) \tag{4.21}$$
and, from (4.6b) and (4.20),
$$\eta = Ay/(3Aa_2\, vx)^{2/3} \tag{4.22}$$
Similarly we can now write
$$u = \frac{A^{2/3}}{(3a_2\, vx)^{1/3}} F'(\eta) \tag{4.23a}$$
$$v = \frac{(3Aa_2\, v)^{1/3}}{3x^{2/3}} (2\eta F' - F) \tag{4.23b}$$
and, from (4.14) and (4.17),
$$F''' + a_2(F'^2 + FF'') = 0 \tag{4.24}$$

4.2 Laminar-Free Jets from Infinitesimal Apertures

for which our boundary conditions (4.3) become

$$\text{at} \quad \eta = 0, \quad F = 2\eta F' = 0 \quad \text{and} \quad F'' = 0 \quad \text{as} \quad \eta \to \infty, \quad F' \to 0 \quad (4.25)$$

Equation (4.24) can be written

$$\frac{d}{d\eta}(F'' + a_2 FF') = 0$$

This expression is integrable if we choose

$$a_2 = 2 \tag{4.26}$$

Whence

$$\frac{d}{d\eta}(F' + F^2) = 0, \quad F' + F^2 = \text{const} = K^2 \tag{4.27}$$

This constant introduces no additional arbitrariness, and at this point may be chosen to be equal to unity; however, we shall carry it along in general form.
From (4.27)

$$\eta = \int \frac{dF}{K^2 - F^2}$$

$$\eta = (1/K)\tanh^{-1}(F/K)$$

$$F = K \tanh K\eta \tag{4.28}$$

$$F'(\eta) = K^2 \operatorname{sech}^2 K\eta \tag{4.29}$$

Now combining (4.18) and (4.29)

$$A = J \bigg/ \rho \int_{-\infty}^{\infty} F'^2(\eta)\, d\eta = J/\rho K^4 \int_{-\infty}^{\infty} \operatorname{sech}^4 K\eta\, d\eta$$

$$= \tfrac{4}{3} J/\rho K^3 \tag{4.30}$$

From (4.23), (4.26), (4.29), and (4.30), we obtain

$$u = \frac{1.5(J/6\rho)^{2/3}}{(vx)^{1/3}} \operatorname{sech}^2 K\eta \tag{4.31a}$$

$$v = \frac{(4.5J/\rho)^{1/3}}{3x^{2/3}}[2K\eta \operatorname{sech}^2 K\eta - \tanh K\eta] \tag{4.31b}$$

where, from (4.22), (4.26), and (4.30),

$$K\eta = \frac{(y/2)(J/6\rho)^{1/3}}{(vx)^{2/3}} \tag{4.31c}$$

Equations (4.31) are the Schlichting–Bickley equations for a two-dimensional laminar jet.

Equation (4.31a) represents the axial velocity profile of the two-dimensional laminar jet. On the jet axis or centerline $\eta = 0$ so that the centerline velocity u_c is

$$u_c = \frac{1.5(J/6\rho)^{2/3}}{(\nu x)^{1/3}} \quad (4.32)$$

Figure 4.2 shows a normalized axial velocity profile (u/u_c versus η) for the two-dimensional jet. The profile is bell shaped with a maximum on the centerline. To demonstrate the effect of momentum flux changes on the profile we may specify a reference value of momentum flux J_f. The reference velocity corresponding to J_f is designated as u_f. Then from Eq. (4.31a) and (4.31c) we may form the ratio

$$\frac{u}{u_f} = \left(\frac{J}{J_f}\right)^{2/3} \frac{\text{sech}^2[\eta(J/J_f)^{1/3}]}{\text{sech}^2 \eta} \quad (4.33)$$

Figure 4.3 shows the effects of momentum flux changes on the axial velocity. An increase in momentum flux causes the velocity in the central region to increase and in the outer region to decrease. A decrease in momentum flux brings about the inverse result. The value of η for which there is no change in velocity depends upon the momentum flux ratio. For example, when momentum flux is doubled $\eta = 1.05$, where $u/u_f = 1.0$. Thus a momentum flux increase narrows the jet and a momentum flux decrease broadens it.

To determine the jet streamlines we can insert the values of u and v from Eqs. (4.31) into the defining equation of a streamline:

$$dy/dx = v/u \quad (4.34)$$

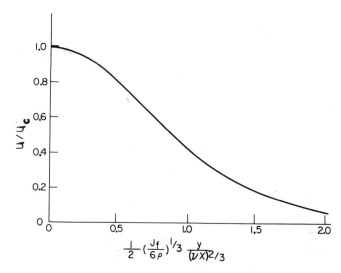

Fig. 4.2 Normalized axial velocity profile for a two-dimensional laminar jet.

4.2 Laminar-Free Jets from Infinitesimal Apertures

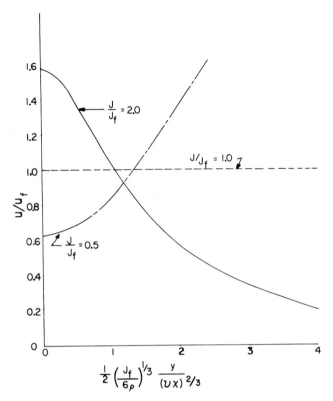

Fig. 4.3 Effect of change in momentum flux on the laminar two-dimensional jet profile.

A simpler procedure, however, is to take advantage of the fact that, for steady flow, the streamlines bound regions of constant mass flow; i.e., the mass flow across a streamline is zero. We may find the lines of constant mass flow by integrating across the axial velocity profile. The formulation for mass flow is therefore

$$\dot{m} = \rho h \int_0^{y_1} u \, dy \tag{4.35}$$

where $\dot{m}$ is the mass flow between $y = 0$ and $y = y_1$, and h is the depth. The substitution of Eq. (4.31a) into (4.35) yields

$$\frac{\dot{m}}{\rho h} = 3 \left(\frac{J}{6\rho}\right)^{1/3} (vx)^{1/3} \tanh \frac{0.5(J/6\rho)^{1/3} y_1}{(vx)^{2/3}} \tag{4.36}$$

Now for each specific constant value of $\dot{m}$ the relation between y_1 and x represents the equation of a streamline. We may place Eq. (4.36) into a more convenient form by assuming reference values for mass flow and momentum flux. Suppose that the reference mass flow is $\dot{m}_f$, where b_f and u_f are, respectively, the jet reference width and velocity so that

$$2\dot{m}_f = \rho h b_f u_f$$

and that the reference momentum flux per unit depth is J_f, where

$$J_f = \rho b_f u_f^2$$

With these definitions Eq. (4.36) becomes

$$\frac{\dot{m}}{\dot{m}_f} = 3.30 \left(\frac{x}{b_f N_R}\right)^{1/3} \tanh \frac{0.275(y_1/b_f)}{(x/b_f N_R)^{2/3}} \qquad (4.37)$$

where N_R is the Reynolds number defined as

$$N_R = u_f b_f / \nu$$

Figure 4.4 shows the streamlines described in Eq. (4.37) for various values of the mass flow ratio $\dot{m}/\dot{m}_f$. Since the defining equations allow no mass flow to issue from an infinitesimal slit, all the mass flow in the jet is the result of entrainment. The sharp change in direction at small values of $x/(b_f N_R)$ is evidence of

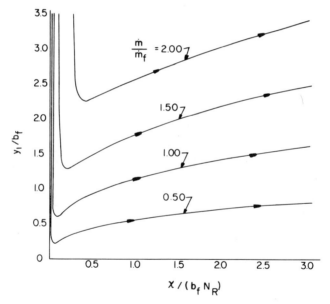

Fig. 4.4 Streamlines for a two-dimensional laminar jet.

the entrained flow. The parameter $x/(b_f N_R)$ reveals that an increase in Reynolds number N_R moves the streamlines closer to the centerline. This is a further result of the fact that laminar jets of increased velocity are narrower.

The lines of constant momentum flux can be determined by a similar procedure, namely that

$$J = 2\rho \int_0^{y_2} u^2 \, dy \qquad (4.38)$$

4.2 Laminar-Free Jets from Infinitesimal Apertures

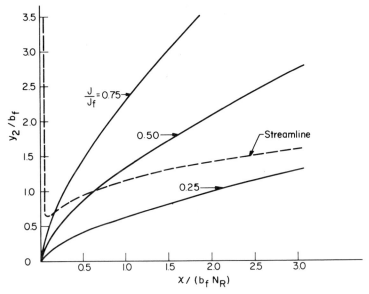

Fig. 4.5 Lines of constant momentum flux for laminar two-dimensional jet.

where J is the momentum flux per unit depth between $y = -y_2$, and $y = y_2$. If the axial velocity from Eq. (4.31a) is employed in Eq. (4.38), the result of the integration and some algebraic manipulation is

$$\frac{J}{J_f} = \frac{3}{2}\left(\tanh\frac{0.275 y_2/b_f}{(x/b_f N_R)^{2/3}} - \frac{1}{3}\tanh^3\frac{0.275 y_2/b_f}{(x/b_f N_R)^{2/3}}\right) \tag{4.39}$$

Figure 4.5 shows the lines of constant momentum flux for $J/J_f = 0.25$, 0.50, and 0.75. These lines emanate from the slit position. A typical streamline is superimposed on the figure to show that the lines of constant momentum flux intersect the streamlines. All the lines of constant momentum flux may be represented by an equation of the type $y/x^{2/3} = \text{const}$.

4.2.2 The Axisymmetric Laminar Jet

Analysis of the axisymmetric incompressible laminar jet involves a similar procedure. The momentum equation in this case is

$$u\frac{\partial u}{\partial x} + u_r\frac{\partial u}{\partial r} = \frac{v}{r}\frac{\partial}{\partial r}\left(r\frac{\partial u}{\partial r}\right) \tag{4.40}$$

and the continuity equation is

$$\frac{\partial}{\partial x}(ru) + \frac{\partial}{\partial r}(ru_r) = 0 \tag{4.41}$$

where u_r is the radial velocity, and the boundary conditions are

$$\text{at} \quad r = 0, \quad u_r = 0, \quad \partial u/\partial r = 0 \quad \text{as} \quad r \to \infty, \quad u \to 0 \tag{4.42}$$

We let

$$ru = \partial\psi/\partial r \tag{4.43a}$$

$$ru_r = -\partial\psi/\partial x \tag{4.43b}$$

$$\psi = G(x) F(\eta) \tag{4.44a}$$

$$\eta = r g(x) \tag{4.44b}$$

Using (4.43) and (4.44), after the coordinate transformation, Eq. (4.40) becomes

$$(gG' + 2g'G)\frac{F'^2}{\eta^2} - gG'\frac{FF''}{\eta^2} + gG'\frac{FF'}{\eta^3} = \frac{vg}{\eta}\frac{d}{d\eta}\left(F'' - \frac{F'}{\eta}\right) \tag{4.45}$$

Then let

$$gG' + 2g'G = mgG' \tag{4.46}$$

and

$$gG' = kvg \tag{4.47}$$

where k and m are constants. From (4.47)

$$G = kvx \tag{4.48}$$

Equation (4.46) can be written as

$$G'/G + 2g'/g = m(G'/G)$$

Whence using (4.48)

$$g = Ax^{(m-1)/2} \tag{4.49a}$$

$$g' = (A/2)(m-1)x^{(m-3)/2} \tag{4.49b}$$

Hence

$$u = (A^2 k v x^m/\eta)F'(\eta) \tag{4.50a}$$

$$u_r = \frac{-Akvx^{(m-1)/2}}{2}\left(\frac{2F}{\eta} + (m-1)F'\right) \tag{4.50b}$$

$$\eta = Arx^{(m-1)/2} \tag{4.50c}$$

and

$$mk\frac{F'^2}{\eta^2} - k\left(\frac{FF''}{\eta^2} - \frac{FF'}{\eta^3}\right) = \frac{1}{\eta}\frac{d}{d\eta}\left(F'' - \frac{F'}{\eta}\right) \tag{4.51}$$

4.2 Laminar-Free Jets from Infinitesimal Apertures

Multiplying by η and choosing $k = -m = 1$,

$$-\frac{F'^2}{\eta} - \frac{FF''}{\eta} + \frac{FF'}{\eta^2} = \frac{d}{d\eta}\left(F'' - \frac{F'}{\eta}\right)$$

$$-\frac{d}{d\eta}\left(\frac{FF'}{\eta}\right) = \frac{d}{d\eta}\left(F'' - \frac{F'}{\eta}\right) \quad (4.52)$$

For $k = -m = 1$, we also obtain from (4.50)

$$u = (A^2 v/x)[F'(\eta)/\eta] \quad (4.53a)$$
$$u_r = (Av/x)(F' - F/\eta) \quad (4.53b)$$
$$\eta = Ar/x \quad (4.53c)$$

$$\frac{\partial u}{\partial r} = \frac{d\eta}{dr}\frac{\partial u}{\partial \eta} = \frac{A}{x}\frac{\partial u}{\partial \eta} = \frac{A^3 v}{x^2}\left(\frac{1}{\eta}F''(\eta) - \frac{F'(\eta)}{\eta^2}\right) \quad (4.54)$$

The boundary conditions (4.42) become

$$\lim_{\eta \to 0} F'(\eta) - F(\eta)/\eta = 0 \quad (4.55a)$$

$$\lim_{\eta \to 0} F''(\eta) - F'(\eta)/\eta = 0 \quad (4.55b)$$

$$\lim_{\eta \to \infty} F'(\eta) < \infty \quad (4.55c)$$

Integrating (4.52) we obtain

$$F'' - F'/\eta + FF'/\eta + C = 0$$

where C is a constant of integration. From (4.50a)

$$\lim_{\eta \to 0} F'(\eta)/\eta < \infty$$

that is, $F'(0) = 0$. Consequently (4.55a) yields

$$F(0) = 0 \quad (4.55d)$$

In this case $C = 0$, and

$$FF' - F' + \eta F'' = 0, \quad \frac{d}{d\eta}\left(\frac{F^2}{2} - 2F + \eta F'\right) = 0$$

or, since the constant of integration is again zero,

$$F^2/2 - 2F + \eta F' = 0$$

This equation separates into the form

$$\frac{dF}{2F(1 - F/4)} = \frac{d\eta}{\eta}$$

$$-\frac{1}{2}\ln\left|\frac{1 - F/4}{F}\right| = \ln \eta + \ln C_1$$

where C_1 is a constant. The condition $F > 4$ leads to physically meaningless results; therefore, we assume $F < 4$ and obtain

$$F = C_1^2 \eta^2 / (1 + C_1^2 \eta^2 / 4) \tag{4.56}$$

We could at this point choose $C_1 = 1$, because C_1 does not actually introduce additional arbitrariness. We will show this by carrying out the algebra.

Let
$$\eta_1 = C_1 \eta \tag{4.57}$$

Substituting into (4.56) we obtain

$$F = \eta_1^2 / (1 + \eta_1^2 / 4) \tag{4.58}$$

Now
$$dF/d\eta_1 = (1/C_1) \, dF/d\eta \tag{4.59}$$

so that (4.53) becomes

$$u = \frac{A^2 C_1^2 v}{x} \frac{dF/d\eta_1}{\eta_1} \tag{4.60a}$$

$$u_r = (AC_1 v/x)(dF/d\eta_1 - F/\eta_1) \tag{4.60b}$$

$$\eta_1 = C_1 \eta = AC_1 r/x \tag{4.60c}$$

Equations (4.60) reveal that C_1 introduces no additional arbitrariness since it appears in exactly the same way as does A; we have merely replaced A by another arbitrary constant $A_1 = AC_1$.

As before we evaluate the constant in terms of the momentum flux

$$J_c = 2\pi\rho \int_0^\infty u^2 r \, dr$$

Using (4.60a) and (4.56), we find

$$A^2 C_1^2 = \tfrac{3}{16} J_c / \pi \rho v^2$$

Whence (4.60) becomes

$$u = \frac{3}{8} \frac{J_c}{\pi \rho v x} \frac{1}{(1 + \eta_1^2/4)^2} \tag{4.61a}$$

$$u_r = \frac{1}{4x} \left(\frac{3J_c}{\pi\rho}\right)^{1/2} \frac{\eta_1(1 - \eta_1^2/4)}{(1 + \eta_1^2/4)^2} \tag{4.61b}$$

$$\eta_1 = (r/4vx)(3J_c/\pi\rho)^{1/2} \tag{4.61c}$$

Let us compare the axial velocity profiles of the axisymmetric and two-dimensional jets at the same centerline velocity. The centerline velocity for the axisymmetric jet, from Eq. (4.61) with $\eta_1 = 0$, is

$$u_c = \tfrac{3}{8} J_c / \pi \rho v x \tag{4.62}$$

4.2 Laminar-Free Jets from Infinitesimal Apertures

If we equate the centerline velocities given by Eqs. (4.32) and (4.62) we obtain

$$J_c = 4\pi\rho(vxJ/6\rho)^{2/3} \tag{4.63}$$

This is the equation which relates the momentum flux of the axisymmetric jet to the momentum flux per unit depth of the two-dimensional jet for the condition of equal centerline velocities. Now if Eq. (4.63) is substituted into Eq. (4.61c), the result is

$$\eta_1 = \frac{\sqrt{3}r}{2} \frac{[J/6\rho]^{1/3}}{(vx)^{2/3}} \tag{4.64}$$

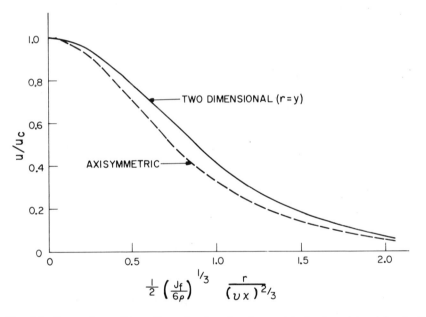

Fig. 4.6 Comparison of two-dimensional and axisymmetric laminar jet axial velocity profile.

and η_1 is in the same form as η in Eq. (4.31c). Figure 4.6 shows the axial velocity profiles of the axisymmetric and two-dimensional jets plotted on the same coordinate axes. Under the condition of equal centerline velocity, the axisymmetric jet is narrower than the two-dimensional jet. The location of the point where the velocity is half the centerline velocity occurs 22% farther from the axis for the two-dimensional jet.

It is difficult to confirm experimentally the distributions obtained for laminar jets because of the assumptions of a hole or slit of infinitesimal width so that, when a finite width aperture is used as must, of course, be true in an actual experiment, similarity is obtained only relatively far downstream. Unfortunately the laminar jet has a very great tendency to become turbulent at downstream distances which are less than those for which similarity may be considered to hold. Thus the distribution which is usually found for a laminar jet in the region which is still laminar is highly dependent on the distribution at the nozzle exit. If the nozzle consists of a fairly long duct such that the flow will be fully developed, then at the nozzle exit there will be a parabolic distribution of velocity and, for an appreciable distance, the distribution will be fairly close to a parabolic distribution. Andrade [3] was able to obtain reasonable agreement with Bickley and Schlichting's results, however, by using very low Reynolds number jets of the order of 30 or less. Sato [4] was able to obtain laminar jets up to Reynolds numbers of about 50, and pointed out that, though the velocity distribution is initially parabolic because of the nozzle, it does approach a modified form of the similarity solution.

The amount of momentum exchange of the jet with its surroundings depends not only on its momentum but also on its viscosity, which in fact is a measure of momentum exchange between layers of fluid. Since laminar viscosity is small compared to turbulent or eddy viscosity, a laminar jet entrains much less of the surrounding fluid than does a turbulent jet.

4.2.3 Jet Width (Laminar Jets)

An important parameter of concern with respect to the jet is its width, which may be defined in a number of ways. One of these is the momentum flux width b_m, which is defined for the two-dimensional jet by

$$J = \rho \int_{-\infty}^{\infty} u^2 \, dy \equiv \rho \int_{-b_m}^{b_m} u_c^2 \, dy = 2\rho b_m u_c^2 \qquad (4.65a)$$

or

$$b_m = J/2\rho u_c^2 \qquad (4.65b)$$

The substitution of the expression for centerline velocity given in Eq. (4.32) into Eq. (4.65) yields

$$b_m = 2.42(vx)^{2/3}(\rho/J)^{1/3} \qquad (4.66)$$

Another way of defining jet width is through the use of the "half" width. This width, $b_{1/2}$, is the distance from the jet axis at which the jet velocity is half of the centerline velocity. From Eqs. (4.31a) and (4.31c) we may calculate the jet half-width as

$$\begin{aligned} b_{1/2} &= 2(6)^{1/3}(vx)^{2/3}(\rho/J)^{1/3} \, \text{sech}^{-1}(0.5)^{1/2} \\ b_{1/2} &= 3.2(vx)^{2/3}(\rho/J)^{1/3} \end{aligned} \qquad (4.67)$$

The jet half-width is 33% larger than its momentum flux width. This is an indication that most of the momentum flux of the jet remains near the jet centerline.

4.3 Turbulent-Free Jets from Infinitesimal Apertures

For an axisymmetric jet the momentum flux width b_{mr} is defined as

$$J_c = 2\pi\rho \int_0^\infty u^2 r\, dr \equiv 2\pi\rho \int_0^{b_{mr}} u_c^2 r\, dr \tag{4.68a}$$

or

$$b_{mr} = (1/u_c)(J_c/\pi\rho)^{1/2} \tag{4.68b}$$

In this case the centerline velocity is given in Eq. (4.62). Thus Eq. (4.68) becomes

$$b_{mr} = (8vx/3)(\pi\rho/J_c)^{1/2} = 4.73\,vx(\rho/J_c)^{1/2} \tag{4.69}$$

and the half-width $b_{(1/2)r}$ for the axisymmetric jet may be obtained from Eqs. (4.61a) and (4.61c) as

$$b_{(1/2)r} = 5.27(vx)(\rho/J_c)^{1/2} \tag{4.70}$$

4.3 TURBULENT-FREE JETS FROM INFINITESIMAL APERTURES

In the case of turbulence the velocity is considered to consist of a mean velocity plus a fluctuation velocity. The mean and fluctuation velocity distributions for turbulent jets are in general considered separately.

In order to obtain a theoretical expression for the mean velocity distribution of a turbulent jet, one must either start from some sort of diffusion equation or make some assumption associating the turbulent diffusion phenomenon with an eddy viscosity. The concept of an eddy viscosity allows one to obtain a similarity solution for a turbulent jet issuing from an infinitesimal nozzle exit.

There are a number of expressions that have been found for the velocity distributions of turbulent jets. We shall here discuss the circular and plane jet distributions based on Prandtl's new mixing theory, and the distribution based on Reichardt's inductive theory.

4.3.1 Distributions Based on Prandtl's Theory

In the case of the distributions obtained by use of Prandtl's new mixing theory, the calculations are very similar to those for the laminar case. The significant difference between the two is that the viscosity cannot be considered constant in the turbulent flow case.

Prandtl showed that the nominal width of a jet b_m is proportional to the downstream distance from the nozzle

$$b_m = kx \tag{4.71}$$

where b_m may be defined in a number of ways, one of which is the momentum flux width.

4.3.1.1 Prandtl's Theory Applied to the Two-Dimensional Jet

If the jet width is taken as the momentum flux width, Eqs. (4.65) and (4.71) can be combined to yield

$$u_c = \frac{1}{x^{1/2}} \left(\frac{J}{2\rho k}\right)^{1/2} \tag{4.72}$$

Using Prandtl's assumption that the eddy viscosity is proportional to the centerline velocity and the width of the jet, we can write

$$\varepsilon = k_1 b_m u_c \tag{4.73}$$

From Eqs. (4.71)–(4.73) we may express the eddy viscosity as a function of x; that is

$$\varepsilon = a_4 x^{1/2} \tag{4.74}$$

where $a_4 = k_1 [Jk/2\rho]^{1/2}$.

Equation (4.74) is the formulation used by Goertler [5] in his solution of the two-dimensional jet.

Consequently, the momentum equation (4.1), with the eddy viscosity replacing the kinematic viscosity, becomes

$$u\,\partial u/\partial x + v\,\partial u/\partial y = a_4 x^{1/2}\,\partial^2 u/\partial y^2 \tag{4.75}$$

We shall solve this equation using another commonly employed procedure, which is a slight variation on the previous method. This procedure is based on the assumption that the functional relationship in x is of the form of x^w.

We assume, therefore,

$$\psi = C_1 x^p F(\eta) \tag{4.76a}$$

$$\eta = \sigma_e x^q y \tag{4.76b}$$

where σ_e is an empirical constant, whence

$$u = \partial \psi/\partial y = C_1 \sigma_e x^{p+q} F'(\eta) \tag{4.77a}$$

$$v = -\partial \psi/\partial x = -[C_1 p x^{p-1} F(\eta) + C_1 \sigma_e y q x^{p+q-1} F'(\eta)] \tag{4.77b}$$

Taking derivatives and inserting them into (4.75),

$$(p+q) C_1^2 \sigma_e^2 x^{2p+2q-1} F'^2 - C_1^2 \sigma_e^2 p x^{2p+2q-1} FF'' = a_4 C_1 \sigma_e^3 x^{p+3q+0.5} FF''' \tag{4.78}$$

so that we must choose

$$(p+q) C_1^2 \sigma_e^2 x^{2p+2q-1} = m C_1^2 \sigma_e^2 p x^{2p+2q-1} \tag{4.79a}$$

and

$$C_1^2 \sigma_e^2 p x^{2p+2q-1} = n a_4 C_1 \sigma_e^3 x^{p+3q+1/2} \tag{4.79b}$$

4.3 Turbulent-Free Jets from Infinitesimal Apertures

From (4.79) we find

$$p = 3/2(2 - m) \tag{4.80a}$$

$$q = 3(m - 1)/2(2 - m) \tag{4.80b}$$

$$C_1 p = na_4 \sigma_e \tag{4.80c}$$

Using (4.79), Eq. (4.78) becomes

$$F''' - n(mF'^2 - FF'') = 0$$

If we choose

$$m = -1 \tag{4.81a}$$

the equation is of the same form as (4.24).
Then choosing

$$n = 2 \tag{4.81b}$$

and, since our boundary conditions are the same, we obtain as before (after choosing the constant equal to unity)

$$F(\eta) = \tanh \eta \tag{4.82a}$$

$$F'(\eta) = \operatorname{sech}^2 \eta \tag{4.82b}$$

But for $m = -1$ Eqs. (4.80) become

$$p = 3/2(2 - m) = \tfrac{1}{2} \tag{4.83a}$$

$$q = -1 \tag{4.83b}$$

so that (4.76b) becomes

$$\eta = \sigma_e y/x \tag{4.84a}$$

Inserting (4.83) into (4.77)

$$u = C_1 \sigma_e x^{-1/2} \operatorname{sech}^2 \eta \tag{4.84b}$$

$$v = (C_1/2x^{1/2})[2\eta \operatorname{sech}^2 \eta - \tanh \eta] \tag{4.84c}$$

We now introduce the momentum flux expression

$$J = \rho \int_{-\infty}^{\infty} u^2 \, dy$$

from which we obtain

$$C_1 = \tfrac{1}{2}(3J/\rho\sigma_e)^{1/2}$$

Then (4.84) becomes

$$u = \tfrac{1}{2}(3J\sigma_e/\rho x)^{1/2} \operatorname{sech}^2 \eta \tag{4.85a}$$

$$v = \tfrac{1}{4}(3J/\rho\sigma_e x)^{1/2}[2\eta \operatorname{sech}^2 \eta - \tanh \eta] \tag{4.85b}$$

$$\eta = \sigma_e y/x \tag{4.85c}$$

where J is the momentum flux per unit length.
Equations (4.85) are Goertler's solutions for the two-dimensional turbulent jet.

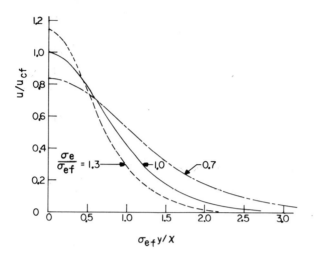

Fig. 4.7 Two-dimensional turbulent jet profiles.

The empirical constant σ_e is sometimes called the spread parameter of the jet. Reichardt [6] in his measurements found that $\sigma_e = 7.67$. However, other investigators have shown that σ_e depends significantly on the shape of the plenum chamber from which the jet issues. To demonstrate the effect σ_e has on the axial velocity profile we define a reference centerline velocity u_{cf} in terms of a reference spread parameter σ_{ef} as

$$u_{cf} = \tfrac{1}{2}(3J\sigma_{ef}/\rho x)^{1/2} \tag{4.86}$$

With this definition Eq. (4.85a) may be written as

$$u/u_{cf} = (\sigma_e/\sigma_{ef})^{1/2} \operatorname{sech}^2(\sigma_e/\sigma_{ef})(\sigma_{ef} y/x) \tag{4.87}$$

Figure 4.7 shows the axial velocity profiles for three values of the ratio σ_e/σ_{ef}. An increase in σ_e narrows the profile while a decrease in σ_e broadens it. When the plenum from which the jet issues encourages the formation of large-scale turbulent eddies the spread parameter will be less than 7.67 and the resulting jet will have an increased spread. On the other hand, if exceptional care is taken to prevent turbulence in the plenum the spread parameter of the jet increases and the jet is narrower.

To determine the streamlines for the turbulent jet we use the axial velocity profile, Eq. (4.85a), in the formulation for mass flow, Eq. (4.35). The result is

$$\dot{m}/\rho h = \tfrac{1}{2}(3Jx/\rho\sigma_e)^{1/2} \tanh \sigma_e y/x \tag{4.88}$$

Now the assumption of a reference momentum flux $J_f = \rho b_f u_f^2$ and a reference mass flow $2\dot{m}_f = \rho h b_f u_f$ reduce Eq. (4.88) to

$$\frac{\dot{m}}{\dot{m}_f} = 1.732 \left(\frac{x}{\sigma_e b_f}\right)^{1/2} \tanh \frac{(y/b_f)}{(x/\sigma_e b_f)} \tag{4.89}$$

4.3 Turbulent-Free Jets from Infinitesimal Apertures

Figure 4.8 shows the streamlines for the two-dimensional turbulent jet. They are similar to those for the laminar jet except that they are not dependent on Reynolds number. For Reynolds numbers above about 20 the laminar jet is narrower than the turbulent jet. In the laminar jet, for example, when $N_R = 100$, the $\dot{m}/\dot{m}_f = 0.50$ streamline at $x/b_f = 50$ passes through $y/b_f = 0.425$. In the turbulent jet with $\sigma_e = 7.67$ the $\dot{m}/\dot{m}_f = 0.50$ streamline at $x/b_f = 50$ passes through $y/b_f = 0.725$.

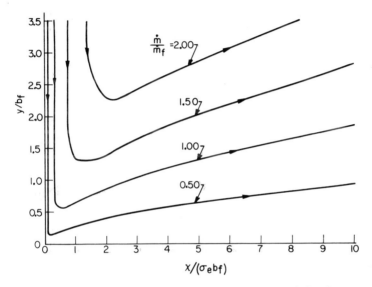

Fig. 4.8 Streamlines for the two-dimensional turbulent jet.

4.3.1.2 Prandtl's Theory Applied to the Axisymmetric Jet

In the case of the axisymmetric jet the momentum flux width is given in Eq. (4.68). The combination of this jet width and Eq. (4.71) yields a centerline velocity of the form

$$u_{cr} = (1/kx)(J_c/\pi\rho)^{1/2} \tag{4.90}$$

Once again we follow Prandtl's assumption that the eddy viscosity is proportional to the centerline velocity and the jet width. The result from Eqs. (4.71) and (4.90) is

$$\varepsilon = \varepsilon_0 = \text{const} \tag{4.91}$$

Thus the eddy viscosity for the axisymmetric jet is not a function of x but is constant throughout the flow field.

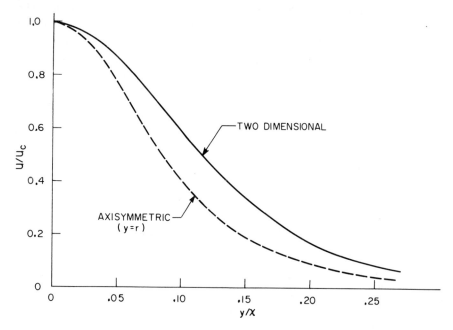

Fig. 4.9 Comparison of two-dimensional and axisymmetric turbulent jet axial velocity profiles.

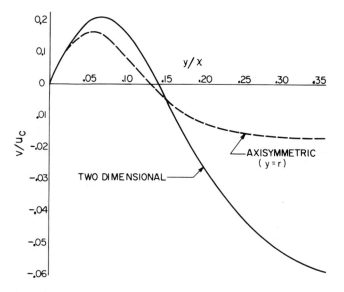

Fig. 4.10 Comparison of two-dimensional and axisymmetric turbulent jet transverse velocity distributions.

4.3 Turbulent-Free Jets from Infinitesimal Apertures

It follows that the turbulent jet momentum equation is identical to the laminar equation previously solved. The solutions are therefore identical, except that ε_0 is substituted for v,

$$u = \frac{3}{8} \frac{J_c}{\pi \rho \varepsilon_0 x} \frac{1}{(1 + \eta_1^2/4)^2} \tag{4.92a}$$

$$u_r = \frac{1}{4x}\left(\frac{3J_c}{\pi\rho}\right)^{1/2} \frac{\eta_1(1 - \eta_1^2/4)}{(1 + \eta_1^2/4)^2} \tag{4.92b}$$

$$\eta_1 = (r/4x\varepsilon_0)(3J_c/\pi\rho)^{1/2} \tag{4.92c}$$

where ε_0 is an empirical parameter.

From measurements on axisymmetric turbulent jets Reichardt [6] determined that $\varepsilon_0(\rho/J)^{1/2} = 0.0161$. With this value of eddy viscosity we may compare the axial and transverse velocities of the axisymmetric jet [Eq. (4.92)] to the corresponding velocities of the two-dimensional jet [Eq. (4.85)]. Figures 4.9 and 4.10 show the axial and transverse velocities of the jets, respectively. For the two-dimensional jet, $\sigma_e = 7.67$. The axial velocity profile of the two-dimensional jet is broader than that of the axisymmetric jet. This is the same relation that was observed for laminar jets (Fig. 4.6). The transverse velocity distributions are considerably different, especially at large distances from the centerline. The velocity of the entrained flow from the jet edges is about four times larger for the two-dimensional jet.

4.3.2 Distributions Based on Reichardt's Inductive Theory

After examining a large amount of data on velocity profiles, Reichardt [7] noted that a normal probability distribution could fit them quite well and, therefore, it should be possible to derive an expression for the velocity profile from a diffusion equation such as, for example, the heat equation

$$\partial T/\partial t = k^2 \, \partial^2 T/\partial y^2 \tag{4.93}$$

where T is temperature, t is time, and k is a constant.

4.3.2.1 Reichardt's Theory Applied to Two-Dimensional Jets

For conditions analogous to that of a jet emerging from a slit of infinitesimal width, the solution to Eq. (4.93) is of the form

$$T = (k/t^{1/2}) \exp(-y^2/4k^2 t) \tag{4.94}$$

Reichardt started with the momentum equation, assuming that the viscosity and pressure are negligible:

$$u \, \partial u/\partial x + v \, \partial u/\partial y = 0 \tag{4.95}$$

where u and v are instantaneous velocities.

Using the equation of continuity, this expression becomes

$$\partial u^2/\partial x + \partial uv/\partial y = 0$$

and averaging over time

$$\overline{\partial u^2}/\partial x + \overline{\partial uv}/\partial y = 0 \tag{4.96}$$

Reichardt then assumed that

$$\overline{uv} = -f(x)\,\overline{\partial u^2}/\partial y \tag{4.97}$$

This assumption is the heart of Reichardt's inductive theory of turbulence. With this assumption (4.96) becomes

$$\overline{\partial u^2}/\partial x = f(x)\,\partial^2 \overline{u^2}/\partial y^2 \tag{4.98}$$

To get this into the form of (4.93), a coordinate transformation is made,

$$(d\zeta/dx)\,\partial \overline{u^2}/\partial \zeta = f(x)\,\partial^2 \overline{u^2}/\partial y^2 \tag{4.99}$$

letting

$$f(x) = d\zeta/dx$$

or

$$\zeta = \int f(x)\,dx \tag{4.100}$$

(4.99) becomes

$$\partial \overline{u^2}/\partial \zeta = \partial^2 \overline{u^2}/\partial y^2$$

Incorporating the pertinent boundary conditions, the solution is

$$\overline{u^2} = (A^2/\zeta^{1/2})\exp(-y^2/4\zeta) \tag{4.101}$$

Now

$$\overline{u^2} = \bar{u}^2 + \overline{u'^2}$$

where $\bar{u}$ is the mean velocity and u' is the fluctuation component; that is

$$u = \bar{u} + u'$$

In general,

$$\overline{u'^2} \ll \bar{u}^2$$

so that

$$\overline{u^2} \cong \bar{u}^2 \tag{4.102}$$

whence (4.101) becomes

$$\bar{u} = (A/\zeta^{1/4})\exp(-y^2/8\zeta) \tag{4.103}$$

4.3 Turbulent-Free Jets from Infinitesimal Apertures

The experimental data conforms closely to

$$\bar{u} = (A_1/x^{1/2}) \exp(-y^2/2C_1^2 x^2) \quad (4.104)$$

We note that for the equivalence of (4.103) and (4.104) it is necessary that

$$8\zeta = 2C_1^2 x^2$$

Whence

$$f(x) = d\zeta/dx = C_1^2 x/2$$

That is, if we assume that

$$-\overline{uv} = (C_1^2 x/2)\, \partial \bar{u}^2/\partial y$$

the desired result is obtained. Then, from the continuity equation,

$$\bar{v} = -\int_0^y \frac{\partial u}{\partial x}\, dy \quad (4.105)$$

From (4.104)

$$\frac{\partial \bar{u}}{\partial x} = \frac{(2y^2 - C_1^2 x^2) A_1}{2 C_1^2 x^{7/2}} \exp(-y^2/2C_1^2 x^2) \quad (4.106)$$

Substituting (4.106) into (4.105) we obtain

$$\bar{v} = \frac{A_1}{2 x^{3/2}} \left[2y \exp(-y^2/2C_1^2 x^2) - \left(\frac{\pi}{2}\right)^{1/2} C_1 x\, \mathrm{erf}\, \frac{y}{\sqrt{2 C_1 x}} \right] \quad (4.107)$$

where

$$\mathrm{erf}\, z \equiv \frac{2}{\sqrt{\pi}} \int_0^z \exp(-y^2)\, dy$$

Then for the momentum flux

$$J = \rho \int_{-\infty}^{\infty} u^2\, dy$$

and using (4.104)

$$A_1^2 = J/\sqrt{\pi \rho C_1}$$

Whence

$$\bar{u} = (J/\rho C_1 x)^{1/2} (1/\pi)^{1/4} \exp(-y^2/2 C_1^2 x^2) \quad (4.108a)$$

$$\bar{v} = -\left(\frac{J}{C_1 \rho}\right)^{1/2} \left(\frac{1}{\pi}\right)^{1/4} \frac{1}{2 x^{3/2}} \left[\left(\frac{\pi}{2}\right)^{1/2} C_1 x\, \mathrm{erf}\left(\frac{y}{\sqrt{2 C_1 x}}\right) \right.$$

$$\left. - 2y \exp(-y^2/2 C_1^2 x^2) \right] \quad (4.108b)$$

It is sometimes convenient to write Eq. (4.108) in terms of the reference velocity u_f and the reference width b_f. If $J = \rho u_f^2 b_f$, Eq. (4.108) becomes

$$\frac{\bar{u}}{u_f} = \frac{0.75}{(x')^{1/2}} \exp(-y'^2/2x'^2) \tag{4.109a}$$

$$\frac{\bar{v}}{u_f} = \frac{0.75 C_1}{(x')^{1/2}} \left\{ 0.63 \, \text{erf}\left(\frac{y'}{\sqrt{2}\, x'}\right) - \frac{y'}{x'} \exp[-(y')^2/2(x')^2] \right\} \tag{4.109b}$$

where

$$x' = C_1 x / b_f, \qquad y' = y / b_f$$

Let us compare the axial velocity distributions from Goertler's theory [Eq. (4.85a)] and from Reichardt's hypothesis [Eq. (4.108a)]. To accomplish this we must first relate the empirical constants σ_e and C_1 from each theory. A relation between σ_e and C_1 can be obtained by equating the centerline velocities. The result is

$$C_1 = 4/3\sqrt{\pi}\sigma_e = 0.752/\sigma_e \tag{4.110}$$

Thus we may now write Reichardt's axial velocity expression in terms of the empirical constant σ_e as

$$\bar{u}/u_c = \exp[-0.889(\sigma_e y/x)^2] \tag{4.111}$$

The equivalent expression from Goertler's theory [Eq. (4.85a)] is

$$u/u_c = \text{sech}^2(\sigma_e y/x) \tag{4.112}$$

Figure 4.11 is a plot of Eqs. (4.111) and (4.112). The agreement between the theories is quite good in the vicinity of the centerline. At $\sigma_e y/x$ between 0.2 and 1.0 Reichardt's theory produces slightly higher velocities. The largest discrepancies occur at the tails of the distributions where $\sigma_e y/x > 1.5$. In this region Goertler's theory predicts higher velocities. Experimental data show good agreement with Reichardt's theory.

4.3.2.2 Reichardt's Theory Applied to Axisymmetric Jets

Alexander et al. [8] applied Reichardt's theory to the axisymmetric jet. Since the method they used follows closely the derivation for two-dimensional jets, we present here only a brief description of it. We begin with the momentum equation (4.40) and neglect viscosity to obtain

$$u \, \partial u/\partial x + u_r \, \partial u/\partial r = 0 \tag{4.113}$$

Now through the application of the continuity equation (4.41), Eq. (4.113) becomes

$$\frac{\partial u^2}{\partial x} + \frac{1}{r}\frac{\partial}{\partial r}(uu_r r) = 0 \tag{4.114}$$

4.3 Turbulent-Free Jets from Infinitesimal Apertures

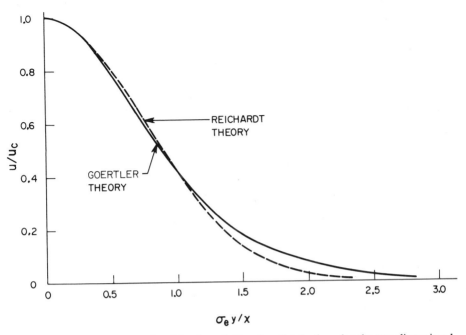

Fig. 4.11 Comparison of Reichardt and Goertler distributions for the two-dimensional turbulent jet.

To obtain an axisymmetric diffusion type equation it is now necessary to assume that

$$\frac{1}{r}\frac{\partial}{\partial r}(uu_r r) = -f(x)\left[\frac{\partial^2 u^2}{\partial r^2} + \frac{1}{r}\frac{\partial u^2}{\partial r}\right] \qquad (4.115)$$

The substitution of Eq. (4.115) into (4.114) yields

$$\frac{\partial u^2}{\partial x} = f(x)\left[\frac{\partial^2 u^2}{\partial r^2} + \frac{1}{r}\frac{\partial u^2}{\partial r}\right] \qquad (4.116)$$

If we assume that

$$f(x) = C_2^2 x/2$$

then

$$u = (J_c/\pi\rho)^{1/2}(1/C_2 x)\exp[-r^2/(2C_2^2 x^2)] \qquad (4.117)$$

is a solution of Eq. (4.116) and Eq. (4.117) is the Reichardt distribution for an axisymmetric jet. We can compare this with the distribution from Goertler's theory if we relate the empirical constant C_2 and the eddy viscosity ε_0. The equality of the centerline velocities in Eqs. (4.92a) and (4.117) yields

$$\varepsilon_0 = (\sqrt{3}C_2/8)(3J_c/\pi\rho)^{1/2} \qquad (4.118)$$

Thus Eq. (4.92) can be placed in the form

$$u/u_c = 1/(1 + \eta_1^2/4)^2 \tag{4.119}$$

where $\eta_1 = [(2/\sqrt{3})r]/(C_2 x)$.

Figure 4.12 compares the velocity distributions from Goertler's theory [Eq. (4.119)] and from Reichardt's theory [Eq. (4.117)] for the axisymmetric jet. The behavior here again is very much like the two-dimensional case, the largest discrepancies occurring at the tails of the distributions.

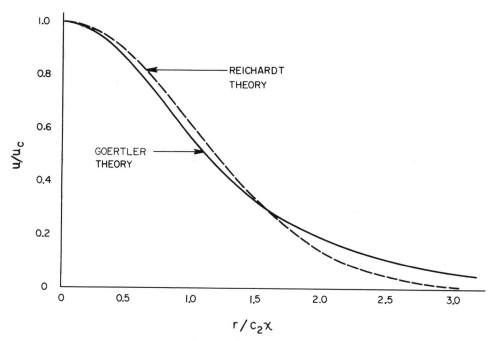

Fig. 4.12 Comparison of Reichardt and Goertler distributions for the axisymmetric turbulent jet.

4.3.3 Jet Width—Turbulent Jets

The momentum flux width defined in Eq. (4.65) for the two-dimensional laminar jet holds also for the two-dimensional turbulent jet. In the turbulent case the centerline velocity may be taken from either Eq. (4.85a) (Goertler's theory) or Eq. (4.108a) (Reichardt's theory). In terms of the empirical constants σ_e and C_1 the momentum flux width of the two-dimensional jet is

$$b_m = \tfrac{2}{3}x/\sigma_e \tag{4.120a}$$

$$b_m = (\sqrt{\pi}/2)C_1 x \tag{4.120b}$$

4.4 Turbulent-Free Jets from Finite Apertures

The jet half-width may also be obtained from Eqs. (4.85a) and (4.108a) and is

$$b_{1/2} = 0.881 x/\sigma_e \qquad (4.121a)$$

$$b_{1/2} = 1.177 C_1 x \qquad (4.121b)$$

As in the laminar two-dimensional jet the half-width of the turbulent two-dimensional jet is 33% larger than its momentum flux width. An important difference, however, is that the laminar jet width depends inversely on the one-third power of the momentum flux per unit depth, whereas the turbulent jet is independent of momentum flux.

The momentum and half-widths of the axisymmetric turbulent jet, based on Goertler's theory, are exactly the same as given for the laminar jet in Eqs. (4.69) and (4.70) except that eddy viscosity must replace kinematic viscosity. Thus for the turbulent jet

$$b_{mr} = 4.73 \varepsilon_0 \, x (\rho/J_c)^{1/2} \qquad (4.122a)$$

$$b_{(1/2)r} = 5.27 \varepsilon_0 \, x (\rho/J_c)^{1/2} \qquad (4.122b)$$

Let us now calculate the axisymmetric jet widths from Reichardt's theory. From Eqs. (4.68) and (4.117) we obtain

$$b_{mr} = C_2 \, x \qquad (4.123a)$$

$$b_{(1/2)r} = 1.177 C_2 \, x \qquad (4.123b)$$

A comparison of Eqs. (4.122) and (4.123) reveals a curious fact. Goertler's theory predicts that axisymmetric jet width depends on momentum flux and Reichardt's theory shows no momentum flux dependency. We must rely on experiments to indicate which formulation is correct. Experiments by Alexander et al. [8] show that the jet width is essentially independent of momentum flux. Thus Prandtl's hypothesis of constant eddy viscosity for the axisymmetric jet [Eq. (4.91)] cannot be correct. Nevertheless, it is interesting to note that widely diverging assumptions on the form of the eddy viscosity do not change the axial velocity profile significantly.

4.4 TURBULENT-FREE JETS FROM FINITE APERTURES

In the previous discussion of the theoretical equations for the velocity distributions, it was assumed that the jet issued from a round hole of zero radius or from an infinitesimal slit. In actual practice, of course, the jet apertures have finite dimensions. As a result the expressions for the distributions near the nozzle exit must be modified.

Equally important to our understanding of the process is that we cannot always assume that the velocity profile is uniform at the nozzle exit. The geometry upstream of the nozzle exit can alter the exit profile and can affect the characteristics of the jet. In addition the upstream geometry may change the tangential and radial components of the mean velocity and often, of more significance, can change the fluctuation velocities.

4.4.1 Albertson's Empirical Distribution

Albertson et al. [9] have considered the velocity distributions from slits of finite width b, and round nozzles of finite diameter d.

Figure 4.13 shows the Albertson model for the two-dimensional turbulent jet. The flow field is divided into two zones: (1) the zone of flow establishment and (2) the zone of established flow. In the zone of flow establishment, turbulence at the jet edges acts to decrease the size of a constant velocity central region

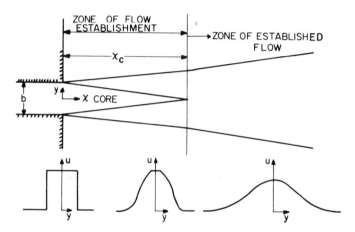

Fig. 4.13 Albertson model for two-dimensional turbulent jet from a finite aperture.

known as the "core" region. In the "core" region the velocity profiles are not similar. They have a flat top which continually decreases in width as x increases. Finally at the end of the core (x_c) the turbulence has penetrated to the jet centerline. Thereafter the profiles are similar and the centerline velocity decays as predicted by the similarity solutions of the previous sections.

Albertson et al. [9] found that they were able to fit their experimental data quite well by the following empirically derived formulas:

$$u/u_0 = \exp[-(y + \sqrt{\pi} C_1 x/2 - b/2)^2/2(C_1 x)^2] \tag{4.124a}$$

for $x \leq x_c$ and $|x/y| \leq 2x_c/b$,

$$u/u_0 = 1 \tag{4.124b}$$

for $x \leq x_c$ and $|x/y| \geq 2x_c/b$, and

$$\frac{u}{u_0} = \left(\frac{1}{\sqrt{\pi} C_1} \frac{b}{x}\right)^{1/2} \exp\left[\frac{-1}{2(C_1)^2} \left(\frac{y}{x}\right)^2\right] \tag{4.124c}$$

4.4 Turbulent-Free Jets from Finite Apertures

for $x \geq x_c$, where u_0 is the velocity at the nozzle exit. From (4.124a), letting $u = u_0$, $y = 0$, $x = x_c$, the core length x_c is related to the slit width b and to C_1 by the expression

$$x_c/b = 1/\sqrt{\pi C_1}$$

or

$$x_c = b/\sqrt{\pi C_1} \tag{4.124d}$$

C_1 is a constant to be experimentally determined.

The above equations correlate well with experimental data for a properly chosen value of C_1.

From Albertson's experiments, $C_1 = 0.109$.

The Albertson formulation for flow from a circular orifice is

$$\frac{u}{u_0} = \exp\left[-\frac{(r + C_2 x - d/2)^2}{2C_2^2 x^2}\right] \tag{4.125a}$$

for $x \leq x_c$ and $(x/r) \leq 2x_c/d$, where d is the nozzle diameter,

$$u/u_0 = 1 \tag{4.125b}$$

for $x < x_c$ and $(x/r) \geq 2x_c/d$, and

$$u/u_0 = (d/2C_2 x) \exp(-r^2/2C_2^2 x^2) \tag{4.125c}$$

for $x \geq x_c$.

For the axisymmetric case the core length is

$$x_c/d = 1/2C_2 \tag{4.125d}$$

and Albertson's data gives $C_2 = 0.081$.

4.4.2 Distribution for Arbitrary Entrance Profile (Two-Dimensional Jets)

Another way of obtaining equations for the case of a finite slit [10], and in fact for any arbitrary distribution at a nozzle, is to use Reichardt's inductive theory. Since the equation that he obtained is linear in $\overline{u^2}$ and the distribution obtained for $\overline{u^2}$ is actually the response of the system to a delta function at $x = 0$, the response for any arbitrary input at $x = 0$ can be found by standard Green's function techniques. Given the velocity distribution at $x = 0$ as

$$u(0, y) = f(y) \tag{4.126}$$

the entire velocity distribution $u(x, y)$ can be found by convolving the input distribution with the previously determined solution for the slit of infinitesimal width. Thus, using (4.108a),

$$\overline{u^2} = \frac{J}{\rho C_1 x} \left(\frac{1}{\pi}\right)^{1/2} \int_{-\infty}^{\infty} f^2(\alpha) \exp\left[-\frac{(y-\alpha)^2}{C_1^2 x^2}\right] d\alpha \tag{4.127}$$

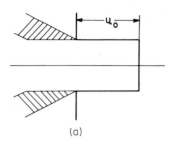

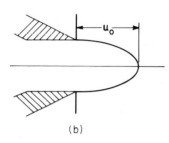

Fig. 4.14 Uniform (a) and parabolic (b) initial velocity distributions.

where the fluctuating component of velocity is assumed negligible and α is a dummy variable.

We consider here the two different entrance velocity profiles shown in Fig. 4.14. For the uniform profile (Fig. 4.14a) the shape function $f(y)$ is

$$f(y) = (1/b)^{1/2}, \quad -b/2 \leq y \leq b/2 \qquad (4.128a)$$

$$f(y) = 0, \quad y \leq -b/2, \quad y \geq b/2 \qquad (4.128b)$$

and the momentum flux per unit depth is $\rho u_0^2 b$. Now the uniform shape factor (4.128) may be substituted into Eq. (4.127) to yield

$$\frac{\overline{u^2}}{u_0^2} = \frac{1}{C_1 x \sqrt{\pi}} \int_{-b/2}^{+b/2} \exp\left[-\left(\frac{y-\alpha}{C_1 x}\right)^2\right] d\alpha \qquad (4.129)$$

where the constant $(1/b)^{1/2}$ was chosen to be consistent with the momentum flux. Equation (4.129) may be placed in the equivalent form

$$\overline{u^2}/u_0^2 = \tfrac{1}{2}[\text{erf } Y_1 + \text{erf } Y_2] \qquad (4.130a)$$

where

$$Y_1 = (0.5 + y')/x', \quad y' = y/b \qquad (4.130b)$$

$$Y_2 = (0.5 - y')/x', \quad x' = C_1 x/b \qquad (4.130c)$$

and thus

$$\bar{u}/u_0 = (\tfrac{1}{2})^{1/2}[\text{erf } Y_1 + \text{erf } Y_2]^{1/2} \qquad (4.131)$$

To the extent that the assumptions are valid, this distribution holds for all values of y and all positive values of x. Figure 4.15 shows the velocity distribu-

4.4 Turbulent-Free Jets from Finite Apertures

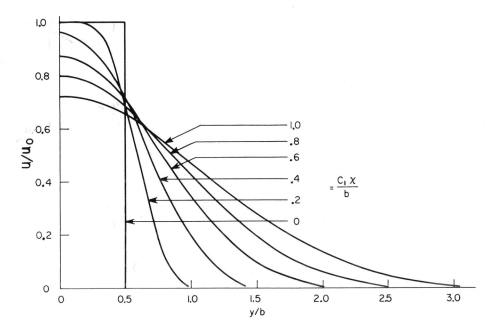

Fig. 4.15 Velocity distributions obtained at various downstream positions for an initially uniform distribution.

tions predicted by Eq. (4.131) for various values of x'. These distributions are not similar. True similarity is approached only asymptotically for large values of x'. However, for practical purposes we may consider that the profiles are similar whenever $x' > 1.0$.

Let us suppose now that the entrance profile is a curve (Fig. 4.14b) which has the shape function

$$f(y) = A_1(1 - 4y^2/b^2)^{1/2}, \quad -b/2 \leq y \leq b/2 \quad (4.132a)$$

$$f(y) = 0, \quad y \leq -b/2, \quad y \geq +b/2 \quad (4.132b)$$

With the shape function of the form given in Eq. (4.132), Eq. (4.127) becomes

$$\frac{\overline{u^2}}{u_0^2} = \frac{1}{C_1 x \sqrt{\pi}} \int_{-b/2}^{+b/2} \left(1 - \frac{4\alpha^2}{b^2}\right) \exp\left[-\left(\frac{y-\alpha}{C_1 x}\right)^2\right] d\alpha \quad (4.133)$$

where in this case A_1 must equal $(3/2b)^{1/2}$ to conserve the momentum flux per unit depth. Integration of Eq. (4.133) yields the result that

$$\frac{\overline{u}}{u_0} = (\tfrac{1}{2} - 2y'^2 - x'^2)^{1/2}(\operatorname{erf} Y_1 + \operatorname{erf} Y_2)^{1/2}$$

$$+ \frac{2x'^2}{\sqrt{\pi}} [Y_2 \exp(-Y_1^2) + Y_1 \exp(-Y_2^2)] \quad (4.134)$$

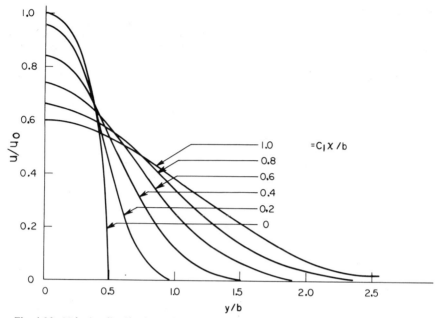

Fig. 4.16 Velocity distributions obtained at various downstream positions for an initially parabolic distribution.

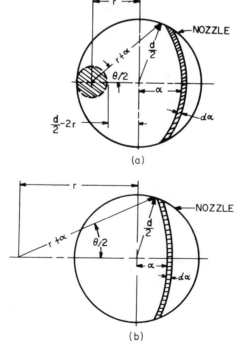

Fig. 4.17 Areas of equal contribution to the velocity: (a) interior points, (b) exterior points.

4.4 Turbulent-Free Jets from Finite Apertures

This function applies at any position downstream of the nozzle exit when the exit profile has the shape given in (4.132). Figure 4.16 shows the distribution for various values of x'. At a considerable distance downstream the shapes again become similar and are in the familiar Gaussian form. Thus, downstream profile measurements give no indication of the exit profile shape.

4.4.3 Distribution for Arbitrary Entrance Profile (Axisymmetric Jets)

For the case of axisymmetric configurations the convolution integral [8, 11] is more involved. The convolution of the infinitesimal nozzle solution equation [Eq. (4.117)] and an axisymmetric profile has the form

$$\bar{u}^2 = \frac{J_c}{\rho \pi C_2^2 x^2} \int_0^\infty f^2(\alpha) \exp\left[-\left(\frac{r-\alpha}{C_2 x}\right)^2\right] \theta(\alpha, r)(r+\alpha)\, d\alpha \qquad (4.135)$$

where as before $f(r)$ is the shape function at the nozzle exit; $\theta(\alpha, r)$ defines the angles of equal contribution. Figure 4.17 gives a graphical representation of $\theta(\alpha, r)$. The views show the cross section of the jet nozzle. Figure 4.17a indicates the values of $\theta(\alpha, r)$ for the interior points of the flow field ($r \leq d/2$) and Fig. 4.17b indicates the exterior points ($r \geq d/2$). The angle of equal contribution for interior points is

$$\theta(\alpha, r) = 2\pi, \qquad -r \leq \alpha \leq (d/2 - 2r) \qquad (4.136a)$$

$$\theta(\alpha, r) = 2\cos^{-1} \frac{r^2 + (r+\alpha)^2 - d^2/4}{2r(r+\alpha)}, \qquad (d/2 - 2r) \leq \alpha \leq d/2 \qquad (4.136b)$$

The angle of equal contribution for the exterior points has the same form as Eq. (4.136b) except that the limits change to $-d/2 \leq \alpha \leq d/2$. Due to the complexities introduced by $\theta(\alpha, r)$, Eq. (4.135) must be integrated numerically to obtain the velocity distribution for the arbitrary input. However, it is possible to find the velocity distribution along the jet axis analytically. At the axis ($r = 0$), Eq. (4.135) reduces to

$$\overline{u^2}(x, 0) = \frac{2J_c}{\rho C_2^2 x^2} \int_0^\infty f^2(\alpha)\, \alpha \exp\left[-\left(\frac{\alpha}{C_2 x}\right)^2\right] d\alpha \qquad (4.137)$$

For a uniform velocity profile at the exit, Eq. (4.137) becomes

$$\overline{u^2}(x, 0) = \frac{J_c K_1^2}{\rho C_2^2 x^2} \int_0^{d/2} (2\alpha) \exp\left(-\frac{\alpha^2}{C_2^2 x^2}\right) d\alpha \qquad (4.138)$$

where K_1 is a scale factor chosen so that $u(0, 0) = u_0$. Since $J_c = \rho \pi d^2 u_0^2/4$, $K_1 = 2/(\sqrt{\pi}\, d)$. Thus from Eq. (4.138) we obtain

$$\frac{\overline{u^2}(x, 0)}{u_0^2} = 1 - \exp\left(-\frac{1}{4C_2^2 x^2/d^2}\right) \qquad (4.139)$$

Figure 4.18 shows the centerline distribution given in Eq. (4.139). Albertson's model as described in Eqs. (4.125b) and (4.125c) is also shown. In the Albertson model the centerline velocity is maintained at the exit magnitude throughout the core and decreases thereafter inversely with the downstream distance. The superposition method, however, yields a single continuous curve which approaches the Albertson model asymptotically.

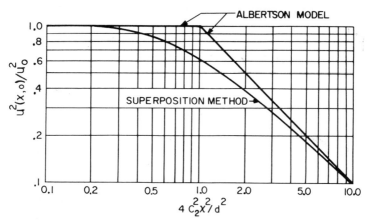

Fig. 4.18 Comparison between Albertson's model and the superposition model.

4.4.4 Schlichting–Abramovich and Simson–Brown Distributions

Schlichting [12] in 1930 obtained the following velocity distribution valid in the wake of a blunt body:

$$u = u_{\max}(1 - \eta^{3/2})^2, \qquad \eta^{3/2} \leq 1$$

where $\eta = y/b_w$ and where b_w is the thickness of the wake.

Abramovich adapted this formula to the boundary layer of a jet in a coflowing stream [13, p. 151 ff] and then showed that it could be used for both the two-dimensional [13, p. 172 ff] and axisymmetric jets [13, p. 176 ff] of finite width in the initial (mixing) region as well as in the main (fully developed) region [13, p. 180 ff] by comparing this theoretical (semiempirical) formula with the theories and experimental results previously obtained. Although derived for the case of a jet in a coflowing stream, the data indicates that the results also hold when the velocity of the coflowing stream is zero.

Brown and Simson [14] point out that, for the two-dimensional jet, Albertson's profile is a better representation than Abramovich's but that the finite profile is easier to work with than Albertson's. They therefore revised the Abramovich distribution to obtain

$$u = u_{\max}[1 - \eta^{7/4}]^2 \tag{4.140}$$

which fits the data very well.

4.4 Turbulent-Free Jets from Finite Apertures

In Eq. (4.140) $\eta = y_e/kx$, where y_e is the distance from the outer edge of the core.

In the initial region $u_{max} = u_0$. By requiring conservation of momentum, one finds

$$k = 1.378b/x_c \qquad (4.141)$$

where b is the width of the slit and x_c is the length of the core.

In the zone of established flow

$$u_{max} = u_0(b/0.726kx)^{1/2} = (x_c/x)^{1/2} \qquad (4.142)$$

Brown and Simson state that, although x_c is actually a weak function of the Reynolds number (varying from about $5.0b$ at $N_R = 10^4$ to about $7.0b$ at $N_R = 10^5$), in Simson's experiments he found that Albertson's result ($x_c = 5.2b$) was also a good fit for his jet.

Figure 4.19 compares the velocity profiles from Eq. (4.85a) (Goertler), Eq. (4.124c) (Albertson), and Eq. (4.140) (Simson). The most significant differences in the profiles occur at the tails of the distribution. Simson's distribution decays most rapidly. In the other regions of the distribution the velocities are essentially the same.

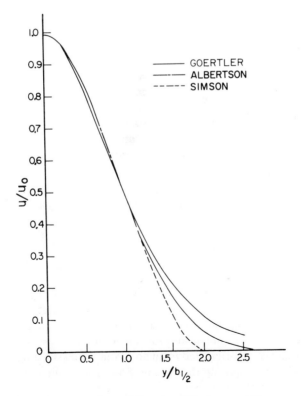

Fig. 4.19 Comparison of Goertler, Albertson, and Simson profiles.

The finite-aperture distributions we have discussed each have certain computational advantages and disadvantages. The erf function distributions obtained from Reichardt's theory have the advantage of a single analytical form all along the jet. Unfortunately integration over y cannot be done in closed form.

Albertson's profile is somewhat easier to work with, but integration over all powers of velocity is simplest using the Simson distribution. The integration may, however, result in an expression having many terms. We will therefore find it advantageous to discuss modifications of the Schlichting–Bickley and Goertler profiles.

4.4.5 The Modified Schlichting–Bickley and Goertler Profiles

The form obtained by Schlichting and by Bickley for the two-dimensional laminar jet issuing from an infinitesimal slit is similar to that obtained by Goertler for a turbulent two-dimensional jet issuing from an infinitesimal slit. We repeat them here:

Schlichting–Bickley Profile

$$u = \frac{1.5(J/6\rho)^{2/3}}{(\nu x)^{1/3}} \operatorname{sech}^2 \frac{(y/2)(J/6\rho)^{1/3}}{(\nu x)^{2/3}} \tag{4.31}$$

Goertler Profile

$$u = \tfrac{1}{2}(3J\sigma_e/\rho x)^{1/2} \operatorname{sech}^2(\sigma_e y/x) \tag{4.85}$$

These can both be written in the form

$$u = k_{30} \operatorname{sech}^2 k_{40} y \tag{4.143a}$$

where

$$k_{30}(x) = \frac{1.5(J/6\rho)^{2/3}}{(\nu x)^{1/3}} \quad \text{for a laminar jet} \tag{4.143b}$$

$$k_{30}(x) = \tfrac{1}{2}(3J\sigma_e/\rho x)^{1/2} \quad \text{for a turbulent jet} \tag{4.143c}$$

$$k_{40}(x) = \frac{(J/6\rho)^{1/3}}{2(\nu x)^{2/3}} \quad \text{for a laminar jet} \tag{4.143d}$$

$$k_{40}(x) = \sigma_e/x \quad \text{for a turbulent jet} \tag{4.143e}$$

This type of distribution shares with the Brown and Simson distribution the attribute that powers of u are easily integrated with respect to y. It has an advantage over the Brown and Simson distribution in the fact that the same form can be used for both laminar and turbulent jets.

By assuming that the jet originated from an infinitesimal slit situated upstream (x_0) of a finite slit and requiring the volume flows to be equal at the finite slit, Bourque and Newman [15] were able to modify the Goertler profile to approximate the flow out of a finite slit.

4.5 Experimental Results on Plane Turbulent Jets

We will find it convenient to use this modified profile for both laminar and turbulent two-dimensional jets throughout this text.

If the jet originates at a point x_0 upstream of the finite slit, its equation will be (if $x = 0$ at the finite slit)

$$u = k_3 \operatorname{sech}^2 k_4 y \qquad (4.144\text{a})$$

where

$$k_3(x) = 1.5(J/6\rho)^{2/3}/v^{1/3}(x + x_0)^{1/3} \quad \text{for a laminar jet} \qquad (4.144\text{b})$$
$$k_3(x) = \tfrac{1}{2}(3J\sigma_e)^{1/2}/\rho^{1/2}(x + x_0)^{1/2} \quad \text{for a turbulent jet} \qquad (4.144\text{c})$$
$$k_4(x) = (J/6\rho)^{1/3}/2v^{2/3}(x + x_0)^{2/3} \quad \text{for a laminar jet} \qquad (4.144\text{d})$$
$$k_4(x) = \sigma_e/(x + x_0) \quad \text{for a turbulent jet} \qquad (4.144\text{e})$$

The virtual origin is determined by forcing the volume flow obtained from the distribution above to be equal to the volume flow out of a slit having a uniform velocity distribution; i.e., the volume flow due to the uniform distribution is

$$q_s = u_0 \, bh \qquad (4.145)$$

where q_s is the flow, b the slit width, and h the slit height, but from (4.144a) at $x = 0$

$$q_s = 2h \int_0^\infty k_3(0) \operatorname{sech}^2 k_4(0) \, y \, dy$$
$$= 2h k_3(0)/k_4(0) \qquad (4.146)$$

whence for the turbulent jet [Eqs. (4.144c) and (4.144e)]

$$u_0 b = (3J x_0/\rho \sigma_e)^{1/2} \qquad (4.147)$$

but

$$J = \rho u_0^2 b \qquad (4.148)$$

hence,

$$x_0 = b\sigma_e/3 \qquad (4.149)$$

In a similar manner we find for the laminar jet

$$x_0 = u_0 b^2/36v \qquad (4.150\text{a})$$

or, in terms of the Reynolds number N_R,

$$x_0 = b N_R/36 \qquad (4.150\text{b})$$

4.5 EXPERIMENTAL RESULTS ON PLANE TURBULENT JETS

As has already been mentioned, Albertson and his colleagues obtained data on plane turbulent jets for which they found the exponential distribution to be a good fit.

Many others have obtained data on the two-dimensional turbulent jet. Most of them found that their data could be fitted about equally well by either the Goertler distribution or the exponential distribution. Some, on the other hand, have claimed that one or the other of these curves is a better fit near the tail ends; for example, Miller and Comings [16] indicate that, if the raw data at the tail of the curve are corrected for the component of transverse velocity that is present, the data fit the exponential curve much better than the Goertler curve (Fig. 4.19).

Miller and Comings [16] have also measured turbulent velocity and static pressure within the jet (Fig. 4.20). Their data show that once past the core the

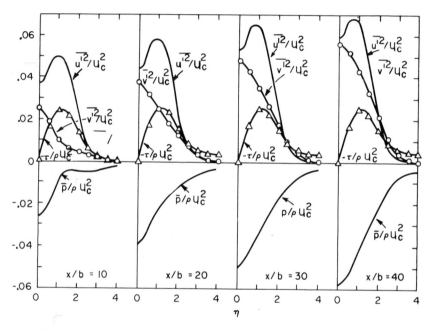

Fig. 4.20 Mean square turbulent velocity profiles and static pressure profiles.

pressure within the jet is below ambient by several percent in places and that the mean square of the turbulent velocity may be as high as 6 or 7% of the mean velocity squared; however, the effects of the negative static pressures and the turbulent velocity tend to cancel each other with respect to their effects on the mean velocity distribution and therefore do not appreciably negate the assumptions used by Goertler and those used by Reichardt.

It should also be noted that the turbulence tends to be greatest where the velocity gradient is greatest. This is particularly evident close to the nozzle.

The pressure distribution (Fig. 4.21) within the transition region (or zone of establishment) [16] is interesting in that, for a given downstream distance x, the pressure is positive near the jet axis and dips sharply to its lowest value on each side of the core region.

4.5 Experimental Results on Plane Turbulent Jets

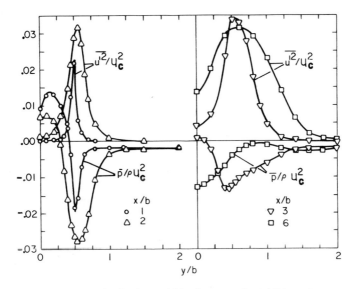

Fig. 4.21 Distributions within the zone of establishment.

4.5.1 Stability

Although a jet of sufficiently low Reynolds number (about 10 or so) can in principle remain laminar until all of its energy is dissipated in mixing with the surrounding fluid, in actual practice the Reynolds numbers used are such that nearly all jets break into turbulence at some point downstream. The region where the jet changes from laminar to turbulent is called the transition region. A number of theoretical and experimental investigations have been made of the process of transition and stability of jets. It is found that in general vortices are shed from the shear layers of the jet. These result in sinusoidal oscillations both in the longitudinal and transverse directions with respect to the direction of flow. These transverse and longitudinal oscillations correspond, respectively, to antisymmetric and symmetric shedding of eddies. This eddy shedding and the associated sinusoidal oscillations are associated with the natural frequencies of the jet. If the jet is excited by a sinusoidal disturbance of some particular frequency, then it is found in general that frequencies are amplified if they correspond with some of these natural frequencies. Studies by Roffman and Toda [17] have shown that maximum amplification is obtained if the sound input is at the base of the jet.

The vortices or eddies at the boundaries of the jet are caused by the shear between adjacent layers of fluid moving at different velocities, and the strength of these eddies are therefore dependent upon the relative velocities between layers, i.e., on the velocity gradient. For example, a nozzle can be shaped so as to cause an approximately uniform profile at its exit or it can be shaped to give an approximately parabolic distribution. The uniform distribution will result in a high velocity gradient as it moves into the stagnant fluid, whereas the gradient for the parabolic distribution will be appreciably less for the same total momentum and mass flow. The parabolic distribution will therefore result in much weaker eddies and can remain laminar longer.

The eddy viscosity of a turbulent stream is due in large part to the eddies at the boundaries penetrating the entire stream. Additional disturbances at the stream boundaries, particularly near the nozzle exit, can interact with the eddies so as to increase the fluctuation energy, thereby effectively increasing the eddy viscosity of the jet even when the jet is already turbulent. Since, as the equations for the width of a jet indicate, the width increases with the viscosity, one expects that the turbulent jet width should increase when disturbed if the disturbance increases the eddy viscosity.

This increase in jet width has been noted for Reynolds numbers up to almost 10,000. The experiments indicate that in general the spreading of an axisymmetric turbulent jet is appreciably more affected by disturbances than is a two-dimensional turbulent jet at comparable Reynolds numbers.

The evidence seems to indicate that the dimension that determines the sensitive frequency is the boundary layer thickness. It is of interest to note that, for both the axisymmetric and the two-dimensional laminar jets, the momentum thickness is inversely proportional to the square root of the axial velocity.

The nondimensional frequency called the Strouhal number is defined as

$$N_s = fd/u \tag{4.151}$$

where d is a characteristic diameter, f is the frequency, and u is the velocity. For a two-dimensional jet it is usual to choose the nozzle width (b) as d in Eq. (4.151).

Most of the data on jet stability indicates that the Strouhal number for maximum jet sensitivity is proportional to the square root of the Reynolds number for Reynolds numbers greater than about 150.

Sato [4] points out and Sato and Sakao [18] reconfirm the significance of the boundary layer width on the sensitive frequency. They show that, by redefining the Strouhal number and the Reynolds number in terms of the boundary layer thickness, the Strouhal number is a constant, i.e., independent of Reynolds number; for example, for the two-dimensional laminar jet we have, from Eq. (4.32),

$$u_c = 1.5(J/6\rho)^{2/3}/(\nu x)^{1/3} \tag{4.152}$$

And, from Eq. (4.66),

$$b_m = 2.42(\nu x)^{2/3}(\rho/J)^{1/3} \tag{4.153}$$

whence

$$u_c b_m^2 = \tfrac{8}{3}\nu x \tag{4.154}$$

4.5 Experimental Results on Plane Turbulent Jets

The Strouhal number using $2b_m$ and u_c as the characteristic diameter and velocity, respectively, is

$$N_{sb} = 2fb_m/u_c \tag{4.155}$$

From Eq. (4.154) at any given x, $b_m \sim u_c^{-1/2}$; hence,

$$N_{sb} \sim fu_c^{-3/2} \tag{4.156}$$

For the higher Reynolds numbers, experiments indicate that the frequency f is proportional to the velocity to the three-halves power, so that the Strouhal number defined in terms of the momentum thickness is a constant.

4.5.2 The Bounded Jet

Most fluidic devices have a design which is etched or cut into some solid or through the solid after which top and bottom plates are added. The result is that the jet nozzles are rectangular in shape and that the jets exiting from these nozzles are always bounded by the top and bottom plates and can spread only in the other directions. The jets within these devices are, therefore, appreciably different from the unbounded jets previously discussed. Let us, then, consider how the presence of solid boundaries may affect a jet.

Figure 4.22 illustrates a bounded two-dimensional jet, i.e., a finite section of a two-dimensional jet bounded by two parallel plates. The geometry is similar to

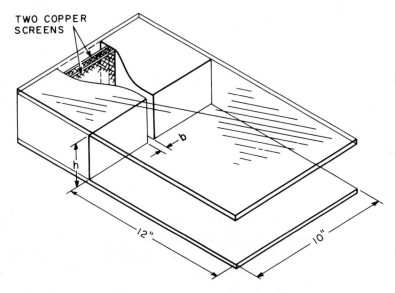

Fig. 4.22 Bounded two-dimensional jet.

that found in proportional fluid amplifiers with large open vents. The nozzle exit is rectangular and has an aspect ratio defined as the nozzle height (which is, of course, the distance between the parallel plates) divided by its width. If the aspect ratio is large, then one expects, and indeed finds, that in the vicinity of the midplane the velocity distribution should be similar to that of the free (unbounded) two-dimensional jet. One also expects, but does not find, that the velocity distribution in the z-direction (perpendicular to the plates) should have its maximum in the midplane since the velocity at both plates bounding the jet must be zero. Instead most investigators [19, 20] have found a double maximum off the axis of the jet—one maximum on each side of the midplane.

These maxima are caused by secondary flow vortices resulting from the fact that the velocity must be zero at the top and bottom plates. This effect can be appreciably increased by vortices formed upstream of the nozzle and can be minimized by flow straighteners and by bringing the top and bottom plates close together. As the top and bottom plates are brought together these vortices are appreciably damped out.

A more important effect of the bounding plates are that they also damp out turbulence in their immediate vicinities and tend to increase it farther away because of the velocity gradient. Consequently, when solid boundaries are present, the eddy viscosity is a function of distance from the plates and approaches zero at the plates, so that if the plates are sufficiently close together (small aspect ratio) the turbulence is greatly attenuated.

Finally, there is a variation of pressure along the plates instead of the approximately uniform pressure distribution present in the case of free jets.

Because of the absence of side walls (such as exist in binary devices, for example) the pressure gradients are quite small, although somewhat larger than those in the free two-dimensional jet.

Trapani's data [20], as well as that of Raju and Kar [21], indicates that the distribution is not self-similar throughout.

This has been determined by the calculation of σ_e using the modified Goertler distribution and measurements of the velocity along a constant velocity ratio (u/u_c) line; i.e., since

$$u = u_c \operatorname{sech}^2[\sigma_e y/(x + x_0)]$$

then

$$\sigma_e = [(x + x_0)/y] \operatorname{sech}^{-1}(u/u_c)^{1/2}$$

For a free jet, σ_e should be the same for all values of u/u_c and for all values of x and y. Trapani [20] and Raju and Kar [21], however, find that, for a given value of u/u_c, σ_e becomes constant beyond the core, but that the value of σ_e differs for different values of u/u_c measured in the midplane. Furthermore, the results depend on aspect ratio. The differences between the values of σ_e measured along the $u/u_c = 0.8$ and 0.4 lines, for example, is shown by Raju and Kar to be appreciably greater for an aspect ratio of 2 than for an aspect ratio of 10. At an aspect ratio of 10, $\sigma_e \sim 8$ and hence does not differ greatly from that of a free jet (7.67).

The increased σ_e for bounded jets indicates less entrainment than for free jets. Because the jet entrains less flow than a free jet, the centerline velocity at the midplane decays less rapidly than that of a free jet. The jet entrains less because of the drag of the plates on the flow that must be brought in from outside the plates to the jet. This drag also results in a greater pressure drop between the axis of the jet and the surrounding atmosphere than for the free jet.

PROBLEMS

4.1 If $J = \rho u_f^2 b_f$, show that the axial velocity from the Schlichting–Bickley equation for a two-dimensional laminar jet becomes

$$\frac{u}{u_f} = \frac{0.454}{B^{1/3}} \text{sech}^2 \frac{0.275 y/b_f}{B^{2/3}}$$

where $B = x/(b_f N_R)$ and $N_R = u_f b_f/\nu$.

4.2 The reference conditions in a two-dimensional laminar jet are related by $J = \rho u_f^2 b_f$. Consider a point in the jet field where $y/b_f = 1.0$ and $x/b_f = 20$. Find the ratio of the axial velocities at this point when the Reynolds number $(u_f b_f/\nu)$ changes from 100 to 200.

4.3 If $J_c = \pi d_f^2 u_f^2/4$ for an axisymmetric laminar jet, show that the axial velocity profile may be expressed as

$$\frac{u}{u_f} = \frac{0.094}{B_1} \left[\frac{1}{(1 + 0.012 r^2/d_f^2 B_1^2)^2} \right]$$

where $B_1 = x/(d_f N_R)$ and $N_R = u_f d_f/\nu$.

4.4 Find the spread parameter σ_e of a two-dimensional turbulent jet from the data measured at two points far downstream of the nozzle exit. The data show only that the velocity ratio between the points is $u_1/u_2 = 1.052$, where the points are at $y_1 = 1.0$ cm, $x_1 = 10.0$ cm, and $y_2 = 1.5$ cm, $x_2 = 20.0$ cm.

4.5 Find the velocity distribution downstream that results when two finite-width two-dimensional parallel jets, one nozzle width apart, issue into a medium at rest. (HINT: Choose an arbitrary profile at the entrance position that has a zero-velocity region and then use the convolution integral.)

4.6 An axisymmetric turbulent jet issues from a finite-diameter nozzle with uniform velocity u_0. If the experimental constant C_2 of Albertson's distribution is 0.070 and the velocity at a point in the established flow region is $0.3 u_0$ at $x/d = 15$, find the radial distance r/d to the point. What is the velocity at this point if C_2 is changed to 0.080? In each case what is the core length?

4.7 In Eqs. (4.50a) and (4.50b) the virtual origins of two-dimensional laminar and turbulent jets were found based on a match of the mass flow at the entrance of the finite nozzle. If instead of this condition we restrict the transverse velocity v at $x = 0$ and $y = b/2$ to zero, show that

(a) for the laminar jet

$$x_0/b = 0.0448 N_R$$

(b) for the turbulent jet

$$x_0/b = 0.459 \sigma_e$$

4.8 Find the mass flow width and the momentum flux width for (a) the Simson–Brown velocity distribution, (b) Reichardt's profile for two-dimensional flow from an infinitesimal slit.

4.9 Using the momentum flux width as the reference diameter in the definition of the Reynolds number and the centerline velocity as the reference velocity, find the Reynolds number as a function of downstream distance for the Simson–Brown and the Albertson distributions.

NOMENCLATURE

b	Slit width
b_f	Reference width of jet
b_m	Momentum flux width of the two-dimensional jet
b_{mr}	Momentum flux width for axisymmetric jet
C_1	Spread factor for the two-dimensional turbulent jet (Reichardt theory)
C_2	Spread factor for the axisymmetric turbulent jet (Reichardt theory)
d	Nozzle diameter
h	Depth
J	Momentum flux per unit depth
J_c	Momentum flux of axisymmetric jet
J_f	Reference momentum flux
$\dot{m}$	Mass flow
$\dot{m}_f$	Reference mass flow for one half the jet width
N_R	Reynolds Number
r	Radial coordinate
u	Axial velocity (x direction)
u'	Fluctuation velocity
$\bar{u}$	Mean velocity
u_c	Jet centerline velocity
u_0	Velocity at the nozzle exit
u_f	Reference velocity
u_r	Radial velocity
v	Transverse velocity (y direction)
x	Position coordinate
x_c	Core length

y Position coordinate
ε Eddy viscosity
ν μ/ρ, kinematic viscosity
ρ Density
σ_e Spread factor for the two-dimensional turbulent jet (Goertler theory)

REFERENCES

1. H. Schlichting, Laminare strahlausbreitung. *Z. Angew. Math. Mech.* **13**, 260 (1933); H. Schlichting, "Boundary Layer Theory." pp. 164–168, 181–184. McGraw-Hill, New York, 1960.
2. W. Bickley, The plane jet. *Phil. Mag.* **23**, Ser. 7, 727 (1939).
3. E. N. da C. Andrade, Velocity distribution in a liquid-into-liquid jet, Part II. The plane jet. *Proc. Phys. Soc.* 784 (Sept. 1939).
4. H. Sato, The stability and transition of a two-dimensional jet. *J. Fluid Mech.* **7**, 53 (1960).
5. H. Goertler, Berechnung von aufgaben der freien turbulenz auf grund eines neuen naherungsansatzes. *Z. Angew. Math. Mech.* **22** (1942); also H. Schlichting, "Boundary Layer Theory," p. 605. McGraw-Hill, New York, 1960.
6. H. Reichardt, "Gesetzmabigkeiten der freien Turbulenz," VDI-Forschungsheft 414, 1942. See also G. N. Abramovich, "The Theory of Turbulent Jets," pp. 113–120. MIT Press, Cambridge, Massachusetts, 1963.
7. H. Reichardt, Uber eine neue theorie der freien turbulenz. *Z. Angew. Math. Mech.* **21** (1941).
8. L. G. Alexander, T. Baron, and E. W. Comings, "Transport of Momentum, Mass and Heat in Turbulent Jets." Univ. of Illinois Eng. Exp. Sta., Bull. No. 413, 1953.
9. M. S. Albertson, Y. B. Dai, R. A. Jensen, and H. Rouse, Diffusion of submerged jets. *Proc. ASCE* **74**, 571 (1948).
10. J. M. Kirshner, Jet flows. *Fluidics Quart.* **1**, No. 3 (1968).
11. K. N. Reid and S. Katz, "Characterization of Free and Impinging Axisymmetric Jets With and Without Auxiliary Flows," ASME paper 70-WA/Flcs-6, November 1970.
12. H. Schlichting, "Boundary Layer Theory," p. 600. McGraw-Hill, New York, 1960.
13. G. N. Abramovich, "The Theory of Turbulent Jets," MIT Press, Cambridge, Massachusetts, 1963 (Russian printing, Fizmatgiz Press, 1960).
14. F. T. Brown and A. K. Simson, "Research in Pressure-Controlled Fluid Jet Amplifiers," Rep. #9213-1, Dept. of Mech. Eng., MIT, Contract DA-19-020-ORD-5650, Nov. 1963; A. K. Simson, "A Theoretical Study of the Design Parameters of Subsonic Pressure-Controlled, Fluid Jet Amplifiers," Ph.D. Thesis, MIT, 1963.
15. C. Bourque and B. G. Newman, Reattachment of a two-dimensional incompressible jet to an adjacent flat plate. The Aeronautical Quarterly, Vol. XI, p. 201ff, August 1960.
16. D. E. Miller and E. W. Comings, Static pressure distribution in the free turbulent jet. *J. Fluid Mech.* **3**, part 1 (Oct. 1957).
17. G. L. Roffman and K. Toda, "A Discussion of the Effects of Sound on Jets and Flueric Devices," ASME paper #69-Vibr-3; *J. Eng. Ind. Ser.* B, **91**, No. 4, 1161 (1969).
18. H. Sato and F. Sakao, An experimental investigation of the instability of a two-dimensional jet at low Reynolds numbers. *J. Fluid Mech.* Vol. **10**, part 2, 337ff (1964).
19. J. F. Foss and J. B. Jones, "A Study of Incompressible Turbulent Bounded Jets," Purdue Res. Foundation Project #3728, 16 Oct. 64. Prepared for Harry Diamond Laboratories.
20. R. D. Trapani, "An Experimental Study of Bounded and Confined Jets," Fluerics-22, Harry Diamond Labs and Advances in Fluidics (Nov. 66) (*Proc. May 1967 ASME-HDL Fluidics Symp.*).
21. V. C. Raju and S. Kar, Studies on bounded jets. *Proc. Int. JSME Symp. Fluid Mach. Fluidics* **3**, 27 (1972).

Chapter 5

JET DYNAMICS

5.1 GENERAL DEVELOPMENT OF JET DYNAMICS*

Figure 5.1 shows an incompressible two-dimensional jet that has a velocity distribution $u(x, y)$ in the axial direction x. Let us consider the dynamics of a particular jet particle at a transverse position y_0 from the jet axis. A transverse pressure gradient, $dp/dy = -g(x, t)$, initiated at $t = 0$, exerts a force on the particle and the force produces a particle acceleration that is given by

$$\rho \, d^2 y_0/dt^2 = g(x, t) \qquad (5.1)$$

where ρ is the density of the fluid.

Rather than follow the motion of a particular particle, we will choose to consider instead the motion of particles passing through a given point. We therefore write

$$dy_0/dt = \partial y_0/\partial t + u \, \partial y_0/\partial x + v \, \partial y_0/\partial y \qquad (5.2)$$

where v is the velocity in the transverse direction y. Now we assume that

$$v \, \partial y_0/\partial y \ll u \, \partial y_0/\partial x$$

so that Eq. (5.2) becomes

$$dy_0/dt = \partial y_0/\partial t + u \, \partial y_0/\partial x \qquad (5.3)$$

---

* Reference [1].

5.1 General Development of Jet Dynamics

Since particles of different axial velocities will be affected differently by the pressure gradient, Eq. (5.3) is meaningful only if the deflection of a particular particle is relatively unaffected by interaction with other particles during the interval of interest. Since the velocity profile is relatively flat in the vicinity of the axis, we expect that the interaction with other particles will be minimum

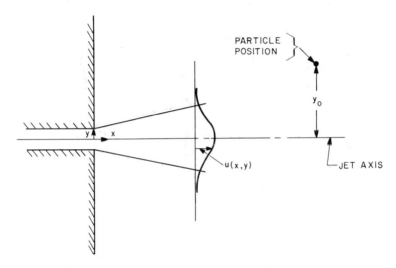

Fig. 5.1 Two-dimensional jet.

in that region. Thus we will restrict the discussion to the motion of the axis of the jet and use the results to qualitatively indicate the effect of the pressure gradient on the entire jet. For the jet axis $u = u_c$, so that Eq. (5.3) changes to

$$dy_0/dt = \partial y_0/\partial t + u_c \, \partial y_0/\partial x \tag{5.4}$$

If Eq. (5.4) is substituted into Eq. (5.1) the result is

$$\frac{\partial^2 y_0}{\partial t^2} + 2u_c \frac{\partial^2 y_0}{\partial t \, \partial x} + u_c^2 \frac{\partial^2 y_0}{\partial x^2} = \frac{g(x, t)}{\rho} \tag{5.5}$$

To place Eq. (5.5) into a more convenient form we define the axial transport time of the particles τ as

$$\tau \equiv \int_0^x \frac{dx'}{u_c} \tag{5.6}$$

where x' is a variable of integration. The application of Eq. (5.6) to Eq. (5.5) yields

$$\frac{\partial^2 y_0}{\partial t^2} + 2 \frac{\partial^2 y_0}{\partial t \, \partial \tau} + \frac{\partial^2 y_0}{\partial \tau^2} = \frac{g[x(\tau), t]}{\rho} \tag{5.7}$$

To obtain the general solution of Eq. (5.7) we employ two separate Laplace transformations. The first Laplace transformation from the t to the s plane gives

$$s^2 Y + 2s\, dY/d\tau + d^2 Y/d\tau^2 = (1/\rho)G[x(\tau), s] \tag{5.8}$$

where Y and G are the transforms, respectively, of y_0 and g, and the initial conditions at $t = 0$ are $y_0 = \partial y_0/\partial t = 0$. The second Laplace transformation from the τ to the σ plane results in

$$s^2 V + 2s\sigma V + \sigma^2 V = (1/\rho)G^1[f(\sigma), s] \tag{5.9}$$

where V and G^1 are the σ transforms of Y and G, and the initial conditions at $\tau = 0$ are $Y = dY/d\tau = 0$. Equation (5.9) is now merely an algebraic equation that has the solution

$$V(\sigma, s) = \frac{G^1[f(\sigma), s]/\rho}{(s+\sigma)^2} \tag{5.10}$$

The inversion of Eq. (5.10) with respect to σ may take either of two equivalent forms of the convolution integral. Thus

$$Y(\tau_l, s) = \frac{1}{\rho} \int_0^{\tau_l} \tau\, e^{-s\tau} G[l - x(\tau), s]\, d\tau \tag{5.11a}$$

$$Y(\tau_l, s) = \frac{1}{\rho} \int_0^{\tau_l} (\tau_l - \tau) \exp[-s(\tau_l - \tau)]\, G[x(\tau), s]\, d\tau \tag{5.11b}$$

where $\tau_l = \int_0^l dx/u_c$ is the transport time for a particle to move the distance l. If the two equivalent forms of Eq. (5.11) are now inverted with respect to s the result is

$$y_0(\tau_l, t) = \frac{1}{\rho} \int_0^{\tau_l} \tau\, g\,[l - x(\tau), t - \tau]\, H(t - \tau)\, d\tau \tag{5.12a}$$

$$y_0(\tau_l, t) = \frac{1}{\rho} \int_0^{\tau_l} (\tau_l - \tau)\, g[x(\tau), t - \tau_l + \tau]\, H(t - \tau_l + \tau)\, d\tau \tag{5.12b}$$

where $H(t)$ is the unit step function. Let us first consider Eq. (5.12a). The unit step function $H(t - \tau)$ is zero for $t < \tau$ and unity for $t > \tau$. Since τ takes on values from 0 to τ_l during integration, we may be sure that the step function equals unity throughout the entire range of integration if t is greater than τ_l. What happens, however, when t is less than τ_l? In this case the step function is zero during integration from t to τ_l, and unity during integration from 0 to t. As a result we may express Eq. (5.12a) as

$$y_0(\tau_l, t) = \frac{1}{\rho} \int_0^{\tau_l} \tau\, g[l - x(\tau), t - \tau]\, d\tau, \qquad t \geq \tau_l \tag{5.13a}$$

$$y_0(\tau_l, t) = \frac{1}{\rho} \int_0^{t} \tau\, g[l - x(\tau), t - \tau]\, d\tau, \qquad 0 \leq t \leq \tau_l \tag{5.13b}$$

5.2 Response of Jet to an Impulse Function

From a physical viewpoint, Eq. (5.13a) represents the deflection of particles that leave the jet nozzle and enter the flow field after the pressure gradient is applied. Equation (5.13b), on the other hand, holds for the deflection of those particles that had already left the nozzle before the application of the pressure gradient.

We may obtain equivalent results from the other convolution integral given in Eq. (5.12b). Thus

$$y_0(\tau_l, t) = \frac{1}{\rho} \int_0^{\tau_l} (\tau_l - \tau) \, g[x(\tau), t - \tau_l + \tau] \, d\tau, \qquad t \geq \tau_l \qquad (5.14a)$$

$$y_0(\tau_l, t) = \frac{1}{\rho} \int_{\tau_l - t}^{\tau_l} (\tau_l - \tau) \, g[x(\tau), t - \tau_l + \tau] \, d\tau, \qquad 0 \leq t \leq \tau_l \qquad (5.14b)$$

A word of caution is in order on the use of the two different forms valid for the same condition, for example, (5.13a) and (5.14a). The form of g in the two expressions corresponds to a shift in coordinates and a reversal in sign. Thus, in Eq. (5.13a), $g(0, t - \tau)$ corresponds to τ taking on the value τ_l, i.e., the upper limit, whereas in (5.14a) $g(0, t - \tau_l + \tau)$ corresponds to τ taking on the value zero, i.e., the lower limit. This point is of importance, for example, when the form of the function changes between 0 and τ_l; e.g.,

$$g(\tau, t) = g_1(\tau, t), \qquad 0 \leq \tau \leq \tau_1$$

$$g(\tau, t) = g_2(\tau, t), \qquad \tau_1 \leq \tau \leq \tau_l$$

thus, from (5.13a),

$$y_0(\tau_l, t) = \frac{1}{\rho} \int_0^{\tau_l - \tau_1} \tau \, g_2(\tau_l - \tau, t - \tau) \, d\tau$$

$$+ \frac{1}{\rho} \int_{\tau_l - t_1}^{\tau_l} \tau \, g_1(\tau_l - \tau, t - \tau) \, d\tau, \qquad t \geq \tau_l \qquad (5.15a)$$

whereas, from (5.14a),

$$y_0(\tau_l, t) = \frac{1}{\rho} \int_0^{\tau_1} (\tau_l - \tau) \, g_1(\tau, t - \tau_l + \tau) \, d\tau$$

$$+ \frac{1}{\rho} \int_{\tau_1}^{\tau_l} (\tau_l - \tau) \, g_2(\tau, t - \tau_l + \tau) \, d\tau, \qquad t \geq \tau_l \qquad (5.15b)$$

5.2 RESPONSE OF JET TO AN IMPULSE FUNCTION

It is of interest to obtain the transfer function of the jet and its time response for a delta function pressure gradient. Figure 5.2 shows the physical configuration. On each side of the jet there is a control nozzle of width x_1. The delta function input only acts on the jet in the control region, i.e., from zero to x_1.

In terms of the transport time variable τ, the control lasts from zero to τ_1, where $\tau_1 = \int_0^{x_1} dx/u_c$. The corresponding transform of the pressure gradient is therefore

$$G[x(\tau), s] = 1, \qquad 0 \leq \tau \leq \tau_1$$

$$G[x(\tau), s] = 0, \qquad \tau_1 \leq \tau$$

Equation (5.11b) then becomes

$$Y(\tau_l, s) = \frac{1}{\rho} \int_0^{\tau_1} (\tau_l - \tau) \exp[-s(\tau_l - \tau)]\, d\tau \tag{5.16}$$

Equation (5.16) may be readily integrated with the result that

$$Y(\tau_l, s) = \frac{1}{\rho s^2} \{\exp[-s(\tau_l - \tau_1)](1 + s\tau_l - s\tau_1) - \exp(-s\tau_l)(1 + s\tau_l)\} \tag{5.17}$$

Equation (5.17) represents the transfer function of a jet. To obtain the time response of the jet to a delta function we must find the inverse Laplace transform of Eq. (5.17). This leads to the time function

$$\rho y_{0l} = tH(t - \tau_l + \tau_1) - tH(t - \tau_l) \tag{5.18}$$

where $y_{0l} = y_0(\tau_l, t)$. Figure 5.3 shows the impulse time response of the jet for various control widths. When the control width is small (solid line) the response is a small sawtooth. There is no deflection at the distance l downstream until

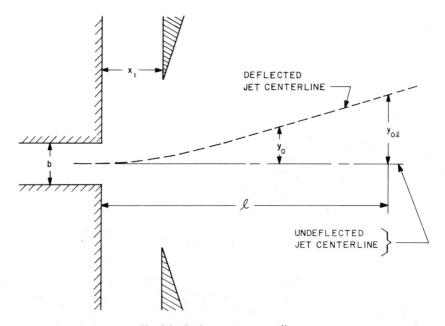

Fig. 5.2 Jet in a pressure gradient.

5.3 Constant Pressure Gradient

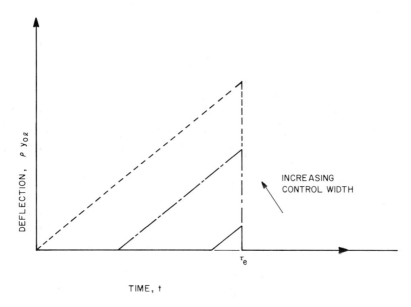

Fig. 5.3 Response to a delta function pressure gradient.

the signal moves through the transport time interval $\tau_l - \tau_1$. As the control width increases (broken line) the sawtooth becomes larger. Finally, when the control width x_1 equals the distance l (dashed line), the jet deflection begins immediately from time $t = 0$. The sawtooth shapes for all the control widths terminate after $t = \tau_l$.

5.3 CONSTANT PRESSURE GRADIENT

Let us now consider the relatively simple case of a two-dimensional jet issuing from a slit of width b into a uniform pressure gradient, constant in time; i.e., $g = g_0 = \text{const} = \Delta p/b$.

Since Eqs. (5.13) do not consider the variation of velocity explicitly, they can be integrated before taking velocity into account, or alternatively the equations can be transformed so as to take the velocity into account before integrating the equations. We will do this calculation the first way. Equation (5.13) gives

$$y_0(\tau_l, t) = \frac{1}{\rho} \int_0^{\tau_l} \tau g_0 \, d\tau, \qquad t \geq \tau_l \tag{5.19a}$$

$$= \frac{1}{\rho} \int_0^{\tau} \tau g_0 \, d\tau, \qquad 0 \leq t \leq \tau_l \tag{5.19b}$$

Integration of Eqs. (5.19) yields

$$y_0(\tau_I, t) = \tau_I^2 \, \Delta p/b2\rho, \qquad t \geq \tau_I \qquad (5.20a)$$

$$y_0(\tau_I, t) = t^2 \, \Delta p/b2\rho, \qquad 0 \leq t \leq \tau_I \qquad (5.20b)$$

Equations (5.20) and (2.80) are equivalent for the condition that the supply pressure p_s equals $\rho u_0^2/2$.

These results give the physically obvious solution that the deflection of the jet in a uniform pressure gradient depends on the time it spends under the influence of that field so that the effect of velocity or distance is important only if if changes the time spent within the field.

To obtain Eq. (5.20a) in terms of velocity and distance, the velocity decay of the jet centerline must be known. For our example, we will assume that the

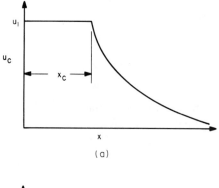

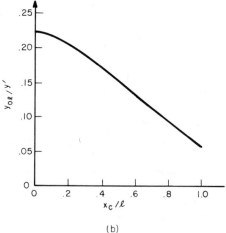

Fig. 5.4 Deflection as a function of core length x_c: (a) centerline velocity, (b) deflection versus decay position.

5.4 Oscillating Pressure Gradient

centerline velocity (Fig. 5.4a) is constant for $0 \leq x \leq x_c$ and then varies with x for $x \geq x_c$ as follows for a turbulent jet [see Eq. (4.85a)]

$$u_c = (k/x_c)^{1/2} = u_1 = \text{const}, \qquad 0 \leq x \leq x_c$$

$$u_c = (k/x)^{1/2}, \qquad x_c \leq x \leq l$$

where $k = \frac{1}{4}(3J\sigma_e/\rho)$, J is the momentum per unit length, and σ_e the empirical spread factor.

The time it takes a particle to go from zero to x_c is τ_c, where

$$\tau_c = \int_0^{x_c} \frac{dx}{u_1} = \frac{x_c}{u_1} = \frac{x_c^{3/2}}{k^{1/2}}$$

and the time to go from x_c to l is τ_2, where

$$\tau_2 = \int_{x_c}^l \frac{x^{1/2} dx}{k^{1/2}} = \frac{2x^{3/2}}{3k^{1/2}}\bigg|_{x_c}^l = \frac{2}{3k^{1/2}}(l^{3/2} - x_c^{3/2})$$

Then, since $\tau_l = \tau_c + \tau_2$, it follows from Eq. (5.20a) that

$$y_{0l}/y' = (1/18)[2 - (x_c/l)^{3/2}]^2, \qquad t > \tau_l \tag{5.21}$$

where $y' = \Delta p l^3/(\rho k b)$.

An interesting fact disclosed by Eq. (5.21) and shown in Fig. 5.4b is that the longer the jet remains at constant velocity the less it is deflected in a uniform pressure gradient. In fact, if the jet is of constant velocity over the entire distance, i.e., if $x_c = l$, Eq. (5.21) gives

$$(y_{0l})_{\text{const}} = l^3 \Delta p/18\rho k b \tag{5.22a}$$

Whereas if the centerline velocity decay begins at the nozzle, i.e., if $x_c = 0$,

$$(y_{0l})_{\text{decay}} = 4l^3 \Delta p/18\rho k b \tag{5.22b}$$

5.4 OSCILLATING PRESSURE GRADIENT

We next consider the effect of an oscillating pressure field on the jet. In the first case we will discuss, the control nozzles are of width x_1 and the distance between power nozzle and receivers is l. We will assume that the pressure field is of constant amplitude between 0 and x_1 and is zero between x_1 and l (Fig. 5.2). For the case of interest

$$g(x, t) = B \cos \omega t, \qquad 0 \leq x \leq x_1$$

$$g(x, t) = 0, \qquad x_1 \leq x \leq l$$

where B is a constant.

Inserting this into Eq. (5.15b)

$$y_0(\tau_l, t) = \frac{B}{\rho} \int_0^{\tau_1} (\tau_l - \tau) \cos \omega(t - \tau_l + \tau) \, d\tau \qquad (5.23)$$

where

$$\tau_1 = \int_0^{x_1} \frac{dx}{u_c}$$

The integration of Eq. (5.23) yields

$$y_0(\tau_l, t) = \frac{B}{\rho\omega} \bigg\{ (\tau_l - \tau_1) \sin \omega(t - \tau_l + \tau_1) - \tau_l \sin \omega(t - \tau_l)$$

$$- \frac{1}{\omega} [\cos \omega(t - \tau_l + \tau_1) - \cos \omega(t - \tau_l)] \bigg\} \qquad (5.24)$$

Now, expanding the trigonometric functions we obtain

$$y_{0l} = \frac{B(m^2 + n^2)^{1/2}}{\rho\omega^2} \bigg\{ \frac{m}{(m^2 + n^2)^{1/2}} \sin \omega t + \frac{n}{(m^2 + n^2)^{1/2}} \cos \omega t \bigg\} \qquad (5.25a)$$

where

$$m \equiv (\beta - \alpha) \cos(\beta - \alpha) - \beta \cos \beta - \sin(\beta - \alpha) + \sin \beta$$

$$n \equiv \beta \sin \beta + \cos \beta - (\beta - \alpha) \sin(\beta - \alpha) - \cos(\beta - \alpha) \qquad (5.25b)$$

$$\alpha \equiv \omega\tau_1, \qquad \beta = \omega\tau_l, \qquad y_{0l} = y_0(\tau_l, t)$$

We may rewrite Eq. (5.25) as

$$y_{0l} = M \cos(\omega t - \phi) \qquad (5.26a)$$

where

$$M = \frac{B\tau_l^2}{\rho\beta^2} \{2[1 + \beta^2(1 - k_1)](1 - \cos k_1 \beta) + k_1^2 \beta^2 - 2k_1 \beta \sin k_1 \beta\}^{1/2}$$

$$\phi = \tan^{-1} \frac{\beta(1 - k_1) \cos \beta(1 - k_1) - \sin \beta(1 - k_1) + \sin \beta - \beta \cos \beta}{\cos \beta + \beta \sin \beta - \beta(1 - k_1) \sin \beta(1 - k_1) - \cos \beta(1 - k_1)} \qquad (5.26b)$$

$$k_1 = \alpha/\beta = \tau_1/\tau_l$$

This result can of course also be obtained from the system response function [Eq. (5.17)] by letting $s = j\omega$.

The maximum deflection amplitude M as a function of β is plotted in Fig. 5.5 for several values of k_1 where $B\tau_l^2/\rho$ has been set equal to unity.

The phase shift ϕ as a function of β is plotted in Fig. 5.6 for the same values of k_1.

5.4 Oscillating Pressure Gradient

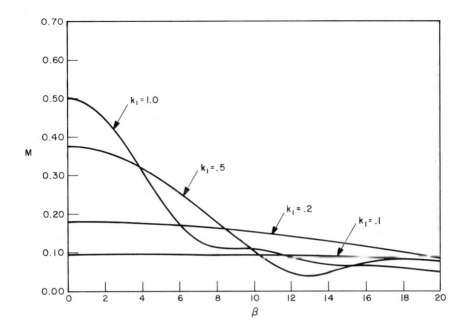

Fig. 5.5 Maximum deflection M as a function of β.

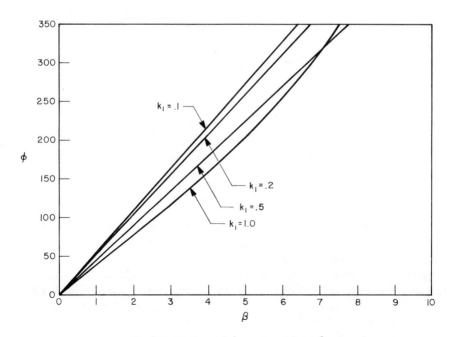

Fig. 5.6 Phase shift ϕ as a function of β.

For a given transport time τ_l (the time it takes a particle to move from the nozzle to the splitter), at low frequencies the deflection increases as τ_1 (the time it takes a particle to cross the control width x_1) increases. At very high frequencies the deflection decreases as τ_1 increases. However, the amplitude of the deflection increases more at low frequencies than it decreases at high frequencies. As a result the bandwidth (where the amplitude is down to 70% of the maximum amplitude and $\beta = \beta_c$) decreases as the control width increases.

In particular we note that, for k_1 small, the magnitude is approximately constant for β as large as 20, thus lending justification to the often used assumption that the transfer function of the jet may be treated as a pure delay at least for those cases where the control width is small compared to the nozzle-splitter distance. It is, however, possible to also show this from the transfer function [Eq. (5.17)], where for $s\tau_1$ small we can expand $\exp(s\tau_1)$ as

$$\exp(s\tau_1) \cong 1 + s\tau_1 + (s^2\tau_1^2)/2 \tag{5.27}$$

The substitution of Eq. (5.27) into Eq. (5.17) yields

$$Y(\tau_l, s) \cong \frac{\exp(-s\tau_l)}{\rho}\left[\tau_1\tau_l - \frac{\tau_1^2}{2} + \frac{s\tau_1^2}{2}(\tau_l - \tau_1)\right] \tag{5.28}$$

For small values of τ_1 Eq. (5.28) reduces to

$$Y(\tau_l, s) \cong \frac{\exp(-s\tau_l)}{\rho}(\tau_1\tau_l) \tag{5.29}$$

and we have confirmed the fact that the jet behaves as a pure time delay for small control widths.

From Fig. 5.5 we see that even for $k_1 = 1$ the deflection amplitude is reasonably constant for small β, dropping by a factor of $\sqrt{2}$ at approximately $\beta_c = 3.5$.

Since the gain of a fluid amplifier is approximately proportional to the jet deflection, and since it is usually desirable that the gain be relatively flat over the frequency range of interest, it is extremely fortunate that this flat region exists.

Figure 5.5 shows, moreover, that the bandwidth associated with β_c is dependent on τ_l since $\omega = \beta/\tau_l$. Thus the bandwidth ω_c is determined by β_c. Since $\beta_c = 3.5$, for $k_1 = 1$, therefore

$$\omega_c = \beta_c/\tau_l = 3.5/\tau_l$$

It follows that to achieve greater bandwidths the transport time must be reduced either by reducing the distance l or by increasing the velocity u_c, or the control width must be decreased.

In order to minimize phase distortion, it is also desirable that the phase shift be approximately proportional to the frequency. Figure 5.6 shows that even for $k_1 = 0.2$ this is true for $\omega\tau_l$ equal to 3 or less. A phase shift proportional to the frequency means that the velocity of propagation is independent of frequency so that there is no dispersion.

If in Eq. (5.26) we write $\beta = \omega \tau_l$, we can see that for a fixed value of ω the deflection magnitude increases as τ_l increases. The phase shift also increases with τ_l for a fixed ω.

Now a larger value of transport time τ is obtained by either a longer path length l or a smaller velocity u_c, or both. Consider, therefore, a jet issuing from a slit with a uniform distribution as it leaves the nozzle. Immediately upon leaving the nozzle the jet boundaries will interact with its surroundings to form a shear layer on each side. Over the width of this shear layer the jet velocity goes from approximately its maximum value to approximately zero. Thus the transport time for particles in this shear layer range from the transport time of the main jet particles to infinity, so that at any given distance from the nozzle there will be some particles that are 180 degrees out of phase with the main stream. These particles, which are being deflected through the shear layer in opposite directions to those of the main stream, will therefore tend to cause mixing. It is apparent that, although very low velocity particles will get 180 degrees out of phase closer to the nozzle than the higher velocity particles, their effect is relatively negligible because of their small momenta. Thus one expects that most of the important mixing effects will be caused by particles whose velocities are of the order of one quarter to one half of those in the main stream.

These qualitative arguments are very difficult to phrase quantitatively because of the difficulties previously alluded to in connection with Eq. (5.3). It is obvious that, if particles are deflected in opposite direction from one y value to another, neither one can have its velocity specified in any simple manner in terms of the original $u(x, y)$; however, this problem is minimal near the jet axis where the velocity changes only gradually for different y values, so that most of the jet does deflect as a whole making the equations used approximately correct.

5.5 PROPAGATING PRESSURE GRADIENT

Finally we consider the case of a propagating and oscillating transverse pressure gradient. The pressure field is assumed to travel in the same direction as the jet axis with sonic velocity a (which may be positive or negative) and to have a strength which is independent of position. Since the integral obtained for the variable centerline velocity jet cannot be evaluated in closed form for this case, we shall assume a constant centerline-velocity jet in order to better illustrate the effect of the propagating field. It should be noted, however, that, particularly for the laminar jet emanating from a finite nozzle, these results can be expected to hold quite well since the laminar jet spreads relatively little for an appreciable distance downstream of the finite nozzle.

For this case the pressure gradient is

$$g(x, t) = B \cos \omega(t - x/a) \qquad (5.30)$$

Now, since u_c is assumed constant and equal to u_0, the transport time τ from Eq. (5.6) is simply x/u_0. If we substitute the propagating pressure gradient [Eq. (5.30)] into Eq. (5.14a) the result is

$$y_0(l, t) = \frac{B}{\rho u_0^2} \int_0^l (l - x) \cos \omega \left(t - \frac{x}{a} - \frac{l - x}{u_0} \right) dx, \quad t \geq \frac{l}{u_0} \quad (5.31)$$

The integration of Eq. (5.31) leads to

$$y_0(l, t) = \frac{Ba}{\rho \omega^2 (a - u_0)^2 u_0} \{ \omega l(u_0 - a) \sin \omega(t - l/u_0)$$
$$- au_0 [\cos \omega(t - l/a) - \cos \omega(t - l/u_0)] \} \quad (5.32)$$

As before we may expand the trigonometric functions and obtain, after some algebraic manipulations,

$$y_0(l, t) = M_1 \cos(\omega t - \phi_1) \quad (5.33)$$

where

$$M_1 = [Bb_1^2 / \rho \omega^2 (b_1 - 1)^2][A_1^2 + 2 - 2A_1 \sin A_1 - 2 \cos A_1]^{1/2}$$

$$\phi_1 = \tan^{-1} \left[\frac{\sin \beta + A_1 \cos \beta - \sin \beta / b_1}{\cos \beta - A_1 \sin \beta - \cos \beta / b_1} \right]$$

$$b_1 = a/u_0, \quad \beta = \omega \tau_l = \omega l / u_0, \quad A_1 = \beta(1/b_1 - 1)$$

Figure 5.7 shows the normalized deflection $(\rho M_1 u_0^2 / Bl^2)$ plotted against the ratio u_0/a for several values of β. When the centerline velocity equals the sonic velocity $(u_0/a = 1)$ the normalized deflection equals 0.5 for all values of β. This

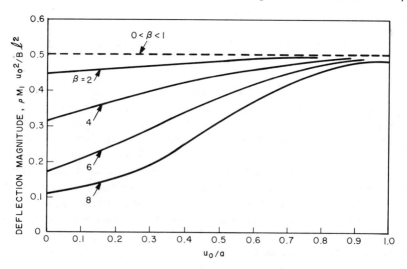

Fig. 5.7 Normalized deflection as a function of u_0/a.

result is essentially identical with expression (5.20a) which was determined for a field unvarying in time, a result that we should expect since a particle traveling at the same speed as the wave is always in phase with it and therefore moves as if it is in a constant gradient.

The important result we have obtained is that the deflection is no longer frequency limited so that the bandwidth would be infinite if the wave propagation and the particle velocities can be kept the same. As the velocity ratio decreases, the deflection becomes a function of β also. For example, when $u_0/a = 0.5$ the normalized deflection is 0.487 for $\beta = 2.0$ but only 0.315 for $\beta = 8.0$. This merely reemphasizes what we have already determined, namely that the deflection decreases as frequency increases. For values of β between zero and unity the deflection is independent of the velocity ratio u_0/a. This is the range of β values that usually occurs in fluid amplifiers. For example, an amplifier, with a power nozzle to output distance of 10 mm, and a nozzle velocity of 30 m/sec, and that operates at 100 Hz, has a β of only 0.207.

Since the phase shift ϕ_1 of Eq. (5.33) is not the phase shift with respect to the propagating wave, but is the phase shift with respect to the phase at the nozzle, evaluation of Eq. (5.33) for $b_1 = 1$ yields $\phi_1 = \beta$. Thus for $u_0/a = 1$ the phase shift is proportional to frequency and the phase distortion is, therefore, minimized.

5.6 TRANSVERSE IMPEDANCE OF A JET

The impedance of the jet in the vicinity of the controls (Fig. 5.2) is derived by finding the change in transverse volume flow δq_1 for a given pressure difference change $\delta(p_2 - p_1)$ across the jet. The impedance is then given by

$$Z_1(s) = \frac{\delta \, \Delta P(s)}{\delta \, Q_1(s)} \tag{5.34}$$

where $\Delta P(s)$ and $Q_1(s)$ are the Laplace transforms of $(p_2 - p_1)$ and q_1, respectively. If we assume that the deflections are small and that they are symmetrical about the jet centerline, then the change in pressure difference $\delta(p_2 - p_1)$ is the same as the pressure difference, and the deflection from zero is y_0. Now the volume flow δq_1 is merely the time derivative of the volume displaced by the jet. In terms of Laplace transforms this is

$$\delta \, Q_1(s) = h \int_0^{x_1} s Y(\tau, s) \, dx \tag{5.35}$$

where x_1 is the control width and h is the distance between top and bottom plates. If we further assume that the pressure gradient is uniform across the jet and that p is a function of t only, then $g(t) = \Delta P/b$ and

$$G(s) = \Delta P(s)/b \tag{5.36}$$

For the uniform pressure gradient given in Eq. (5.36), the deflection of the jet described in Eq. (5.11a) reduces to

$$Y(\tau, s) = (h\,\Delta P(s)/\rho b s^2)[1 - s\tau\,e^{-s\tau} - e^{-s\tau}] \tag{5.37}$$

In the vicinity of the nozzle the jet width and velocity are approximately constant, hence we let $\tau = x/u_0$. Now when we substitute Eq. (5.37) into Eq. (5.35) and perform the integration, the result is

$$Z_1(s) = \frac{\Delta P}{\delta Q_1} = \frac{\rho b s}{h[x_1 + (2u_0/s)\exp(-sx_1/u_0) - (2u_0/s) + x_1\exp(-sx_1/u_0)]} \tag{5.38}$$

After considerable algebraic manipulations the transverse jet impedance in the frequency domain $(s = j\omega)$ is

$$Z_1(j\omega) = \frac{\rho b \omega^2}{4hu_0} \frac{(\sin \beta/2 - j\cos \beta/2)}{(\sin \beta/2 - \beta/2 \cos \beta/2)} \tag{5.39}$$

where $\beta = \omega x_1/u_0$. For $\beta \ll 1$ we can model Eq. (5.39) as $Z_{11} = R_{11} + 1/(j\omega C_1)$. Then by Taylor series approximation the resistance and capacitance of the jet at low frequencies are

$$R_{11} = 3\rho b u_0/h x_1^2 \tag{5.40a}$$

$$C_1 = h x_1^3/6 b \rho u_0^2 \tag{5.40b}$$

Equation (5.40b) presents the same result that was obtained in a simplified manner in Eq. (2.82) with $p_s = \rho u_0^2/2$. For $\beta \gg 1$ the impedance model of Eq. (5.39) is $Z_{12} = R_{12} + j\omega L_1$ and we may determine that

$$R_{12} = -(\rho b\omega/2hx_1)\tan(\omega x_1/2u_0) \tag{5.41a}$$

$$L_1 = \rho b/2hx_1 \tag{5.41b}$$

The jet acts as a resistance and capacitance in series when β is small and as a resistance and an inertance in series when β is large. For intermediate values, the impedance is partially capacitive, partially inertive, and partially resistive. This becomes more apparent if we note that the phase of the jet impedance described in Eq. (5.39) is $\beta/2 - 90°$. Thus at low frequencies (β small) the volume flow leads the pressure difference. When $\beta = 3.14$ the capacitive and inertive reactances are equal and there is no phase difference. For large values of β (high frequencies) the phase becomes positive and inertive reactance dominates. In most cases of interest β is small. This is fortunate since one cannot place any great reliance on the results (quantitatively) at high frequencies because the jet profile and in particular the jet centerline velocity undoubtedly change appreciably (a point we discuss in Section 5.8).

The loss of energy implied by a resistive component needs to be explained because viscosity has not been taken into account. This loss is due to power-jet particles being deflected and thus acquiring energy within the control region and then moving out of that region carrying the energy with them. At large values of β, the particles may make cycles of oscillation within the field and may leave the field moving in either the same or opposite sense as the field.

5.7 The Effects of Feedback (Edgetones)

Consequently, as β varies the resistance goes through positive and negative values.

In Section 2.9.4.4 we briefly mentioned the possibility of obtaining a point-to-point capacitance from a jet barrier and two side cavities. The jet impedance derived here does exhibit capacitive properties at low frequencies. Unfortunately, however, there are some practical difficulties. Foremost among these is the nature of the jet itself. Actually the jet consists of all the fluid particles that emanate from the power nozzle. These particles exchange momentum with fluid in the side chambers (both controls and vents). When the jet deflects because of a side pressure differential, some of the jet particles may leave the jet and enter the side chambers. Conversely, side cavity fluid may be entrained by the jet and flow out of the vents or controls. Thus, from a circuit viewpoint there is a leakage resistance in parallel with the variable volume jet capacitance. The leakage resistance is nonlinear and may sometimes be negative. At present it has not been calculated but it appears capable of effectively shorting out the jet capacitance. The leakage problem is much more serious when the jet is turbulent. The laminar jet spreads less and should produce less leakage. However, the signal levels that can be used with a laminar jet must be very small to avoid disturbances.

Although the construction of a jet barrier capacitive component is difficult, the analyses of proportional amplifiers by Brown and Humphrey [2] and Manion and Mon [3] and others show that the jet capacitance must be taken into account to predict the amplifier dynamics.

5.7 THE EFFECTS OF FEEDBACK (EDGETONES)

We turn now to the application of these results to feedback from the receivers and splitter.

When a jet impacts against a surface, part of its dynamic pressure is converted to static pressure, so that we expect this effect to occur when the jet strikes the splitter. When the jet strikes an opening the conversion of dynamic pressure to static pressure is dependent on the impedance of the opening. Consequently if the impedance varies with frequency, then the amount of conversion of dynamic to static pressure will be frequency dependent.

The pressure gradient resulting is thus a function of the deflection and impedance characteristics of the splitter and receivers.

5.7.1 Feedback from a Wedge

Let us consider the effect of a splitter or wedge alone. A two-dimensional jet striking a wedge (Fig. 5.8) goes into oscillation resulting in the phenomenon known as edgetones. These edgetones have various stages of oscillation, and there are discrete jumps from one stage to another as the velocity or the nozzle-to-wedge distance is changed.

Since the time edgetones were reported by Sondhaus in 1854, a great deal of work has gone into experimental and theoretical investigation of the edgetone effect.

We shall not attempt here to survey this field. Certainly this has already been done more than adequately by a number of others [4–6].

We shall instead restrict ourselves to several papers which in essence propose two opposing theories and to several other papers giving some important experimental results.

There seems to be little doubt that the edgetone results from feedback from the wedge. The major difference between the two theories that we shall discuss is the importance of the jet sensitivity or instability.

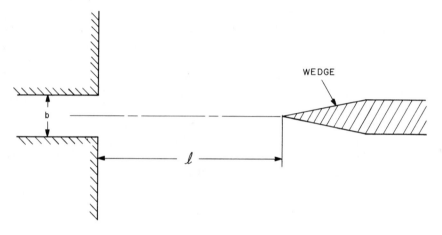

Fig. 5.8 Jet–wedge configuration.

Nyborg [7] in 1954 assumed that the oscillating jet striking the wedge results in a square wave in time reflected from the wedge. This square wave acting all along the sides of the jet in turn causes the jet to oscillate, thus completing the loop. His theory omits any mention of jet instability, yet he was able to obtain very good agreement with the results of Brown [8] on the ratio of the frequencies excited in the various stages. He was also able to calculate the motion of the jet centerline obtaining results which were experimentally verified by Bouyoucos [9]. His theory does not attempt, however, to explain the jumps from one stage to another.

Powell [5] in 1961 amplified on a theory proposed by Curle [10] in 1953, which assumes that feedback from the wedge causes vortices to be shed at the nozzle. These disturbances are amplified, resulting in oscillation as the jet proceeds toward the wedge.

The theory implicitly assumes that the direct effect on the jet of the pressure wave reflected from the wedge is negligible compared to the effect of the growth of the disturbances originating from vortex shedding at the nozzle.

5.7 The Effects of Feedback (Edgetones)

The jumps in frequency which occur between the various stages can then be explained, according to Powell, by the fact that changes in velocity or distance eventually cause the frequency of oscillation to deviate appreciably from the frequencies to which the jet is most sensitive. When this occurs the edgetone will jump to another stage in which its frequency of oscillation does lie within the sensitive region.

In obtaining his results Nyborg used an equation similar to Eq. (5.13a) and assumed that $g(x, t)$ was a square wave function in time alone and did not depend on x; i.e.,

$$g(t) = B\left[\sum_{n=1}^{\infty} H(t) + 2(-1)^n H(t - nT)\right] \quad (5.42)$$

where T is the half-period of the square wave, B is a constant, n is an integer, and $H(t)$ is the unit step function. The waveform given assumes that the wave is initiated at $t = 0$ with an amplitude $+B$, and that at T, the half-period, its value changes by two units to become $-B$. The deflection caused by this waveform is obtained by substituting Eq. (5.42) into Eq. (5.13a), so that

$$y_0(\tau_l, t) = \frac{B}{\rho}\left[\int_0^{\tau_l} \tau H(t - \tau)\, d\tau \right.$$
$$\left. + 2\sum_{n=1}^{\infty}(-1)^n \int_0^{\tau_l} \tau H(t - nT - \tau)\, d\tau\right], \quad t > \tau_l \quad (5.43)$$

The step wave originates as a result of the motion of the jet past the wedge; i.e., the wave changes direction the instant the jet passes the center. Thus, when $t = mT$, where m is any integer, $y_0(\tau_l, t)$ must be equal to zero. For these conditions Eq. (5.43) becomes

$$\int_0^{\tau_l} \tau H(mT - \tau)\, d\tau + 2\sum_{n=1}^{m}(-1)^n \int_0^{\tau_l} \tau H([m-n]T - \tau)\, d\tau = 0, \quad mT > \tau_l \quad (5.44)$$

Equation (5.44) relates the period of oscillation T to the jet transport time τ_l. To simplify the relation we must carry out the integration. As a first step, note that all the step functions in the summation are zero unless the integrating variable τ is less than $(m - n)T$. Thus the upper limit of integration for the integrals under the summation cannot exceed $(m - n)T$. Furthermore, since the upper limit must be positive, only those values of n are permitted for which $m > n$. Thus, Eq. (5.44) may be written as

$$\int_0^{\tau_l} \tau\, d\tau + 2\sum_{n=1}^{m}(-1)^n \int_0^{(m-n)T} \tau\, d\tau = 0 \quad (5.45)$$

Now the integration of Eq. (5.45) yields

$$\frac{\tau_l^2}{2} + T^2 \sum_{n=1}^{m}(-1)^n(m - n)^2 = 0 \quad (5.46)$$

The terms in the summation are merely the square of decreasing integers which alternate in sign. If we let $i = m - n$ we can write the same summation for increasing integers as

$$\frac{\tau_l^2}{2} + T^2 \sum_{i=1}^{m-1}(-1)^i i^2 = 0 \qquad (5.47)$$

Evaluation of the summation in Eq. (5.47) leads to

$$T^2\left[(-1)^{m-1}\frac{(m-1)(m)}{2}\right] = -\frac{\tau_l^2}{2} \qquad (5.48)$$

To solve Eq. (5.48) for T, we recognize that real values are obtained only if m is even. This corresponds to an even number of half-periods. Even values of m restrict the solution to those cases where the pressure gradient and the deflection are 180 degrees out of phase. Thus the period of oscillation is $2T$, where

$$T = \tau_l[1/m(m-1)]^{1/2}, \qquad m = 2, 4, 6, \ldots \qquad (5.49)$$

The possible oscillation frequencies are then given by

$$f = 1/2T = [m(m-1)]^{1/2}/2\tau_l, \qquad m = 2, 4, 6, \ldots \qquad (5.50)$$

Brown [8] found experimentally that

$$f \cong 0.466(u_0/l)j_n \qquad (5.51)$$

where his experimental results gave values to j_n of 1, 2.3, 3.8, and 5.4 for the first four stages of oscillation. If we rewrite Eq. (5.50) as

$$f = (\sqrt{2}/2\tau_l)[m(m-1)/2]^{1/2} \qquad (5.52)$$

we see that for $m = 2, 4, 6, 8$ the radical takes on the respective values 1, 2.45, 3.87, and 5.29. Thus the radical gives good agreement with the j_n of Brown.

Now τ_l depends on the velocity decay of the jet, which in turn depends on the width of the nozzle and conditions upstream of the nozzle. As a result the coefficient of the radical in Eq. (5.52) may be different under different conditions. Although Brown's results [Eq. (5.51)] are most often quoted, other investigations have found values different from the 0.466 given by Brown. In particular, the experiments of Bouyoucos and Nyborg [9] show good agreement with Eq. (5.52).

5.7.1.1 The Frequency Jumps

For a given nozzle-to-wedge distance there is a threshold velocity below which the edgetones do not occur. Similarly, for a given velocity there is a nozzle-to-wedge distance below which the system does not oscillate. Above these thresholds, one finds that if, for example, the velocity is kept constant and the nozzle-to-wedge distance l is varied, the frequency will shift in accordance with a relation of the form of (5.51) for certain values of l, but at some point when l

5.7 The Effects of Feedback (Edgetones)

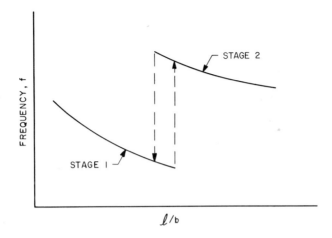

Fig. 5.9 Stages 1 and 2.

is increased the frequency will suddenly jump from one stage to the next. A similar effect occurs if l is fixed and the velocity is allowed to vary.

Figure 5.9 is a schematic drawing that demonstrates the transition from stage 1 to stage 2. The figure represents the results for a jet of fixed width and velocity when the wedge position is changed. At small distances between the wedge and nozzle the frequency is in stage 1. As the wedge distance increases the stage 1 frequency decreases. Ultimately the frequency jumps to stage 2. Now a reduction of wedge distance increases the stage 2 frequency until the frequency jumps back to stage 1.

Thus there is a hysteresis loop in the frequency–wedge distance relation.

Nyborg's theory fits the frequency relation sufficiently well that it seems worthwhile to consider whether it is possible to adapt it so as to take into account the frequency jumps. Let us therefore examine some of Nyborg's assumptions and attempt to provide a physical reason for the feedback phenomenon assumed by Nyborg.

One can surmise that the pressure wave arises from the wedge because every part of the wedge against which some of the jet stagnates becomes a source of pressure waves. Measurements made by Unfried for Powell [5] have shown that the wedge acts very much like a dipole source originating slightly behind the vertex of the wedge.

A jet having a uniform velocity (and hence momentum) distribution would be expected to set up a pressure gradient which depends on distance from the vertex and which is essentially proportional to the jet deflection. Even though the particle velocity gradually falls off away from the jet axis, for small deflections it will still be approximately true that the strength of the pressure gradient resulting from stagnation against the wedge is proportional to the jet deflection.

We have seen that Nyborg assumed that the pressure gradient caused by the jet deflection was a square wave which was not a function of distance. In other words, when the jet is on one side of the splitter the pressure gradient has some constant value independent of deflection, and the instant the jet crosses the splitter it takes on the negative of the same constant value.

By forcing the phase of the deflection to match up properly around the loop, Nyborg was able to show that only certain frequencies were allowed, these frequencies being associated with the various stages of the edgetone frequency as experimentally found by Brown. Nyborg's results, however, fail to explain a number of other effects associated with the edgetone. One of these is the experimental fact that there is a minimum nozzle-to-wedge distance below which the edgetone does not occur. Nyborg realized the assumptions he had used made the gain around the loop independent of this distance, so he then changed his assumption that the deflection resulted in a constant gradient with a sign change when the jet crossed the splitter and replaced it with a similar assumption except that, for deflections less than some particular value, the gradient was assumed to be zero. This ad hoc assumption does result in a minimum nozzle-to-wedge distance but is difficult to justify and does not explain another important phenomenon that occurs, namely, the fact that as the velocity or the nozzle-to-wedge distance is increased a point will be reached at which the edgetone frequency will jump to the next higher stage.

Powell unified what a number of other individuals had been saying occurred in the edgetone. A major hypothesis is that the jet gain is due to the jet instability instead of to the jet deflection in the pressure gradient. According to this hypothesis the important effect of the pressure gradient formed at the edge is to give transverse impulses to the jet just as it leaves the nozzle thereby causing vortices to be shed. These vortices grow because of the jet instability as they travel toward the wedge, thereby resulting in a deflection increase (or gain), in turn giving rise to the formation of a pressure gradient at the wedge.

Since the jet is more sensitive to some frequencies than to others, the jet gain will decrease when the frequency varies appreciably from one of the more sensitive frequencies. (The frequency of oscillation varies as the transport time is changed by changing either the nozzle-wedge distance or the jet velocity.) The frequency will then jump to another stage in order to again oscillate at a frequency to which it is more sensitive.

Powell's theory thus seems to explain the jumps, although much of the argument is qualitative rather than quantitative. Furthermore, it neglects the dynamics of the jet, implicitly assuming that no gain is due to deflection of the jet in the pressure gradient.

Now a fundamental difference between Nyborg's and Powell's theories is that. he first-stage phase shift between nozzle-and-wedge predicted by Nyborg is about 0.6 cycle [Eq. (5.51) shows 0.466 cycle and Eq. (5.52) shows 0.707 cycle] whereas Powell requires a phase shift of 1.25 cycles. Karamacheti [6, 11, 12] and his students made a very thorough study of the phase and velocity at many points of the oscillating jet for various Reynolds numbers and found in all cases

5.7 The Effects of Feedback (Edgetones)

that the phase shift at the wedge was approximately 0.6 cycle for the centerline velocity.

We are thus faced with a dilemma; Nyborg's theory, which completely neglects stability effects, predicts correctly phase shift and frequency ratio but fails to predict frequency jumps and is physically not properly justified.

Powell's theory, on the other hand, which is based on jet instability, can explain the jumps in frequency but is quantitatively incorrect.

5.7.2 Feedback from Receivers

In the following we shall try to expand on Nyborg's results and simultaneously show that similar results to those obtained in the case of the wedge should also be obtained from the receivers.

It is apparent from Fig. 5.10 that for small deflections the pressure reflections from the receivers will be approximately proportional to the deflection and to the

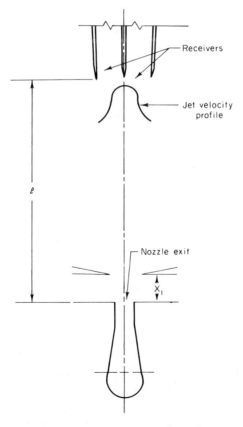

Fig. 5.10 Jet receiver configuration.

impedance of the apertures; that is, the transverse pressure gradient in the immediate vicinity of the receivers will be of the form

$$dp/dy = k_2 Z(\omega) y_0(\tau_l, t) \tag{5.53}$$

where $Z(\omega)$ is the receiver impedance and k_2 is a constant. The pressure gradient at a given point will depend on the distance of that point from the origins of the reflections and on the propagation time of the jet. We will, however, neglect these effects and assume that the pressure gradient is a function of time only. As a further simplification we will restrict our solution to receiver openings designed to be primarily resistive over the frequency range of interest, in which case Eq. (5.53) becomes

$$-dp/dy = g(t) = -H_1 y_0(\tau_l, t) \tag{5.54}$$

where H_1 is the product of k_2 and the resistance of the receiver openings. The negative sign arises from the fact that the static pressure must be 180 degrees out of phase with the dynamic pressure that gave rise to it. If Eq. (5.54) is substituted into (5.14a) we obtain

$$y_0(\tau_l, t) = -\frac{1}{\rho} \int_0^{\tau_l} (\tau_l - \tau) H_1 y_0(\tau_l, t - \tau_l + \tau) \, d\tau \tag{5.55}$$

This is an eigenvalue equation to which we fortunately have the solution, for we have already found the value of the integral of Eq. (5.55) for the particular case when $y_0(\tau, t) = (B/H_1) \cos \omega t$ [see Eq. (5.23)]. The integral in Eq. (5.55) is identical to that of Eq. (5.23) with the limit $\tau_1 = \tau_l$ and the result is therefore given by Eq. (5.26) with $k_1 = 1$; hence,

$$y_0(\tau, t) = (B/H_1) \cos \omega t = -M \cos(\omega t - \phi) \tag{5.56a}$$

where

$$M = (B/\rho\omega^2)[2(1 - \cos \beta) + \beta^2 - 2\beta \sin \beta]^{1/2} \tag{5.56b}$$

$$\phi = \tan^{-1}\left[\frac{\sin \beta - \beta \cos \beta}{\cos \beta + \beta \sin \beta - 1}\right] \tag{5.56c}$$

where

$$\beta = \omega \tau_l$$

If Eq. (5.56a) is a solution, then the following two conditions must be satisfied:

$$B/H_1 = M \tag{5.57a}$$

$$\phi = (2n - 1)\pi, \quad n = 1, 2, 3, \ldots \tag{5.57b}$$

The conditions specify the amplitude and phase of the deflection which will permit oscillations to occur. We can find the frequency of the oscillations by equating Eqs. (5.56c) and (5.57b). Thus

$$\tan^{-1}\left[\frac{\sin \beta - \beta \cos \beta}{\cos \beta + \beta \sin \beta - 1}\right] = (2n - 1)\pi, \quad n = 1, 2, 3, \ldots \tag{5.58}$$

5.7 The Effects of Feedback (Edgetones)

Equation (5.58) is equivalent to finding the alternate roots of the equation, $\tan \beta = \beta$. The first four alternate roots are

$$\begin{aligned} \beta_1 &= 4.493, \\ \beta_2 &= 10.904 = 2.43\beta_1 \\ \beta_3 &= 17.220 = 3.83\beta_1 \\ \beta_4 &= 23.519 = 5.23\beta_1 \end{aligned} \quad (5.59)$$

These values are compared with the experimental values of G. B. Brown and the theoretical ones obtained by Nyborg in Table 5.1. From Eq. (5.59) we may

TABLE 5.1

	Brown	Nyborg square wave	Sine wave
β_2/β_1	2.3	2.44	2.43
β_3/β_1	3.8	3.86	3.83
β_4/β_1	5.4	5.29	5.23

obtain the oscillating frequencies by replacing β by $\omega\tau_l$; i.e., for the first stage the frequency $f = 4.493/(2\pi\tau_l) = 0.705/\tau_l$, which agrees with Eq. (5.52). The fact that the results for the sinusoid agree so well with Nyborg's is not at all surprising because a square wave (as used by Nyborg) contains a large proportion of the fundamental frequency, and the other assumptions are the same.

The similarity in results for these two cases indicates that the actual wave shape can be anything between these extremes (or even possibly of some other shape) without greatly affecting the frequency. As a matter of fact the results of Bouyoucos and Nyborg [9] as confirmed by Shields and Karamcheti [12] show that "the motion of the center line from its position of extreme deflection until it crosses the edge is more rapid than its motion from the edge to the opposite position of deflection." This behavior is what one would expect if the driving field were of square shape such as that suggested by Nyborg rather than simply a sinusoidal oscillation.

The above values of β are the values that β must take on if oscillation occurs. The amplitude condition that determines whether oscillation will occur comes from Eq. (5.57a) and simply states that the gain around a closed loop must be unity. Now the open loop gain G_0 for this system (Fig. 5.11) may be expressed with the aid of Eq. (5.56b) as

$$G_0 = H_1(M/B) = (H_1/\rho\omega^2)[2(1 - \cos \beta) + \beta^2 - 2\beta \sin \beta]^{1/2} \quad (5.60)$$

In order for oscillation to be initiated, the open loop gain must be greater than or equal to unity. Since the closed loop gain must be unity, H_1 cannot actually be constant but must depend on the amplitude of deflection, and indeed we expect that H_1 depends on the jet momentum, which is, of course, not actually uniform, so that H_1 decreases as the deflection increases. Furthermore, as the jet deflection increases the apparent origin of the dipole shifts farther downstream of the vertex, thereby also decreasing the effective value of H_1; H_1 will then stabilize at a value that causes the closed loop gain to equal unity.

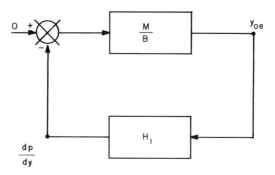

Fig. 5.11 Block diagram for receiver feedback system.

Equation (5.59) indicates that there exists a minimum nozzle-to-wedge distance for oscillation because, in any particular stage of operation, $\beta = \omega \tau_l$ is a constant; consequently, if the distance l is decreased while the velocity and $\omega \tau_l$ are allowed to remain constant, ω must increase. If ω gets sufficiently large the gain around the loop will become less than unity and oscillation will cease. If we assume H_1 is momentum dependent, then again we see that there is some velocity (momentum) below which oscillation will not occur.

It is apparent, therefore, that the assumption that the jet deflection (and therefore the loop gain) depends on the pressure gradient arising from the wedge agrees with a number of experimental facts.

5.7.3 Frequency Jumps Related to Nonlinearity

Although the above agreement with experiment is rather convincing proof that the jet deflection of importance to edgetone generation arises from the pressure gradient directly rather than from amplification of the vortices shed near the base of the jet and which are caused by the pressure gradient, we still cannot rule out the possibility that the instability plays some role in the edgetone phenomenon. We have seen, for example, that the mixing effects are a function of frequency, and indeed experiments have shown that the spreading characteristics and consequently the velocity distribution changes with frequency. This effect is, however, not necessarily related to the jumps in frequency.

Karamcheti and Bauer [6] suggest that the reflected pressure field acts as an input at every point along the jet rather than only at the nozzle. Presumably this input must somehow synchronize the growth of the disturbance caused by

5.7 The Effects of Feedback (Edgetones)

the jet instability in a manner similar to that with which an applied field can order the vortex shedding of a free jet. If this hypothesis could be verified, it would make these two theories (Nyborg's and Powell's) compatible and explain most of the important effects occurring in the edgetones.

It is possible, however, to explain the jumps in a qualitative way from the pressure gradient viewpoint alone. Consider the system shown in Fig. 5.11. The forward loop block M/B represents the open loop relation between a side pressure gradient and the jet deflection. The feedback loop block H_1 determines the magnitude of the pressure gradient that occurs at the wedge due to jet deflection and which is available to act back along the sides of the jet. We know that for oscillation to occur the magnitude of $H_1 M/B$ must equal unity. Let us examine the blocks separately first and then try to understand how they work together.

Figure 5.12 shows a portion of the relation between $M\rho/B$ and frequency that is given in Eq. (5.56b). We have plotted only the first and second stages, and have used different scales for each along the abscissa. Suppose, for example, that the geometry and jet velocity place the first-stage frequency at 500 Hz. Then the corresponding second-stage frequency is 1215 Hz. The first-stage value of $M\rho/B$ is equal to 0.079×10^{-5} sec^2, whereas the second-stage value is 0.030×10^{-5} sec^2. Since stage 1 values of $M\rho/B$ are always larger than stage 2 values, it would appear that if oscillation were possible it would always take place in stage 1. To see why this is not true we must examine the characteristic of the H_1 block.

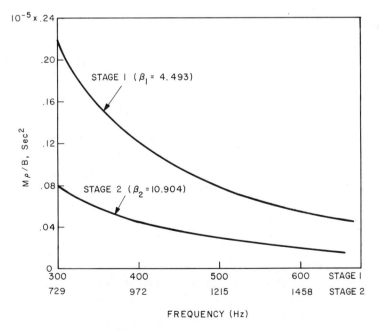

Fig. 5.12 $M\rho/\beta$ as a function of frequency.

Throughout the jet dynamics development we have concentrated on the deflection of the jet centerline and we have omitted any consideration of the velocity profile. To demonstrate the nonlinear characteristic of the H_1 block we must assume a velocity profile. For a two-dimensional laminar jet and nozzle momentum flux of $\rho u_0^2 b$, Eq. (4.144) can be rearranged as

$$u = 0.454 k_N u_0 \operatorname{sech}^2 0.275(y_{0l} k_N^2 / b) \tag{5.61}$$

where

$$k_N = [bN_R/(l + x_0)]^{1/3} \quad \text{and} \quad x_0 = bN_R/36$$

Now the dynamic pressure at the wedge is $p = \rho u^2 / 2$, so that $dp/dy = \rho u \, du/dy$. Thus from Eq. (5.61) we obtain

$$dp/dy = 0.227 k_N^4 (\rho u_0^2 / 2b) \operatorname{sech}^4 0.275(y_{0l} k_N^2 / b) \tanh 0.275(y_{0l} k_N^2 / b) \tag{5.62}$$

If we include the effect of the apparent point of emanation x_0 upstream of the jet nozzle, then k_N is about 2.0 for the sizes and Reynolds numbers that are usually used in edgetone configurations. The magnitude of H_1 $[=(1/y_{0l}) \, dp/dy]$ is therefore

$$H_1 = (3.632) \left(\frac{\rho u_0^2}{2b^2} \right) \frac{\operatorname{sech}^4 1.1 y_{0l}/b \tanh 1.1 y_{0l}/b}{y_{0l}/b} \tag{5.63}$$

Equations (5.62) and (5.63) are shown plotted in Fig. 5.13. The relation between pressure gradient dp/dy and deflection is almost linear for small deflections. However, as the deflection increases the relation becomes nonlinear and exhibits a maximum at $y_{0l}/b = 0.45$. The value of H_1, on the other hand, is monotonically decreasing as the deflection increases.

To visualize the effect of the H_1 block nonlinearities, we apply a sinusoidal input of deflection $M_2 \sin \pi(t/T)$, where M_2 is the maximum value of the deflection for a particular case. Figure 5.14 shows the waveform of the output pressure gradient from the H_1 block that corresponds to sinusoidal inputs of several different amplitudes. Only a half-cycle of the output waveform is shown. When M_2 is small (0.1) the output waveform remains approximately sinusoidal, as we would expect for a linear system. As M_2 increases to 0.4 the waveform squares up appreciably. However, a Fourier analysis of this shape would still produce a large-amplitude wave at the applied frequency of the input. Thus the effective value of H_1 (at $M_2 = 0.4$) would be only slightly less than the values indicated on Fig. 5.13 at $y_{0l}/b = 0.4$. When the amplitude is still larger, at $M_2 = 0.8$, the waveform has two peaks for each half-cycle. Now the magnitude of H_1 at the fundamental frequency is considerably less than predicted by the dc values (Fig. 5.13). The situation becomes even more severe for $M_2 = 1.2$ where a Fourier analysis of the waveform will show only a small amplitude at the applied frequency, and therefore H_1 will be small for this case.

As a result of the foregoing discussion we realize that $H_1 M/B$ may become less than unity when the deflections are large. This is a situation that can happen in stage 1 operation where, as we have seen in Fig. 5.12, larger deflections are

5.7 The Effects of Feedback (Edgetones)

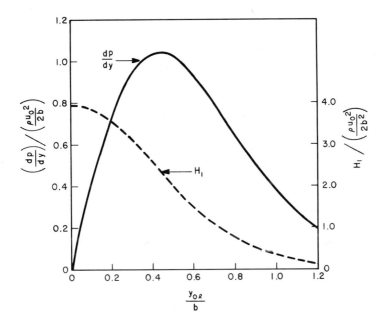

Fig. 5.13 Pressure gradient and H_1 as a function of distance.

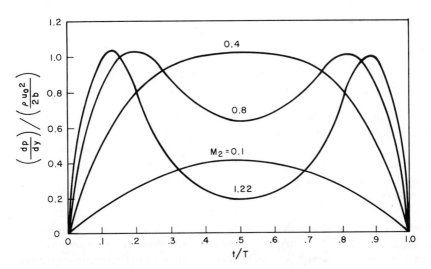

Fig. 5.14 Form of pressure gradient during a half period.

produced. When stage 1 operation becomes impossible because the gain around the loop is less than unity, the system will jump to the second stage. Before this occurs, however, two first-stage frequencies will be present simultaneously, the fundamental first-stage frequency and the third harmonic, as implied by the double peak of Fig. 5.14. The simultaneous existence of the first stage and the third harmonic has been found by Karamcheti and Bauer [6]. The third harmonic of the first stage helps trigger the second stage into oscillation.

The above arguments indicate that the jumps can be explained by modification of Nyborg's theory. However, up to this point the effect of the velocity variation in causing these jumps has not been pointed out.

The stagnation pressure gradient formed at the wedge depends on the slope of the velocity profile. The edgetone-producing jet is laminar for most of its downstream distance and a laminar jet becomes narrower as the velocity is increased, resulting in a greater slope and consequently in an increased pressure gradient formed at the wedge. Thus an increase in velocity causes an increased jet deflection.

Finally, a word with respect to the hysteresis:

As we point out in the next section, the profile of the jet is distorted during edgetone oscillation. This, of course, changes the slope of the profile and consequently the loop gain. When the jet is oscillating, for example, in stage 1, it has some profile. When oscillating in stage 2, the profile could easily be different so that, under the same conditions of velocity and nozzle-to-wedge distance, the gain for stage 1 and for stage 2 depends on the profile; that is, the stage 1 gain is different when the jet has a stage 1 profile than when it has a stage 2 profile.

This phenomenon could explain the hysteresis effect.

5.8 VELOCITY PROFILE OF OSCILLATING JET

In the analysis of the jet dynamics, it is usually assumed that the jet while in motion has essentially the same profile as the stationary jet. Unfortunately the profile actually changes for two reasons: (1) different velocity particles of the jet are apparently not deflected by the same amount, and (2) the Reynolds stresses change, thereby increasing the turbulence and the entrainment characteristics.

We will consider two types of dynamic velocity profile. The first is the *instantaneous mean* jet profile. This is the profile that would be measured by a transverse multiprobe array that was read simultaneously. Averaging of the measurements over several cycles at specific values of deflection in a given direction would then remove the effect of any turbulent or statistical fluctuations that may have existed. The second type of profile, the *effective* profile, is obtained analytically from the apparent profile, which in turn is found by time averaging

5.8 Velocity Profile of Oscillating Jet

the velocity measured by a single probe while it is fixed at different transverse positions.

Measurements of the instantaneous mean profile have been made by Shields and Karamcheti [12] as follows:

A reference probe (hot wire) was placed at a fixed position in the jet and the signal obtained as the jet oscillated in the field was used to trigger an oscilloscope. This established a common time base for velocity measurements that were made by use of a second hot wire. In this manner phase and amplitude could be determined as a function of position and time.

Profiles of the form shown in Fig. 5.15 were obtained. It is seen that the profiles are not symmetric. The profile tends to steepen in the direction of motion and to trail out along the rear edge. An interesting aspect is the presence of the steplike structure on the trailing edge. Shields and Karamcheti's work was done with jets set into oscillation by edgetones for which the frequencies involved are relatively high. The jets they worked with were of relatively low Reynolds number and were therefore laminar over an appreciable distance.

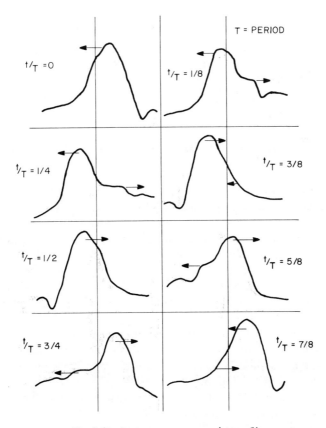

Fig. 5.15 Instantaneous mean jet profiles.

Low-frequency measurements (100–800 Hz) of the *apparent* profile were made by Stiffler [13] on an externally excited turbulent jet. He time averaged the velocity at a point and then moved the probe transversely to obtain measurements at other points. He then showed analytically that the time average (apparent) profile was equal to the effective profile plus a term equal to the product of the amplitude of the second harmonic of the excitation frequency times the distance of the probe from the axis of the unexcited jet. The amplitude of the second harmonic can be determined by use of a narrow-band filter so that the effective profile can readily be found once the apparent profile has been measured. The method implicitly assumes that the effective profile is symmetric, so that it is not necessarily the same as the instantaneous mean profile which (at least under edgetone conditions) is asymmetric, but the two profiles conceivably could be quite similar at low frequencies.

Stiffler's [13] measurements and semiempirical theory indicate that the jet spread does not increase appreciably for frequencies such that

$$\omega(x - x_0)/u_0 < 0.7 \tag{5.64}$$

where x_0 is the virtual origin of Albertson's profile. Thus, in its more usual applications in a proportional amplifier, the jet does not spread additionally because of frequency. At sufficiently high frequencies, however, the slower particles may get 180 degrees or more out of phase with the higher-velocity particles, as indicated by the steps of Fig. 5.15.

5.9 EFFECT OF THE JET ON THE PRESSURE FIELD

In all the foregoing discussions the pressure gradient has been assumed constant across the nominal width of the jet. However, this assumption would only be correct if no jet were present. Then the two sources of pressure (such as the two controls of an amplifier) result in an approximately constant pressure gradient (Fig. 5.16a). The presence of a uniform jet causes most of the pressure drop to occur across the jet. The gradient is now steeper but still approximately constant across the jet (Fig. 5.16b), although it seems obvious that the ratio of pressure drop occurring across the jet to that occurring outside the jet boundaries should depend on the jet velocity.

Our assumption throughout has been that no pressure gradient occurs outside of the jet and that the gradient across the jet is constant (Fig. 5.16c); nevertheless, the actual gradient should be greatest where the velocity is greatest and least where the velocity is least (Fig. 5.16d).

The pressure gradient distribution will, of course, affect the dynamic velocity profile. If, for example, the gradient were proportional to the velocity, the jet

5.9 Effect of the Jet on the Pressure Field

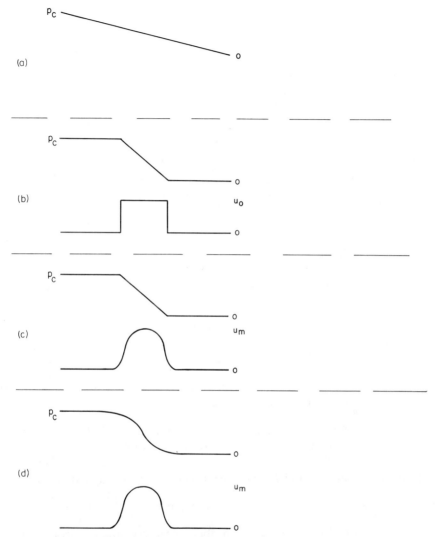

Fig. 5.16 Effect of jet on pressure field: (a) $u = 0$, (b) $u = u_0 = $ const, (c) $u = u(x, y)$ (assumed pressure field), (d) $u = u(x, y)$, probable actual pressure field.

profile would have little tendency to be altered (except for turbulence effects) in the field. In this case particles of all velocities would be deflected by essentially the same amount.

On the other hand, if the gradient were proportional to the momentum, higher-velocity particles would deflect more than lower-velocity particles. The steepening of the front edge of the instantaneous mean profiles measured by Shields and Karamcheti [12] seems to indicate that this is indeed the case.

PROBLEMS

5.1 Assume that a constant pressure Δp exists across the jet from zero to x_1 and a second constant pressure Δp_v exists across the jet from x_1 to l. Show that the deflection is given by

$$y_0(l) = (\Delta p/2b\rho)\{2\tau_1\tau_l - \tau_1^2\} + (\Delta p_v/4b\rho)(\tau_l - \tau_1)^2$$

5.2

(a) Find the time that it takes a centerline particle of a laminar jet to reach a point l that is four nozzle widths downstream of a nozzle of width b.

(b) If the velocity at the nozzle exit is u_0, write the edgetone frequency in terms of u_0 and l, where l is the nozzle-to-wedge distance.

(c) Find numerical results for a and b if $b = 0.2$ mm, $l = 0.8$ mm, and $u_0 = 10$ m/sec.

5.3 A laminar jet of air issues from a slit of width 0.5 mm from a pressure source of 0.7 kN/m². Assume that the profile out of the nozzle is uniform, that the jet maintains constant width and velocity, and that the relation between pressure supply and nozzle velocity is given by $p_s = \frac{1}{2}\rho u_0^2$.

(a) A square pulse of pressure (0.05 kN/m²) lasting for 1 msec is applied over the region $0 \leq x \leq 2$ mm.

(b) Plot the shape of the jet centerline ($0 \leq x \leq 1$ cm) at $t = 1.0$ msec.

5.4 Assume the same conditions as for Problem **5.3** except that an edge at 2.0 mm causes edgetones to be produced.

(a) Find the shape of the jet centerline at the instant the deflection is maximum in stages 1 and 2. (First define maximum deflection for stage 2.)

(b) Discuss the difference in type of oscillation between the third harmonic of first-stage oscillation and second-stage oscillation at the same frequency.

5.5 Find the frequency response of the jet from its transfer function.

5.6 Discuss the problems involved in obtaining a broad-bandwidth amplifier by producing an input pressure configuration as a function of time and position such that the input pressure change apparently travels downstream at the same rate as the jet. (For example, discuss an amplifier having a series of inputs, the second pair downstream of the first pair, the third pair downstream of the second pair, etc. Delayed versions of the same signal are fed to these controls timed to the amplifier jet velocity.)

5.7 Edgetone data are often plotted in terms of the Strouhal number N_S, defined as

$$N_S = fb/u_0$$

where b is the nozzle width. Karamcheti and Bauer's [6] results show that, when two edgetone frequencies are present simultaneously, the Strouhal number corresponding to the first-stage fundamental decreases. Discuss a possible reason for this. (Hint: Consider the effect of additional jet spreading on the transport time.)

NOMENCLATURE

$g(x, t)$	$= -dp/dy$, Negative pressure graident
$G(x, s)$	$= \mathscr{L}\{g(x, t)\}$
$H(t)$	Unit step function
H_1	Feedback coefficient
k_1	$= \alpha/\beta = \tau_1/\tau_l$
M	Magnitude of jet deflection
s	Complex transform variable
t	Time
u_0	Constant velocity
u_c	Centerline velocity
v	Inherent jet transverse velocity
x	Axial distance variable
x_1	Width of field acting on jet
y_0	Instantaneous transverse jet deflection
Y	$= \mathscr{L}\{y_0\}$, Laplace transform of y_0 with respect to time
α	$= \omega\tau_1$
β	$= \omega\tau_l$
δ	Small change in a variable
σ	Complex transfunctional variable
τ	$= \int_0^x dx/u_c(x)$, Transport time
τ_l	$= \int_0^l dx/u_c(x)$
τ_1	$= \int_0^{x_1} dx/u_c(x)$
ϕ	Phase lag

REFERENCES

1. J. M. Kirshner, Response of a jet to a pressure gradient and its relation to edgetones. *Int. JSME Symp. Fluid Mach. Fluidics, 2nd Tokyo* (Sept. 1972).
2. F. T. Brown and R. A. Humphrey, Dynamics of a proportional amplifier—Part 2. *ASME Trans. J. Basic Eng.* **92**, 303–312 (1970).
3. F. M. Manion and G. Mon, "Fluerics 33. Design and Staging of Laminar Proportional Amplifiers," HDL-TR-1608, September 1972.
4. G. B. Brown, The mechanism of edgetone production. *Proc. Phys. Soc. London* **49**, 508 (1937).
5. A. Powell, On the edgetone. *J. Acoust. Soc. Amer.* **33**, 395 (1961).
6. K. Karamcheti and A. B. Bauer, "Edgetone Generation," SUDAAR No. 162. Stanford Univ., July 1963.
7. W. S. Nyborg, Self-maintained oscillations of the jet in a jet-edge system, 1. *J. Acoust. Soc. Amer.* **26**, 174 (1954).

8. G. B. Brown, The vortex motion causing edgetones. *Proc. Phys. Soc. London* **49**, 493 (1937).
9. J. V. Bouyoucos and W. S. Nyborg, *J. Acoust. Soc. Amer.* **26**, 511 (1954).
10. N. Curle, The mechanics of edgetones. *Proc. Roy. Soc. London* **A216**, 412 (1953).
11. G. R. Stegen and K. Karamcheti, "On the Structure of an Edgetone Flow Field," SUDAAR No. 303. Stanford Univ., February 1967.
12. W. L. Shields and K. Karamcheti, "An Experimental Investigation of the Edgetone Flow Field," SUDAAR No. 304. Stanford Univ., February 1967.
13. A. K. Stiffler, "Sinusoidal Excitation of a Free Turbulent Jet." Ph.D. Thesis, Pennsylvania State Univ., September 1971.

Chapter 6

STATIC CHARACTERISTIC CURVES

6.1 INTRODUCTION

Although there has been considerable progress in the analysis of fluidic components, it is still not possible to derive their exact characteristics theoretically. Since the combination of fluidic components into circuits and systems requires a prior knowledge of the component characteristics, the fluidic circuit designer must often rely on information provided by characteristic curves.

6.2 CONCEPT OF SOURCE AND LOAD

The concept of a source and a load is fundamental to the use of characteristic curves. In the usual case the source supplies energy ("active" source) and the load receives energy ("passive" load). Although their roles may be reversed in special situations (i.e., "passive" source and "active" load), this does not happen too often. The rate at which the source delivers energy to the load depends upon the impedance associated with each. For the static characteristics the impedances are purely resistive. Thus in this presentation we refer only to source resistance and load resistance.

Figure 6.1 shows the typical arrangement of a fluid circuit for which the source and load characteristics are to be measured. In the physical configuration (Fig. 6.1a) the source and load are both connected to opposite sides of an enlarged measuring section. The enlargement is necessary because the across signal

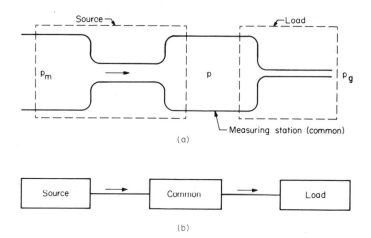

Fig. 6.1 Source and load connection: (a) typical source–load physical configuration, (b) schematic of source–load configuration.

variable is the total pressure and this is most conveniently measured when the velocity is negligible. This brings up a point, which we will return to shortly, concerning the difference between characteristics measured with and without the enlargement. For the present, however, let us assume that the configuration is the one shown physically in Fig. 6.1a and schematically in Fig. 6.1b.

6.2.1 Source Characteristics

The source characteristic is the locus of all possible combinations of total pressure and volume flow that the source can deliver to the measuring station. Fig. 6.2 shows several typical source characteristics. To understand the reason

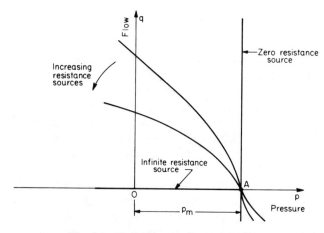

Fig. 6.2 Typical source characteristics.

6.2 Concept of Source and Load

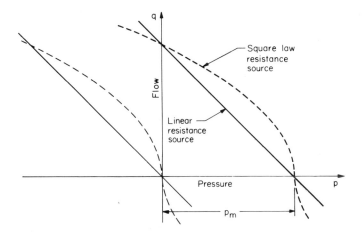

Fig. 6.3 Linear and square law sources.

for the particular shape of source curves, imagine that it is possible to create, by means that need not be specified, any desired pressure at the measuring station. Thus, for example, when the measuring station pressure equals the source pressure p_m, there will be no flow from the source. This condition is represented by point A on Fig. 6.2. At each measuring station pressure less than p_m, there is a corresponding volume flow from the source and the source is termed "active." When the measuring station pressure exceeds p_m, flow passes into the source and the source is termed "passive." The magnitude of the volume flow at each pressure level depends on the source resistance. Low-resistance active sources supply more flow at each measuring station pressure than higher-resistance active sources. At the extremes are the zero-resistance and infinite-resistance sources. The zero-resistance or "ideal" source can deliver flow with no pressure decrease. Although ideal sources are not realizable without moving parts, they may be approximated closely enough for many practical applications. At the other extreme, the infinite-resistance source delivers no flow.

Figure 6.3 shows the effects of a change in p_m, on the q versus p source characteristics. The characteristics are merely displaced along the p axis. Thus, even when there is no source pressure ($p_m = 0$), a source characteristic exists. The characteristic on the left of the axis represents the flow–pressure relation that would occur if the pressure at the measuring station were in the vacuum region ($p < 0$). In addition, Fig. 6.3 shows linear and square law source characteristics. Actual fluidic devices have source characteristics that may be linear over a range of pressures and nonlinear over another adjacent pressure range. The equations of the linear and square law source characteristics are

$$p = p_m - Rq \quad \text{(linear)} \tag{6.1a}$$

$$p = p_m - Kq^2 \quad \text{(square law)} \tag{6.1b}$$

where R is the linear resistance and K is the resistive coefficient.

The power W delivered by a source is calculated from $W = pq$. If we define a reference power $W_m = p_m q_m$, where q_m is the flow when $p = 0$, then the power delivered by linear and square law sources is

$$W/W_m = q/q_m - |q/q_m|^2 \quad \text{(linear)} \qquad (6.2a)$$

$$W/W_m = q/q_m - |q/q_m|^3 \quad \text{(square law)} \qquad (6.2b)$$

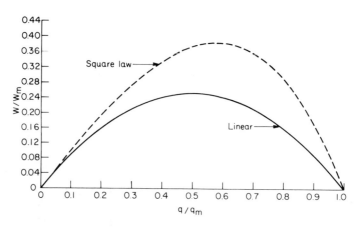

Fig. 6.4 Power delivered by sources.

Figure 6.4 shows the relation between delivered power and volume flow that is given by Eqs. (6.2a) and (6.2b). A source with a square law resistance can supply more power than a linear resistance source. The maximum power from a linear source occurs when $q/q_m = 0.500$, whereas from a square law source the maximum shifts to $q/q_m = 0.577$.

6.2.2 Load Characteristics

The load characteristic is the locus of all possible combinations of total pressure at the measuring station and volume flow that the load will accept or allow to pass. When the downstream end of the load (Fig. 6.1) is connected to an atmospheric pressure reference, the typical load characteristics are as shown in Fig. 6.5. In this case the zero flow condition occurs when the measuring station pressure is also at atmospheric pressure. Now an increase in measuring station pressure causes an increase in flow to the load. This is the region of "passive" load. When the measuring station pressure drops below the reference pressure the flow reverses and the load becomes "active". It should be re-emphasized that the general concepts of characteristic curves need not be concerned with the

6.2 Concept of Source and Load

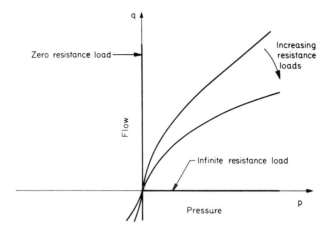

Fig. 6.5 Typical load characteristics.

manner in which the measuring station is maintained at each pressure level. All that is necessary, is that we recognize that a corresponding flow exists for every possible measuring station pressure.

Figure 6.6 shows the effects of changing the reference pressure p_g. In the figure the change is shown in the negative direction for demonstration purposes. When the reference pressure is above atmospheric pressure the load characteristic is translated to the right. The equations that represent the linear and square law load characteristics are

$$p = p_g + Rq \quad \text{(linear)} \tag{6.3a}$$

$$p = p_g + Kq^2 \quad \text{(square law)} \tag{6.3b}$$

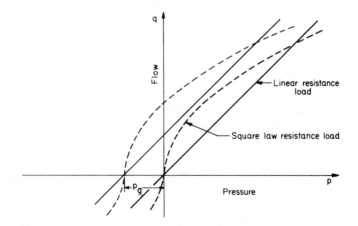

Fig. 6.6 Linear and square law loads.

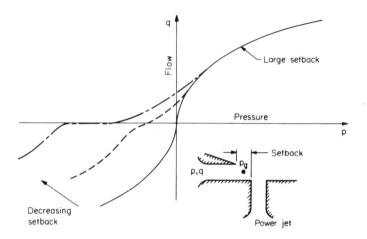

Fig. 6.7 Typical proportional amplifier load characteristics.

When the load is the input of a fluidic component it is possible to obtain load characteristics that appear vastly different from those shown in Figs. 6.5 and 6.6. Figure 6.7 shows some typical relations between p and q at the control of a beam deflection fluid amplifier as a function of setback (or offset). There are two effects taking place to obtain loads of this type: (1) a change in reference pressure level within the amplifier, and (2) a change in the area of the minimum cross section due to the deflection of the power jet. A theoretical formulation for loads that result in inflection points is presented in Chapter 9.

6.2.3 The Operating Point

When the source and load characteristics represent pressures and flows measured at the same location in the fluid circuit, they may be plotted on the same graph. Figure 6.8 shows a typical source characteristic and a typical load characteristic drawn on the same p, q axes. Each characteristic represents the locus of all possible circuit conditions. Obviously when this particular source is connected to this particular load the point of intersection becomes the only point that satisfies both loci. This point of intersection is called the operating point and it is analogous to the point determined by the solution of two simultaneous equations. The operating point does not necessarily indicate that the source and load are matched. The concept of matched impedance is used to describe the conditions under which maximum power is transferred from source to load. Depending on the particular load and source, the operating point may be at maximum power or at any lesser power.

Consider now that an enlarged section was used to obtain the source–load characteristics and that in an actual circuit there is no enlarged section between the source and load. This situation is shown in Fig. 6.9a. We should recognize

6.2 Concept of Source and Load

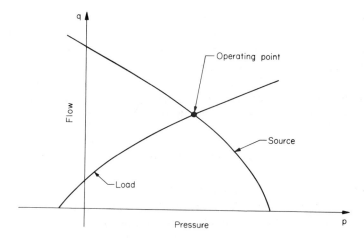

Fig. 6.8 Superposition of source and load.

that the enlargement will probably have less effect on the load characteristic than on the source characteristic. The reason for this is that the enlargement adds a nozzle to the load resistance whereas it adds a diffuser to the source resistance. Thus source curves measured with an enlarged section tend to produce larger source resistances than actually exist in the device. Figure 6.9b shows some typical source and load curves with and without the enlargement. In this case we assume that the load curve is unaffected by the enlargement. The measured source characteristic (solid line), however, is shown with larger resistance than the actual source characteristic (dashed line). Thus there are two points of intersection. The lower intersection is the operating point that would be predicted by measurements with an enlarged section. The upper intersection is the operating point that would actually exist in the circuit.

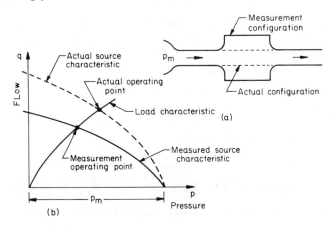

Fig. 6.9 The effect of measuring station enlargement: (a) actual and measurement configurations and (b) corresponding source–load curves.

The foregoing discussion does not present measured data. It is meant only as a caution to the user of characteristic curves. Accurate determination of operating points requires that the measuring technique must not alter the fluid circuit. Thus, in circuits that have appreciable velocity and where enlargements do not appear naturally it may be necessary to compute the total pressure from a combination of static pressure measurements, flow measurements, and some assumption regarding the velocity distribution in the circuit.

6.3 THE TWO-TERMINAL PAIR

The characteristic curve approach treats the fluidic component as a "black box." In general, the box may have any number of inputs and outputs. However, when there are more than one input and one output set of signal variables, the number of curves required becomes impractical. Fortunately, a fluidic component can usually be modeled as a two-terminal pair.

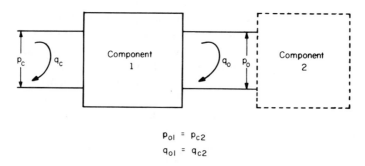

$p_{o1} = p_{c2}$
$q_{o1} = q_{c2}$

Fig. 6.10 The two-terminal pair.

Figure 6.10 shows a fluidic component as a two-terminal pair. At the input terminal the signal variables are p_c, q_c, and at the output terminal they are p_o, q_o. If the pressures at the terminals are assumed to be functions of the input and output flow, the functional relations are

$$p_o = p_o(q_o, q_c) \tag{6.4a}$$

$$p_c = p_c(q_o, q_c) \tag{6.4b}$$

In differential form Eq. (6.4) becomes

$$\Delta p_o = [\partial p_o/\partial q_o]_{q_c=C_1} \Delta q_o + [\partial p_o/\partial q_c]_{q_o=C_2} \Delta q_c \tag{6.5a}$$

$$\Delta p_c = [\partial p_c/\partial q_o]_{q_c=C_1} \Delta q_o + [\partial p_c/\partial q_c]_{q_o=C_2} \Delta q_c \tag{6.5b}$$

where C_1 and C_2 are constants that may take on any value in the range of flows used by the component. The partial derivatives in Eq. (6.5) represent the characteristics that are required to describe the component. Thus $[\partial p_o/\partial q_o]_{q_c=C_1}$

6.4 Proportional Amplifier Characteristics

is the output source characteristic and $[\partial p_c/\partial q_c]_{q_o=c_2}$ is the input load characteristic. The partial derivatives $[\partial p_o/\partial q_c]_{q_o=c_2}$ and $[\partial p_c/\partial q_o]_{q_o=c_1}$ are the transfer characteristics of the component. For components with adequate venting ports the latter partial derivative $[\partial p_c/\partial q_o]$ is negligible and Eq. (6.5) reduces to

$$\Delta p_o = [\partial p_o/\partial q_o]_{q_c=c_1} \Delta q_o + [\partial p_o/\partial q_c]_{q_o=c_2} \Delta q_c \qquad (6.6a)$$

$$\Delta p_c = [\partial p_c/\partial q_c]_{q_o=c_2} \Delta q_c \qquad (6.6b)$$

To describe the component the partial derivatives given in Eqs. (6.5) and (6.6) must be determined. This may be accomplished graphically through the following three characteristic relations (where subscript 1 refers to driving amplifier):

(1) input characteristics (p_{c1} versus q_{c1})
(2) output source characteristics (p_{o1} versus q_{o1})
(3) transfer characteristics (p_{o1} versus q_{c1} and p_{c1} versus q_{o1})

In addition, the performance of a component in a circuit requires the fourth characteristic relation (where subscript 2 refers to driven amplifier):

(4) load characteristic (p_{c2} versus q_{c2})

All the fluidic components that follow in this chapter are represented by the four characteristic relations given above.

6.4 PROPORTIONAL AMPLIFIER CHARACTERISTICS

To be useful, component characteristic curves must provide information about the important performance criteria of the component. For proportional amplifiers the measures of static performance are input resistance, output resistance, and pressure, flow, or power gain. Figure 6.11 shows two beam deflection proportional fluid amplifiers in cascade. The first amplifier in the cascade is called the driving amplifier and the second amplifier is called the driven amplifier. When one amplifier is connected to another the gain of the first amplifier can be obtained if the characteristics (partial derivatives) are known.

For example, the pressure gain of the driving amplifier G_{p1} can be formulated from Eq. (6.6). The procedure is to apply Eq. (6.6a) to amplifier 1 with the result

$$\Delta p_{o1} = [\partial p_{o1}/\partial q_{o1}] \Delta q_{o1} + [\partial p_{o1}/\partial q_{c1}] \Delta q_{c1} \qquad (6.7)$$

Now since $G_{p1} = \Delta p_{o1}/\Delta p_{c1}$, the incremental flows Δq_{o1} and Δq_{c1} in Eq. (6.7) must be expressed in terms of the incremental pressures Δp_{o1} and Δp_{c1}. The application of Eq. (6.6b) to amplifiers 1 and 2 yields

$$\Delta p_{c1} = [\partial p_{c1}/\partial q_{c1}] \Delta q_{c1} \qquad (6.8a)$$

$$\Delta p_{c2} = [\partial p_{c2}/\partial q_{c2}] \Delta q_{c2} \qquad (6.8b)$$

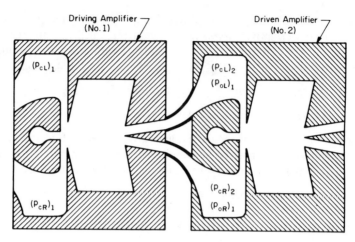

Fig. 6.11 Proportional amplifiers in cascade.

From the circuit arrangement shown in Fig. 6.11, we recognize that $\Delta p_{c2} = \Delta p_{o1}$ and $\Delta q_{c2} = \Delta q_{o1}$. Thus Eq. (6.8b) may be written as

$$\Delta p_{o1} = [\partial p_{c2}/\partial q_{c2}]\, \Delta q_{o1} \tag{6.9}$$

The substitution of Eqs. (6.8a) and (6.9) into Eq. (6.7) yields

$$\Delta p_{o1} = \left[\frac{\partial p_{o1}/\partial q_{o1}}{\partial p_{c2}/\partial q_{c2}}\right] \Delta p_{o1} + \left[\frac{\partial p_{o1}/\partial q_{c1}}{\partial p_{c1}/\partial q_{c1}}\right] \Delta p_{c1} \tag{6.10}$$

When Eq. (6.10) is rearranged the pressure gain of amplifier 1 takes the form

$$G_{p1} \equiv \frac{\Delta p_{o1}}{\Delta p_{c1}} = \frac{[\partial p_{o1}/\partial q_{c1}][\partial p_{c2}/\partial q_{c2}]}{(\partial p_{c1}/\partial q_{c1})[\partial p_{c2}/\partial q_{c2} - \partial p_{o1}/\partial q_{o1}]} \tag{6.11}$$

Note that all of the four types of characteristic relations given in the previous section are required to determine the pressure gain.

To express the pressure gain in terms of flow gain and resistance, note that Eq. (6.7) at zero output pressure is

$$\frac{\partial p_{o1}}{\partial q_{c1}} = \left[-\frac{\partial p_{o1}}{\partial q_{o1}}\right]\frac{\Delta q_{o1}}{\Delta q_{c1}} = (R_{o1})(G_{f1}) \tag{6.12}$$

where R_{o1} is the output resistance and G_{f_1} is the flow gain for unit 1 at zero output pressure. Through the use of Eq. (6.12), Eq. (6.11) may be rewritten as:

$$G_{p1} = \frac{R_{o1} R_{c2}\, G_{f1}}{R_{c1}[R_{c2} + R_{o1}]} \tag{6.13}$$

6.4 Proportional Amplifier Characteristics

where R_{c1} and R_{c2} are the input resistances of unit 1 and 2, respectively. Now if unit 1 is block loaded, $R_{c2} \to \infty$ and the pressure gain for blocked output G_{pb} is

$$G_{pb} = (R_{o1}/R_{c1})G_{f1} \tag{6.14}$$

The pressure gain in Eq. (6.13) may then be written in terms of G_{pb} as

$$G_{p1} = G_{pb}[R_{c2}/(R_{c2} + R_{o1})] \tag{6.15}$$

There is no unique way to present the characteristic curves of proportional amplifiers. Compact presentations are often used. In these, some of the required information appears as parameters on other curves. The following sections indicate two different ways of showing proportional amplifier characteristics.

6.4.1 Compact Presentation

Figure 6.12 shows output source characteristics that have control pressure and control flow as parameters [1]. The source characteristics are actually the graphical equivalents of Eq. (6.6a). For each constant value of control flow (and control pressure), the second term in Eq. (6.6a) is zero and there is a single source output characteristic. Since the amplifier may operate with any of an infinite number of control signals, there are an infinite number of output characteristics. Normally only a few of these, at specific fixed control signal increments, are plotted. Curves for other values of control signals are obtained through a graphical interpolation. To demonstrate that these curves contain all the information in Eq. (6.6a), note that the slope along the source characteristics provides values for the partial derivative $\partial p_o/\partial q_o|_{q_c=C_1}$. The other derivative in Eq. (6.6a), $\partial p_o/\partial q_c|_{q_o=C_2}$, may only be approximated. To obtain this, a line of constant

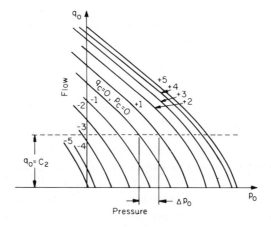

Fig. 6.12 Proportional amplifier output source.

output flow (dashed horizontal line) is drawn on Fig. 6.12. Then the approximate value of $\partial p_o/\partial q_c$ is merely the ratio of the two differentials $\Delta p_o/\Delta q_c$. For example, when q_c goes from 0 to 1, $\Delta q_c = 1$ and the corresponding value of Δp_o is indicated. This method of presentation thus yields source type output curves that contain transfer characteristic information and input characteristic information as parameters on the curves. The superposition of load characteristics then provides a one-quadrant representation of the entire proportional amplifier.

Figure 6.13 shows the superposition of a load characteristic on a typical one-quadrant set of output source curves. In a differential amplifier there are always two source characteristics, one for each output. For example, when $p_c = q_c = 0$

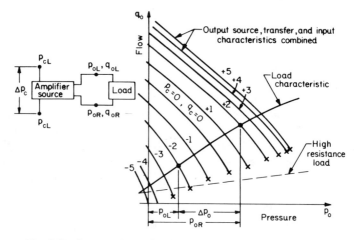

Fig. 6.13 Superposition of proportional amplifier characteristics.

the output characteristic applies equally to the left and right output. When $p_{cL} - p_{cR} = 2$, the right output characteristic is denoted as $+2$ and the left output characteristic as -2. There are also two load characteristics. In the case illustrated the loads are assumed to be identical. The operating points (dots) give the value of pressure and flow at each output. From this, the pressure or flow gain can be calculated. If the right control pressure exceeds the left control pressure by the same amount, i.e., $p_{cR} - p_{cL} = 2$, then in a symmetrical amplifier the operating points will retain their values but will be reversed. The left output pressure will be at the higher-pressure operating point, in this case.

The x symbol on the output source characteristics designates the end of stable amplifier operation. This instability does not occur on all proportional fluid amplifiers. The unstable region is a function of the amplifier vent size, and also depends on whether it has a center dump. If a high-resistance load (shown dashed) were connected to the amplifier, as shown in Fig. 6.13, the source and load curves would not intersect over an appreciable region of normal amplifier performance. This indicates that stable operating points would not then exist over a useful range.

6.4.2 Complete Presentation

Figure 6.14 shows the characteristic curves arranged in three quadrants for the case of a single-sided proportional amplifier such as a direct impact modulator adjusted to give zero output for zero input. This is similar to Verhelst's [2] curves for a NOR element. As before, the output source curves are in the first quadrant. This time, however, there are no parameters on the curves. They are merely spaced evenly at convenient intervals of output pressure. The transfer characteristic is plotted in the fourth quadrant and is based on the output pressure at some known output flow. In this case the output flow is blocked. For any

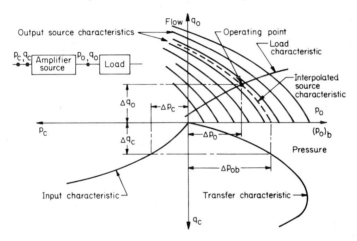

Fig. 6.14 Complete characteristics for single-ended proportional amplifiers.

value of q_c the transfer characteristic yields a value for the blocked output pressure, and from this it is possible to interpolate an output source characteristic (dashed line). The input characteristic appears in the third quadrant and the load characteristic in the first quadrant. Thus, for any pressure or flow input it is possible to enter the transfer characteristic and to construct the pertinent source characteristic on which the operating point must lie. The operating point is the intersection between the load characteristic and the interpolated source characteristic. This point yields the output pressure and flow that result from the changes in input pressure and flow. As a consequence, the pressure, flow, and power gain of the loaded amplifier can be determined.

Figure 6.15 shows the three-quadrant representation for a differential (beam deflection) proportional amplifier. In this case the transfer characteristic provides the blocked output pressure at each output terminal for changes in the difference of the left and right control input flows. The control flow value with which to enter the transfer curve is obtained from the input characteristic, as shown on Fig. 6.15. The transfer curves then provide two values of the blocked output pressure to serve as starting points for interpolated output source characteristics

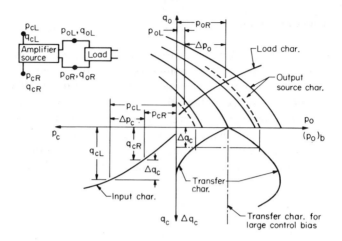

Fig. 6.15 Complete characteristics for differential proportional amplifiers.

(dashed lines). As before, the intersection of the interpolated characteristics and the load characteristic yields the output operating points. One advantage of the three-quadrant method is that it allows the inclusion of control bias level effects. Thus the fourth quadrant may contain a series of transfer characteristics with control bias as a parameter on each characteristic. For example, for very large control bias the output pressure, in beam deflection proportional amplifiers, becomes insensitive to input flow changes. This is reflected in the dot–dash curve in the transfer quadrant.

6.5 BISTABLE SWITCH

In the wall attachment bistable fluid switch the jet issuing from the power nozzle has only two stable positions. This contrasts with the beam deflection proportional fluid amplifier in which the jet has an infinite number of possible positions. The stable positions of the bistable device are designated as RESET and SET on Fig. 6.16a. To represent the bistable switch and predict its performance in a circuit requires four characteristic curves. The curves, which are similar to those already discussed for the proportional amplifier, are (1) input, (2) output, (3) transfer, and (4) load. For the bistable switch, however, the emphasis and certain other details are different.

The most significant difference between bistable and proportional characteristic curves relates to the importance and interpretation of the input (and load) characteristics [1, 3]. Figure 6.16b shows a typical input characteristic (q_c versus p_c) when the bias control is fixed at atmospheric reference conditions. With the jet in the RESET position the pressure within the element is further below the reference pressure than it is when the jet is in the SET position. This occurs

6.5 Bistable Switch

because all the jet entrainment in RESET must come through the signal control port. When the jet is SET, the entrainment on the control port side may also come from the output and vent ports on that side. As a result the input characteristic falls on one of two approximately parallel curves. The appropriate input characteristic then depends on the state of the amplifier. For example, in RESET, the flow–pressure relation (q_c, p_c) follows the upper curve (in Fig. 6.16b) until the control signal is sufficient to switch the amplifier to the SET position. The switching point is marked by an x. Conditions within the element then change

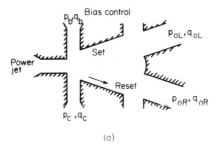

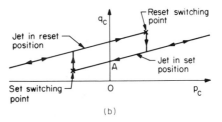

Fig. 6.16 (a) Bistable switch and (b) a typical input characteristic.

abruptly and the input characteristic suddenly changes to the lower curve in Fig. 6.16b. The operating point on the lower curve after the switch depends on the impedance of the source used in making the test. However, the precise path of the input characteristic during the switch from RESET to SET does not represent stable operation and is usually not shown on static characteristics. Instead it is conventional to connect the RESET and SET characteristics by constant pressure lines. With the device now in the SET position, return of the control pressure p_c to atmospheric pressure places the input characteristic at point A. To switch the unit back to RESET requires a suction signal at the signal control port (p_c, q_c). This is obviously the type of signal that would be expected from a physical viewpoint. The suction switching point is designated by x on the SET portion of the characteristic. After sufficient suction has been applied, the jet returns to the RESET position. There is, at the same time, a discontinuous increase in the control flow signal on the characteristic curve. The input characteristic, as viewed from the signal control port, traces out a hysteresis loop. There is a switching point at each end of the positive slope diagonal in the loop. Usually, as we shall see later, it is necessary and convenient to consider the input characteristic as two single characteristic curves, one curve pertaining to each of the two controls of a bistable switch.

The magnitude and shape of the input characteristic depends upon geometric parameters, bias level, and output loading. Figure 6.17 shows some of these effects schematically. For example, a decrease in the distance from the power nozzle to the splitter tends to reduce the size of the hysteresis loop (Fig. 6.17a). Ultimately, if the splitter is in close enough, the element loses its bistability.

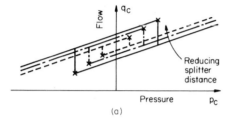

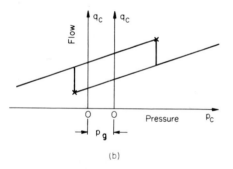

Fig. 6.17 The effect of (a) geometry and (b) bias on the input characteristics of the bistable switches.

Similar effects also occur when the attachment wall setback is changed. Control bias (Fig. 6.17a) acts mainly to displace the flow axis q_c without significant alteration of the hysteresis loop [4]. If the bias is large enough, both RESET and SET switching points take place at positive values of control pressure p_c.

The bistable switch has only two output source characteristics (Fig. 6.18). When the element is in RESET position the SET side output characteristic normally lies in the second quadrant. For most bistable units the SET side output characteristic is almost equivalent to the characteristic from a zero-pressure source. In this case the operating point pressure must be below atmospheric pressure to draw off any flow. When the element switches to the SET position the source characteristics reverse. Now the SET output is in the first quadrant and the RESET output is in the second quadrant. The shape and position of the output characteristics depend on the element geometry and the supply pressure operating range. The observations on source characteristics made previously

6.5 Bistable Switch

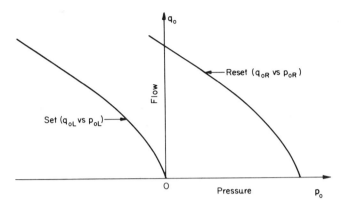

Fig. 6.18 Typical output source characteristics for the bistable switch.

(see Figs. 6.2 and 6.3) also apply generally to the output characteristics of the bistable switch. As in the case of proportional amplifiers, load instabilities can also be indicated on the bistable switch output characteristic that lies in the first quadrant. A bistable switch that is unable to retain state when connected to a high resistance or blocked load has an output characteristic that terminates before it reaches the pressure axis.

Figure 6.19 shows a four-quadrant representation of a bistable unit (no. 1) that is loaded by a bistable unit (no. 2) with different characteristics. The arrangement of the curves is similar to that given for the proportional amplifier in Fig. 6.14. The load characteristics that appear in quadrants 1 and 2, come from unit no. 2. All the other characteristics refer to unit no. 1. The output source characteristics are in quadrants 1 and 2 while the input characteristic is plotted

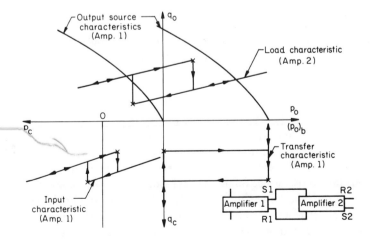

Fig. 6.19 Complete characteristics for bistable switches.

in quadrant 3. The transfer characteristic is again placed in quadrant 4. This characteristic also has a hysteresis loop. At first glance the transfer characteristic looks vastly different from the proportional amplifier transfer characteristic. However, if geometric changes are made in the bistable switch, the hysteresis loop

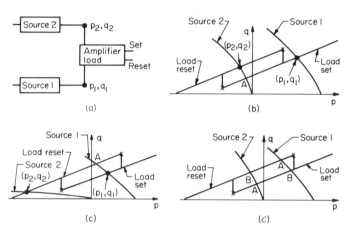

Fig. 6.20 The operation of bistable switches: (a) test arrangement and (b)–(d) characteristic curves.

narrows and the transfer characteristic begins to look more like that of a high-gain proportional amplifier. In this case, of course, all the other characteristic curves would also change and approach proportional amplifier characteristics. Thus the bistable switch is really a special case of the proportional amplifier. This is not surprising since the geometrical similarities allow us to think of the bistable switch as a proportional amplifier with internal feedback.

The complete set of bistable element characteristic curves is rarely used in practice. Instead it is customary to consider only the input characteristics of the bistable switch and the sources to which the unit will be connected in a circuit. A typical test arrangement is shown in Fig. 6.20a, where an element receives signals at its controls from two sources (1 and 2). The purpose of the characteristic curves in this case is to determine the stable operating points (p_1, q_1) and (p_2, q_2), if they exist, and to predict the ability of these sources to change the state of the switch. Suppose that the source and load characteristics are the ones shown in Figs. 6.20b. To find the operating points we must recognize first that the load characteristic actually describes two input relationships. The upper branch pertains to the orifice on the RESET side while at the same time the lower branch pertains to the orifice on the SET side. Thus one operating point must lie on each branch. It will never be possible for both operating points to occur on a single branch. In addition, the source–load concept shows that the same observation holds true for sources. A single source cannot contain two operating

6.5 Bistable Switch

points. With this in mind, point A on Fig. 6.20b cannot be an operating point. If it were, source 1 would have to intersect the RESET branch. Since source 1 has no intersection except on the SET branch, we know that the other operating point must occur at the intersection of source 2 and the RESET branch. As a result, if this switch were originally in RESET and then connected to the sources shown, the state of the element would switch to SET and remain there with the operating points (p_1, q_1) and (p_2, q_2).

Suppose now that the same unit is attached to the sources shown in Fig. 6.20c. In this instance, source 2 has no intersection with the SET branch and the operating point from source 2 lies on the RESET branch. Thus the intersection at A between source 1 and the RESET branch cannot be an operating point. The operating point (p_1, q_1) occurs where source 1 crosses the SET branch. Under these conditions, the unit, in RESET prior to connection of the sources, would switch to SET and remain there.

Finally let us consider the situation shown in Fig. 6.20d. The operating points are either at A—A or B—B, depending on the element state. If the switch is in RESET, the operating points are at A—A. On the other hand, in the SET position the operating points would occur at B—B. These particular sources are not able to switch the element.

In most practical cases the important output source is the one that appears in the first quadrant. Seldom is switching by a second-quadrant characteristic desired. More often, care must actually be exercised to prevent second-quadrant-type characteristics from switching the unit. For this reason we focus special

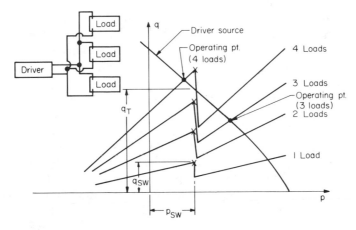

Fig. 6.21 Driving identical parallel loads.

attention on the first-quadrant operation. Figure 6.21 shows a driver source connected to a number of identical units (loads) in parallel. The lowest flow load curve represents a single load characteristic. From the circuit theory for parallel connections, a composite load curve can be obtained by adding the flows at each value of the pressure. This process yields composite load characteristics for 2, 3, and 4 identical loads. The superposition of the load and source curves show that this particular driver can switch up to three units in parallel. If four units are connected to this source, none of them will switch. The operating

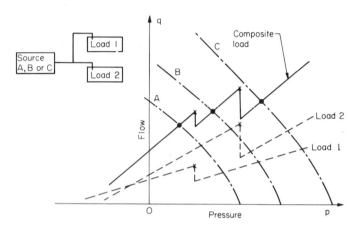

Fig. 6.22 Composite load line procedure.

point for four loads is indicated on the RESET portion of the composite curve. At the operating pressure each element receives only one quarter of the operating point flow. When there are three loads the operating point falls on the SET portion of the three-load composite characteristic. A quick method of determining the number of units that can be operated from a given source is to find the total flow q_T that the source can deliver at the switching pressure p_{sw}. Now the ratio of this total flow to the flow necessary to switch one device q_{sw} is the fan-out n. More precisely, the fan-out is the closest integer number less than q_T/q_{sw}.

Figure 6.22 shows the source–load procedure when a source is connected to different types of elements in parallel. From the individual input characteristics 1 and 2, a composite input characteristic can be constructed as before. In this case, however, the switching points do not occur at the same pressure. Thus the composite curve has two switching points instead of one and three branches instead of two. The leftmost branch represents the condition of both elements in RESET. In the rightmost branch both elements are in SET. The central branch holds during the condition that unit 1 is SET and unit 2 is RESET. In this case the operation depends on the intersection of the source characteristic with the three-branch composite load curve. If the source had the characteristic curve A, the intersection would occur in the leftmost branch and neither unit would switch. A source of the type described by curve B would switch element 1 but not element 2. Source curve C, on the other hand, is sufficient to switch both elements.

6.6 NOR ELEMENTS

NOR elements are active logic components with a number of control input ports and a single output port. These elements may operate discontinuously (wall attachment, turbulent transition) or continuously (beam deflection). Whatever the fluid mechanism, the basic function is the same. The NOR function has an output signal level (ON) only when all the control signals are less than some specific level. If any control or controls exceed some other specific level, there is no output signal (OFF).

The NOR is of particular importance because it is a universal logic element. Thus Boolean combinatorial and sequential functions of any complexity can be implemented entirely with NOR elements. Furthermore, there is the attractive possibility of fabricating fluid circuits with components of a single fixed-geometry design. For this reason the emphasis here is on NOR elements in circuits that consist exclusively of identical elements.

Figure 6.23 shows two stages of a general NOR circuit. The output signal from the element in stage 1 (p_{o1}, q_{o1}) supplies control signals (p_{c2}, q_{c2}) to some number of identical elements in stage 2. The state of any NOR (ON or OFF) is defined by the level of its output signal.

Proper circuit operation requires that when stage 1 is ON all the elements in stage 2 must be OFF, and conversely when stage 1 is OFF all the elements in stage 2 must be ON.

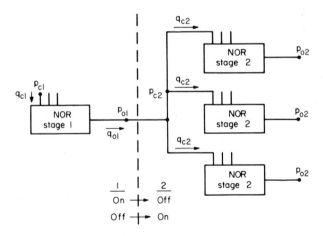

Fig. 6.23 NOR circuitry.

As in the previous cases of proportional amplifiers and bistable switches, the same four important characteristic curves apply to NOR elements. The input, transfer, and output characteristics refer to the element under consideration. The load characteristic comes from the input characteristic of the following

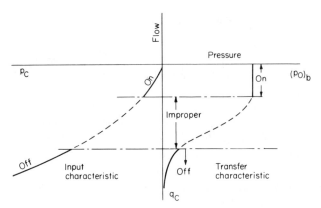

Fig. 6.24 NOR input and transfer characteristics.

stage. Since the elements in each stage are identical, the input and load characteristics are equivalent. They are, however, still plotted in different quadrants. Figure 6.24 shows the typical input and transfer characteristics for an analog-type NOR element. Switching-type NOR's are also possible. There are three distinct regions on the NOR characteristic curves. Only the ON (higher output pressure) and OFF (lower output pressure) states represent proper operation. The pressure signals anywhere in a NOR element network must always fall within the proper regions. The improper region between the ON and OFF states should be avoided or else errors will occur in the circuit function. On the characteristic curves, proper operation is indicated by the solid-line segments. Improper operating conditions may be shown as dashed lines, as in Fig. 6.24, or merely left blank.

Figure 6.25 shows the complete set of NOR characteristic curves. The first

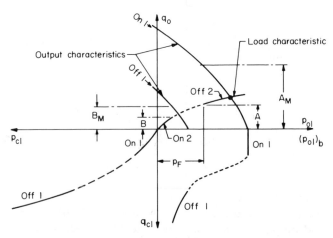

Fig. 6.25 Complete NOR characteristics.

quadrant is of particular interest since it provides a measure of NOR performance. In this quadrant the output characteristics are shown as two discrete lines. However, since there are operating regions instead of single points on the transfer characteristic, the output characteristics should really be shown as output bands. The OFF 1 output characteristic line represents the upper limit of one band which extends into the fourth quadrant. The ON band, on the other hand, is usually quite narrow and more nearly approaches the discrete ON 1 line. The load characteristic (input characteristic of elements in stage 2) is actually the same characteristic that is shown in quadrant 3 except that it is plotted on a different scale. Now the operating regions occur where the load characteristic intersects the source characteristics. There are two intersecting regions that correspond to the Boolean states. The intersection of ON 1 and OFF 2 is indicated by a solid dot on Fig. 6.25. As a result of the output source bands, the intersection may extend over a small line segment of the OFF 2 line. The maximum fan-out can be determined by a flow ratio at the minimum pressure reached along the OFF 2 line (p_F). Thus, with reference to Fig. 6.25, the maximum fan-out is the integer immediately below A_M/A. The other intersection, the one between OFF 1 and ON 2, points up an interesting situation that sometimes arises. When a NOR element is connected to only a single other NOR element, the upper limit of the OFF source characteristic may intersect the load characteristic in the improper region. This condition can be rectified by using a dummy load or by connecting other elements. Thus there is also a minimum fan-out, and in this case it is the integer immediately above the ratio B_M/B.

6.7 PASSIVE LOGIC ELEMENTS

The passive logic elements are those that require no power source. They operate directly on the fluid signals, usually through jet interaction or jet augmentation. The passive logic components, of course, cannot perform inversion. Only active components can produce the logic complement. Nevertheless, passive components are often used in fluidic logic circuits. They are particularly valuable for one-of-a-kind circuits and for applications where power is limited. Two of the most common passive logic components and the ones considered here are the passive AND the passive inclusive OR.

6.7.1 Passive AND

Figure 6.26a is a schematic drawing of one type of passive AND element. A pressure signal applied at one control or the other (p_{c1} or p_{c2}) creates a jet which is oriented toward a vent at reference pressure. When signals are present at both controls (p_{c1} and p_{c2}) the jets interact to produce an output signal p_0 at a centrally located output port.

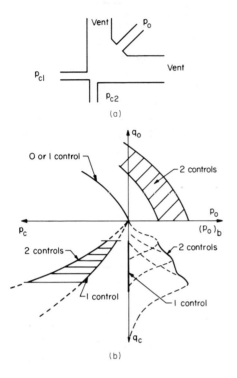

Fig. 6.26 AND characteristics: (a) passive AND schematic, (b) AND characteristic curves.

The performance of passive components is extremely sensitive to the level of the logic control input signals. For example, if 1 kN/m^2 represents the input logic "1", the blocked output logic "1" may be 0.5 kN/m^2. Suppose, however, that one input signal is 1.1 kN/m^2 and the other is 0.9 kN/m^2. Then the output logic "1" may only be 0.3 kN/m^2. The characteristic curves exhibit this sensitivity through the use of operating bands rather than discrete operating lines.

Figure 6.26b shows some typical characteristic curves for the passive AND. The control resistance (third quadrant) depends on the number of control signals (logic "1"s) present. When one control is alone, the resistance flow is merely the orifice resistance. The presence of both controls restricts the flow through each input and increases the input (and load) resistance of the AND. The operating region is bounded by the solid resistance lines. This places a limit on the difference between the actual control signal and the nominal signal for a logic "1". The limit may be ± 10 or 20% of the nominal, depending on the circuit and the component. Discrepancies in excess of 20% usually lead to logic errors and are to be avoided.

Figure 6.26b shows two transfer characteristic curves in the fourth quadrant (solid lines). When one control signal is present the flow passes out through a vent and there is no output signal. For the activated case of two control signals, the output depends on the magnitude of the control signals [5]. This transfer characteristic includes all the possibilities for cases in which the signals may

6.7 Passive Logic Elements

differ from each other by up to some predetermined percentage. The dashed construction lines indicate the sweep (of the resultant jet) by the output port, as the signals go outside their tolerance levels.

As a result of the transfer characteristic extent, the output characteristic in the first quadrant contains an operating band. Thus, when an AND element drives another component, the precise operating point is not determined. If the AND output is required to switch a bistable element, the extremity of the band at small pressure and flow must exceed the switching point. In general, the passive components require greater care and attention than the active components. Also it is bad practice to connect two passive components together. This magnifies the uncertainties and is often unsatisfactory.

6.7.2 Passive Inclusive OR

Though the configuration and function of the passive inclusive OR [6] (Fig. 6.27a) is entirely different from the AND, they are similar in their dependence on signal level. In the OR the output is present if either or both controls are present. The device is basically a junction with vents. The purpose of the vents is to decouple the inputs from one another.

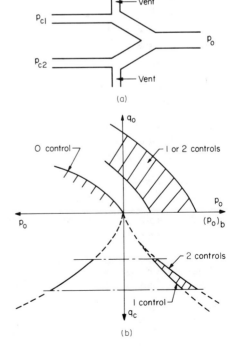

Fig. 6.27 Inclusive "OR" characteristics: (a) passive inclusive OR, (b) OR characteristic curves.

Figure 6.27b shows the input, transfer, and output characteristics of the OR element in their accustomed positions. The input characteristic is a single line, and this indicates that the vents have effectively decoupled the controls. The solid portion of the input characteristic line represents the entire range of allowable signal variables. On the transfer and output characteristic, the linear input range extends over a region. The extension is due to the geometry of the inclusive OR. When two control signals are present there is more energy available at the output.

PROBLEMS

6.1 For a source–load configuration (Fig. 6.1a) the source characteristic is $p = 10 - q$ and the load characteristic is $p = -6 + 3q$.

(a) Draw the load and source characteristics on the same graph.
(b) What is the operating point? (Check by solving source–load equations simultaneously.)
(c) What is the maximum power that the source can deliver?
(d) How would the resistance of the load have to be changed for maximum power transfer?

6.2 A particular source–load configuration has a nonlinear source characteristic $p = 7 - 2q^2$ and a linear load characteristic $p = 2 + 1.5q$.

(a) Plot the source–load characteristics on the same graph.
(b) What is the operating point?
(c) What is the maximum power that the source can deliver?
(d) If the load resistance is fixed, what change in reference pressure is required for maximum power transfer?

6.3 A load with characteristic $p = -2 + 4q^2$ is connected to a source with characteristic $p = -6q$.

(a) Plot the characteristics and find the operating point.
(b) If the source resistance is changed to zero, what is the operating point?
(c) If the source resistance is infinite, what is the operating point?

6.4 Find an expression for the power gain G_w in terms of input, output, transfer, and load characteristics. Is $G_w = G_p G_f$?

6.5 A bistable switch receives input signals from sources 1 and 2 as shown in Fig. 6.20a. The input characteristics are

$$p_{c1} = q_{c1}^2 - 2; \quad p_{c1} = 7, \quad q_{c1} = 3 \quad \text{(at switch point)}$$
$$p_{c2} = q_{c2}^2 - 1; \quad p_{c2} = -0.75, \quad q_{c2} = 0.5 \quad \text{(at switch point)}$$

The source characteristics are

$$p_1 = 10 - k_1 q_1^2, \quad p_2 = -k_2 q_2^2$$

and the switch is always on the reset side before the sources are connected.

(a) If $k_1 = k_2 = 1.0$, what are the operating points at each control?
(b) If $k_1 = 0.1$ and $k_2 = 1.0$, what are the operating points?
(c) What is the largest value of k_1 that will cause the unit to switch? If $k_2 = 1.0$, where will the stable operating points be in this case.
(d) If source p_1 is replaced by $p_1 = 0$, what is the smallest value of k_2 that will cause a switch? What are the operating points?
(e) If $p_1 = 20 - 0.1q^2$, how many units with the identical input characteristics could be switched?

6.6 The output source characteristics of a single-sided proportional amplifier may be expressed as

$$p_o = 3 + 10p_c - 4q_o^2$$

where $0 \leq p_c \leq 0.3$ and $p_c - 2q_c$. The amplifier is loaded with a component that has the characteristic

$$p = -2.0 + 8.0q^2$$

(a) Draw a one-quadrant representation of the output characteristics for $p_c = 0, 0.1, 0.2,$ and 0.3.
(b) Superimpose the load characteristics on this graph.
(c) What is the pressure gain when p_c is changed from 0 to 0.1?
(d) What is the flow gain when p_c is changed from 0.2 to 0.3?
(e) If another load could be selected, what load would produce the maximum pressure gain? What would be the magnitude of the maximum pressure gain?
(f) What load would produce the maximum flow gain and what is the magnitude of the maximum flow gain? For this load what would be the pressure gain?

NOMENCLATURE

G_f	Flow gain
G_p	Pressure gain
G_{pb}	Blocked output pressure gain
K	Resistive coefficient, kN-sec²/m⁸
p	Static pressure, kN/m²
p_c	Input pressure, kN/m²
p_g	Reference pressure, kN/m²
p_m	Source pressure, kN/m²
p_o	Output pressure, kN/m²
q	Volume flow, m³/sec
q_c	Input flow, m³/sec
q_m	Volume flow at $p = 0$, m³/sec
q_o	Output flow, m³/sec
R	Linear resistance, kN-sec/m⁵
R_c	Input resistance, kN-sec/m⁵

R_o Output resistance, kN-sec/m^5
W Power, watts
W_m Reference power, watts

REFERENCES

1. S. Katz and R. J. Dockery, "Fluid Amplification 11. Staging of Proportional and Bistable Fluid Amplifiers," HDL-TR-1165, August 1963.
2. H. A. M. Verhelst, On the design characteristics and production of turbulence amplifiers. *Cranfield Fluidics Conf.*, *2nd* Paper F-2 (January 1967).
3. R. E. Norwood, A performance criterion for fluid jet amplifiers. *Symp. Fluid Jet Contr. Devices ASME* (November 1962).
4. C. P. Wright, "Some Design Techniques for Fluid Jet Amplifiers," IBM Gen. Prod. Div. Tech. Rep. TR 01.758, October 1963.
5. K. Foster and G. A. Parker, "Fluidics—Components and Circuits," pp. 572–589. Wiley, New York, 1970.
6. J. M. Iseman, "Fluerics 9. Fluid Digital Logic Elements and Circuits," HDL-TR-1302, August 1965.

Chapter 7

THE IMPACT MODULATOR

7.1 INTRODUCTION

Impact modulators were first introduced in 1964 by Bjornsen [1] and Lechner and Sorenson [2]. From an operational viewpoint these impacting devices are the no-moving-part equivalent of the conventional flapper–nozzle valve [3] with the flapper replaced by a fluid streamsurface. Although impact modulators are proportional amplifiers, they have relatively small dynamic range and are therefore most often used as NOR elements in logic circuits.

Figure 7.1 shows a schematic drawing of the impact modulator. Here two round nozzles, an *emitter* and a *source*, are located along a common axis.

The source flow moves through the hole in the housing surrounding the source. Partially blocking this hole will cause some of the source flow to enter the housing and raise the pressure within it. Changing the amount of blockage will cause this output pressure to change.

When both the emitter and the source are on, as in normal operation, they impact against each other in some region between the two. The impact point moves as the emitter flow is varied, thus changing the blockage of the source flow and consequently varying the output pressure p_o.

The emitter flow in turn can be controlled by three methods:

(1) The emitter pressure may be varied (emitter modulation).

(2) Flow may be introduced into an annular chamber surrounding the emitter (dashed lines of Fig. 7.1) to modulate the emitter flow (annular modulation).

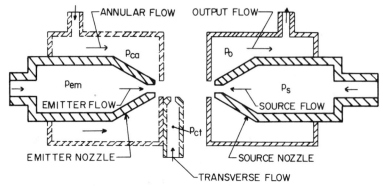

Fig. 7.1 Impact modulator schematic.

(3) Flow from an auxiliary transverse nozzle (dashed lines) may be used to modulate the emitter flow (transverse modulation).

When methods 2 or 3 are used, the device actually consists of two stages of amplification.

To relate the output pressure of an impact modulator p_o to its control pressure p_c requires a knowledge of the fundamental flow processes taking place within the modulator.

To simplify the structure and the resulting analysis, Katz [4] incorporated the output within the source nozzle as in Fig. 7.2 in his investigations. This is the structure that will be discussed in most of this chapter.

The concept of the "dividing streamsurface" [4] assists in the logical separation of fundamental processes from the complex flow field of the modulator. The dividing streamsurface is an apparent barrier that separates emitter flow

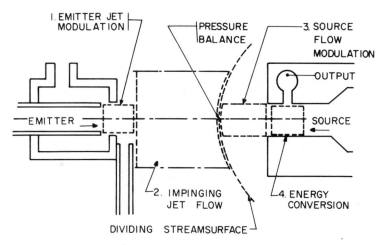

Fig. 7.2 Fundamental processes.

7.1 Introduction

from source flow. Figure 7.3 shows two hypothetical dividing streamsurfaces for the extreme operating conditions of an impact modulator.

If the initial conditions are such that the impact point and the resultant dividing streamsurface are well away from the source nozzle, as in Fig. 7.3a, the source flow is only slightly restricted by the dividing streamsurface. In this condition the source flow is almost independent of the emitter flow. An appropriate change in control signal moves the dividing streamsurface toward the

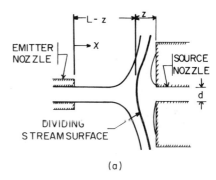
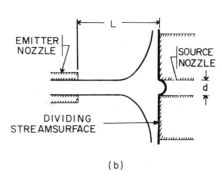

Fig. 7.3 Hypothetical representation of dividing streamsurface: (a) beginning of flow modulation, (b) cutoff.

source and causes a reduction of source flow. At the limiting condition the streamsurface contacts the source nozzle and the source flow is "cut off." When the source is completely blocked ("cut off"), the output is maximum. This is the procedure that occurs in annular and emitter modulation for increasing control signals. For a transverse modulator the source is set at cutoff when there is no control signal. Then the application of control moves the streamsurface in the reverse direction from cutoff to the beginning of flow modulation. At the beginning of flow modulation the pressure at the output of an actual impact modulator may be negative. This is due to the diffuserlike action of the chamber surrounding the source nozzle, and can often be used to advantage.

Figure 7.2 shows the impact modulator separated into four fundamental flow processes. There are two processes on each side of the dividing streamsurface.

On the emitter side of the streamsurface, where the fluid is in the form of a jet, the processes are

(1) emitter jet modulation by control flows;
(2) submerged jet impinging on the streamsurface.

On the source side, where the flow is restricted, the processes are

(3) source flow modulation by the streamsurface;
(4) energy conversion; the output pressure signal is the static pressure portion of the total source pressure near the exit of the source nozzle.

The position and shape of the dividing streamsurface provide the connection between the impinging jet (processes 1 and 2) and the source flow modulation (processes 3 and 4). At present there is no analytical method for determining the streamsurface shape and we must rely on assumptions or experimental measurements. However, the position of the streamsurface on the source–emitter axis lies on the stagnation streamline and can be approximated by a pressure balance once the streamsurface shape is specified.

The concept of a centerline balance is shown schematically in Fig. 7.4. The centerline total pressure distributions for the emitter jet and source flow are superimposed. The centerline position of the streamsurface or "balance" point occurs at the point of intersection. This is the point where the pressures are equal. In the operating region the balance point is always in close proximity to the source nozzle. Thus the balancing pressure is approximately equal to the total

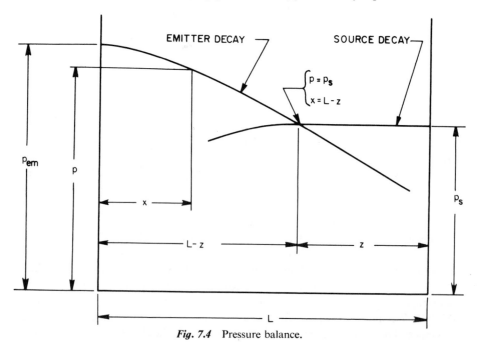

Fig. 7.4 Pressure balance.

pressure of the source p_s. Since the balance point is farther from the emitter nozzle, the emitter jet has sufficient distance in which to decay. As a result the emitter pressure p_{em} must be larger than the source pressure.

We are now in a position to formulate the functional relations that are necessary to determine the connection between control pressure p_c and output pressure p_o. As a starting point, consider that the centerline total pressure p_p of the emitter jet impinging on a plate may be expressed as

$$p_p = f(p_{em}, C_p, x, d, \text{streamsurface shape}) \tag{7.1}$$

where C_p is the jet spread or decay factor for a jet impinging on a plate, x is the distance from the emitter nozzle, and d is the emitter diameter. This particular decay factor and the reason for it are subsequently described in greater detail. Now when the pressure in the emitter jet equals the source pressure ($p_p = p_s$), the balance point is a distance z from the source and a distance $L - z$ from the emitter. Thus, Eq. (7.1) becomes

$$p_s = f(p_{em}, C_p, L - z, d, \text{streamsurface shape}) \tag{7.2}$$

where L is emitter–source spacing.

From the emitter jet flow processes (1 and 2) it is possible to obtain experimentally a relation between decay factor and control pressure p_c of the form

$$C_p = C_p[p_{em}, p_c, \text{streamsurface shape}] \tag{7.3}$$

The source flow modulation (processes 3 and 4) may be described by the relation

$$p_o = p_o[p_s, z, \text{streamsurface shape}] \tag{7.4}$$

If the streamsurface shape is assumed, Eqs. (7.2)–(7.4) contain only the three unknowns, C_p, z, and p_o. Thus, for every control pressure it is possible to calculate the corresponding output pressure. Furthermore, by taking the derivatives of these equations, formulations for the pressure gain of impact modulators are attainable.

In the following three sections (7.2–7.4) some appropriate analytical and empirical expressions that correspond to the functional relations [Eqs. (7.2)–(7.4)] are introduced. Then, in Section 7.5, these expressions are used to determine the pressure gains of the various types of impact modulator.

7.2 CENTERLINE TOTAL PRESSURE DECAY OF FREE AND IMPINGING JETS

To obtain analytical expressions for the functional relations given in Eqs. (7.1) and (7.2) for a jet impinging on a surface requires first the consideration of the free jet. For a uniform velocity profile at the nozzle exit we have found in Chapter 4 [Eq. (4.139)] that the expression for the centerline pressure p of a free axisymmetric jet is

$$p = p_{em}\{1 - \exp[-d^2/4C_2^2 x^2]\} \tag{7.5}$$

where C_2 is the free-jet decay factor and d is the nozzle diameter. Equation (7.5) corresponds to the functional relation given in Eq. (7.1) when the jet does not impinge on a surface.

Le Clerc [5] performed an inviscid analysis for the impingement of an axisymmetric jet on a perpendicular flat plate. The displacement of the free streamlines indicates the distance H from the plate (Fig. 7.5) at which the jet behaves like a free jet. Katz [6] showed that as a result of the displacement distance the jet stagnation pressure recovered on a flat plate (p_p) exceeded the stagnation pressure of the free jet (p) measured at the same downstream distance. To demonstrate this, consider the centerline decay curve shown in Fig. 7.5. A total pressure probe paced at some downstream distance x_p measures the jet stagnation pressure. If a perpendicular flat plate with a small pressure tap in it replaces the probe at the same downstream position x_p, the plate causes the jet particles to change direction. This change takes place over the distance H in front of the plate. The free-jet region then extends only up to the distance $x_p - H$. At this location the normal free-jet entrainment terminates. If no losses occur in the region $(x_p - H) < x < x_p$, the centerline stagnation pressure measured on the flat plate at x_p is equal to the total pressure that occurred in the free jet at $x_p - H$. The centerline total pressure of the impinging jet p_p then becomes from Eq. (7.5)

$$p_p = p_{em}\{1 - \exp[-d^2/4C_2^2(x_p - H)^2]\} \tag{7.6}$$

Since the precise impingement distance H and the amount of stagnation recovery on the plate are uncertain, it is more convenient to utilize a flat-plate decay C_p in conjunction with Eq. (7.6) such that

$$p_p = p_{em}\{1 - \exp[-d^2/4C_p^2 x_p^2]\} \tag{7.7}$$

where $C_p = C_2(1 - H/x_p)$. The plate decay factor C_p may be obtained by fitting measured data of jet impingement on a flat plate to the formulation given in Eq. (7.7). By definition, however, the plate decay factor must always be less than the free-jet decay factor. Experimental results for x_p between 4 and 10 nozzle diameters usually show that C_p is between 80 and 90% of C_2. Figure 7.6 presents some typical test results for a 12.7-mm (0.50-in.) diam jet. For the free jet the best fit to Eq. (7.5) yields a jet decay factor $C_2 = 0.075$. When the jet impinges on a perpendicular flat plate, the best fit of the data to Eq. (7.7) results in a plate decay factor $C_p = 0.066$.

In the case of the impinging jet that occurs in impact modulators, the dividing streamsurface is generally not flat. If we assume that the streamsurface behaves like a solid plate, then we need to recognize the effects of plate curvature on centerline decay. At the present time there is no data available on the influence of plate shape. However, we may still obtain some idea of the effects by interpolation and extrapolation of free-jet and flat-plate data. Figure 7.7 shows some hypothetical centerline decay curves for curved plates. When the plate curvature is convex, the centerline decay should lie between the free-jet and flat-plate

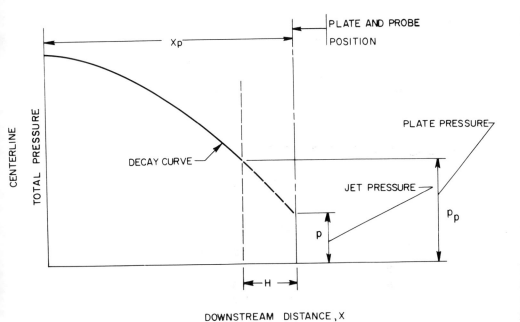

Fig. 7.5 Schematic representation of total pressure recovered in free and impinging jets.

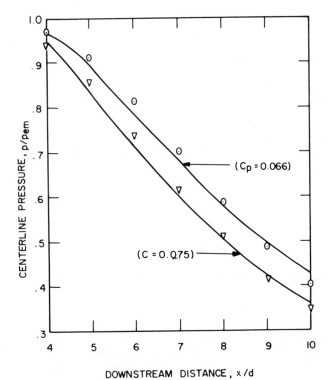

Fig. 7.6 Centerline total pressure on free and impinging jets: $p_{cm} = 7.5$ in. H_2O, $d = 1.27$ cm (0.500 in.), ⊙ impinging jet, ▽ free jet.

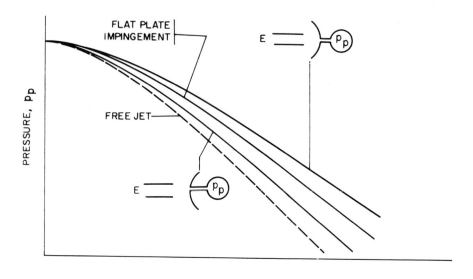

Fig. 7.7 Hypothetical representation of pressure distribution (centerline) on curved surfaces.

decays. When the curvature is concave, the centerline pressures exceed those measured on a flat plate. This seems reasonable because the impingement distance H increases for the concave surface.

Since the centerline pressure distribution depends on streamsurface shape and this shape, in turn, is continually changing as it moves through the modulator operating range, Katz and Reid [7] used an iterative procedure to determine the balance point. For the purpose of demonstrating some typical modulator performance in a straightforward manner, we assume here that the streamsurface remains flat throughout. The consequences of this simplification are discussed in Section 7.5 of this chapter. At the balance point then $p_p = p_s$ and the equation corresponding to the functional relation in Eq. (7.2) can be obtained from Eq. (7.7) as

$$p_s = p_{em}\{1 - \exp[-d^2/4C_p^2(L-z)^2]\} \tag{7.8}$$

7.3 THE EFFECTS OF CONTROL FLOWS ON THE PLATE DECAY FACTOR

The influence of relatively small annular and transverse auxiliary flows on the emitter jet is basic to the analysis of impacting amplifier performance. Equation (7.3) shows that the plate decay factor is a function of control and emitter pressure. Each type of impact modulator (emitter, annular, and transverse) has its

7.3 The Effects of Control Flows on the Plate Decay Factor

own control method. Since more than one control method is never used in the same device, Eq. (7.3) may be conveniently separated into three equations:

$$[C_p]_e = C_p[p_{em}] \qquad \text{emitter modulation} \qquad (7.9a)$$

$$[C_p]_a = C_p[p_{em}, p_{ca}] \qquad \text{annular modulation} \qquad (7.9b)$$

$$[C_p]_t = C_p[p_{em}, p_{ct}] \qquad \text{transverse modulation} \qquad (7.9c)$$

where the subscripts e, a, and t refer to emitter, annular, and transverse modulation, respectively.

The decay factor of a jet is extremely sensitive to the nozzle and supply chamber configuration. Turbulence levels in the supply chamber exert a particularly strong influence on the jet decay factor. Large supply turbulence intensity produces a jet with a large spread and a large decay of centerline pressure. This means that the computed decay factor is significantly more than it would be in a setup with less turbulence. In this section some typical experimental results taken from Katz [6] are given for each modulation method. These results are not meant to be used in the design of impact modulators but rather as an indication of the general effects of control pressure signals.

7.3.1 Effect of Emitter Pressure

The emitter nozzles in the typical cases considered here have diameters of 3.18 mm (0.125 in.) and 12.70 mm (0.500 in.). Figure 7.8 shows the configurations of the nozzles and supply chambers. The smaller nozzle (Fig. 7.8a) has a 12.7-mm (0.500-in.) diameter plenum chamber and a supply tube perpendicular to the nozzle axis. The larger nozzle (Fig. 7.8b) is supplied through an in-line tube at the end of a 44.5-mm (1.75-in.) diameter chamber. This nozzle has a removable screen in the plenum to make the flow more uniform. The emitter pressure is measured by static pressure taps (not shown) in the chamber adjacent to the nozzles.

Figure 7.9 shows the effect of emitter pressure changes on the flat-plate decay factor. The decay factors were computed from Eq. (7.7) using centerline pressure measurements that were made on a flat plate normal to the nozzle axis and at various downstream distances. The results show that the small nozzle has a greater decay factor than the larger nozzle. This signifies that more turbulence exists in the smaller nozzle chamber. For the particular two nozzles under discussion, the plate decay factor at first increases slightly with increasing emitter pressure but eventually decreases. In an actual emitter modulator the emitter pressure need only be changed by 1–5% to sweep through the entire operating range. It appears that emitter momentum change is the major modulating factor and that we may assume that the plate decay factor for emitter modulation $[(C_p)_e]$ is a constant.

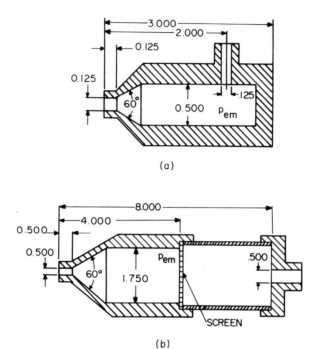

Fig. 7.8 Experimental nozzles: (a) 0.125-in. nozzle, (b) 0.500-in. nozzle.

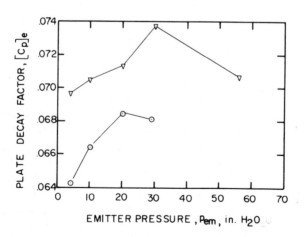

Fig. 7.9 The effect of emitter pressure on plate decay factor: $\triangledown$, $d = 0.125$ in.; $\odot$, $d = 0.500$ in.

7.3.2 Effect of Annular Control Pressure

Figure 7.10a shows the annular jet assembly used to demonstrate some typical effects of annular control. The assembly consists of an orifice plate, an auxiliary housing, a primary nozzle, and a primary housing. As annular control pressure increases from ambient pressure the turbulence intensity decreases in the downstream jet up to a point and thereafter increases. The decreasing region occurs because the velocity gradient at the edge of the jet is becoming less steep due to the supply of a small amount of annular control flow. Ultimately a jet is created, however, that receives considerable flow from both the primary plenum and the perpendicular auxiliary plenum. When this happens the plenum conditions are more turbulent and as a result the turbulent intensities in the jet are increased.

Figure 7.11 shows the effects of annular control on the plate decay factor. The abscissa, in normalized form, is the annular force ratio $p_{ca} A_{ca}/p_{em} A_e$, where $A_e = \pi d^2/4$, $A_{ca} = \pi dS$, and the clearance S is shown in Fig. 7.10. The plate decay factor decreases until the force ratio is about 0.08. This force ratio

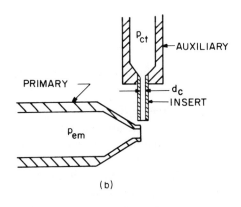

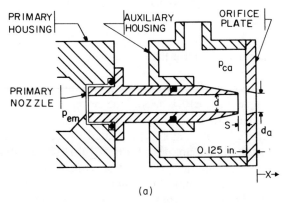

Fig. 7.10 Experimental arrangement for auxiliary flows: (a) annular, (b) transverse.

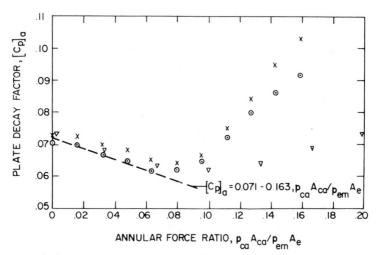

Fig. 7.11 The effect of annular control on plate decay factor: ⊙, $d_a = 0.550$, $S = 0.125$; ×, $d_a = 0.550$, $S = 0.250$; ▽, $d_a = 0.600$ in., $S = 0.125$ in.

corresponds to the minimum jet turbulence intensity. In the region of decreasing plate decay factor the empirical counterpart of Eq. (7.9b) is

$$C_{pa} = 0.071 + K_a(p_{ca} A_{ca}/p_{em} A_e) \tag{7.10}$$

where $K_a = -0.163$.

This equation is shown as the dashed line on Fig. 7.11. Since the plate decay factor versus annular force ratio relation has both decreasing and increasing regions, the annular impact modulator can have either positive or negative pressure gain. That is, in the low control pressure region the application of control tends to move the dividing streamsurface toward the source. In the high control pressure region an increase in control pressure has the opposite effect. This phenomenon may be especially valuable in the generation of special functions.

7.3.3 Effect of Transverse Control Pressure

To demonstrate the typical action of transverse control the experimental test arrangement shown in Fig. 7.10 is used. The primary (emitter) and auxiliary nozzles are perpendicular to each other and their axes are in the same horizontal plane. The transverse control plenum is fabricated to accommodate either 6.35- or 12.70-mm diameter control nozzles (d_c Fig. 7.10b). The emitter nozzle is the one used previously in the emitter pressure tests (Fig. 7.8b).

In an early description of impact modulator operation, Kirshner [8] states that the interaction of the transverse control and emitter jets causes an increase in emitter jet turbulence. As a result the emitter jet spreads out more as it moves

7.3 The Effects of Control Flows on the Plate Decay Factor

contacted by the control. As the emitter jet moves downstream the increase in turbulence intensity propagates throughout the jet. Within four nozzle diameters downstream the emitter jet has larger turbulence intensities at every radial position.

Figure 7.12 is a plot of plate decay factor versus transverse control force ratio $p_{ct} A_{ct}/p_{em} A_e$, where $A_{ct} = \pi d_c^2/4$ and d_c is shown in Fig. 7.10b. There is a considerable, almost linear, increase in plate decay factor with transverse control signal. For modulators with high pressure gain, a large slope is essential. A linear representation of the data leads to an empirical expression of the form

$$C_{pt} = K_t(p_{ct} A_{ct}/p_{em} A_e) + C_p \quad (7.11)$$

where K_t depends on the nozzles and can vary widely. In the test results shown in Fig. 7.12, $K_t = 0.332$ when the diameter ratio $d_c/d = 1.0$, and $K_t = 0.494$ when $d_c/d = 0.5$. The properties of the small control jet are used more effectively in modulating the emitter jet. To illustrate the variability of K_t, other experiments made with two nozzles (3.15-mm diameter) of the type shown in Fig. 7.8a yielded values of 0.660. This is not really surprising since these nozzles are known to produce more turbulence (see Fig. 7.9). The performance of the more turbulent nozzle and plenum geometry emphasizes that the deliberate construction of highly turbulent control arrangements can lead to amplifiers with increased gain.

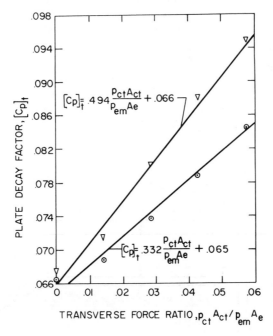

Fig. 7.12 The relation between transverse control and plate decay factor: ▽, $d_c = 0.250$; ⊙, $d_c = 0.500$ in.

downstream. Subsequent experiments [6, 9] confirm this assertion. Initially, at the emitter nozzle exit, the turbulence intensities in the transverse control jet increase the turbulence intensities only in the portion of the emitter jet that is

One further facet of transverse control deserves separate mention. This concerns the deflection of the emitter jet by the transverse control jet. To determine this effect, Katz [6] simulated the jet deflection mechanically in the absence of control flow. This was accomplished by making small angular displacements in the emitter jet axis while the source remained fixed. The results showed that, for angular displacements of the order of magnitude that are caused by the transverse control, there was no source flow modulation. This means that the performance of the transverse impact modulator depends only on the turbulence increase and receives no benefit from jet deflection.

7.4 SOURCE FLOW MODULATION

The source flow modulation process is analogous to the flow phenomenon in the well-known back-pressure sensor (Fig. 7.13). In the sensor the movement of a flat plate in proximity to a pressurized nozzle alters the static pressure near the nozzle exit. The impact modulator has a fluid streamsurface in place of the plate, but the general operating principle is the same.

Kirshner [10] derives a general equation for the pressures at various positions along a converging–diverging source nozzle in the back-pressure sensor. For the special case of a constant diameter source nozzle this equation reduces to

$$p_o/p_s = 1 - 16c_d^2(z/d)^2 \qquad (7.12)$$

where $z/d < 1/4c_d$ and c_d is the discharge coefficient for the annular gap flow.

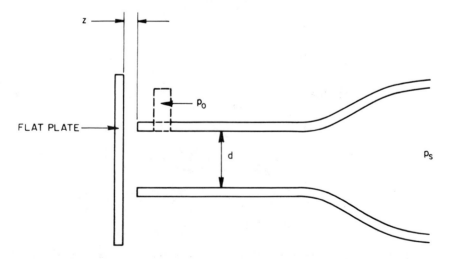

Fig. 7.13 Source flow modulation by flapper nozzle (back-pressure sensor).

7.5 Impact Modulator Pressure Gain

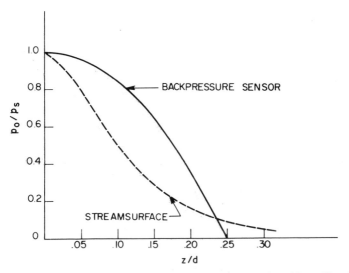

Fig. 7.14 The relation between output pressure and position of barrier.

The solid line in Fig. 7.14 shows the normalized relation between the output (or static) pressure and the plate distance when $c_d = 1.0$. The dashed line represents experimental data from a fluid streamsurface [6]. Almost throughout the entire modulating range, the plate yields higher output pressures than the streamsurface at the same spacing. The reason for this is that the streamsurface has velocity perpendicular to the nozzle whereas the plate does not. As a result the streamsurface offers less resistance to flow. For the purposes of demonstrating the impact modulator performance we assume that Eq. (7.12) with $c_d = 1.0$ is an adequate expression for the functional relation required in Eq. (7.4).

7.5 IMPACT MODULATOR PRESSURE GAIN

The pressure gain of a fluid amplifier (G_p) is defined as

$$G_p \equiv dp_o/dp_c \tag{7.13}$$

In the case of impact modulators the control pressure signal p_c is p_{em} for emitter modulation, p_{ca} for annular modulation, and p_{ct} for transverse modulation. With the above definition and the pressure balance [Eq. (7.8)], the blocked-output source flow modulation [Eq. (7.12)], and a control pressure-decay factor relation [Eqs. (7.10) and (7.11)], the blocked-output pressure gain of impact modulators can be formulated. The following sections demonstrate the derivation of expressions for the pressure gain of emitter, annular, and transverse impact modulators.

7.5.1 Emitter Modulation Pressure Gain

The jet experiments (Fig. 7.9) indicate that the plate decay factor does not change appreciably when the emitter pressure changes. Thus, if the plate decay factor is assumed constant, the emitter modulator pressure gain G_{pe} may be expressed as

$$G_{pe} = \left[\frac{d(p_o/p_s)}{d(z/d)}\right]\left[\frac{d(z/d)}{d(p_{em}/p_s)}\right] \quad (7.14)$$

The first bracketed quantity in Eq. (7.14) originates from Eq. (7.12) and is

$$\frac{d(p_o/p_s)}{d(z/d)} = -32(z/d) \quad (7.15)$$

The second bracketed quantity in Eq. (7.14) comes from the derivative of Eq. (7.8), so that

$$\frac{d(z/d)}{d(p_{em}/p_s)} = -\frac{(1-Y)^2[2C_p{}^2(L/d-z/d)^3]}{Y} \quad (7.16)$$

where $Y = \exp[-1/(4C_p{}^2[L/d-z/d]^2)]$. The substitution of Eqs. (7.15) and (7.16) into Eq. (7.14) yields

$$G_{pe} = 64(z/d)(1-Y)^2[C_p{}^2(L/d-z/d)^3]/Y \quad (7.17)$$

The emitter pressure gain is a function of streamsurface position (z/d), nozzle spacing (L/d), and plate decay factor C_p. With the streamsurface at midposition $(z/d = 0.125)$, Fig. 7.15 shows the emitter pressure gain versus nozzle spacing for various values of plate decay factor. At all values of decay factor the emitter pressure gain increases monotonically as the nozzle spacing decreases. The increase is due to the flattening out of the jet centerline pressure distribution in proximity to the emitter nozzle. Theoretically the emitter pressure gain increases without bound as the spacing decreases. However, the centerline pressure distribution represents only a time average value of pressure. The turbulent fluctuations that are present in the emitter jet result in unstable performance at small nozzle spacings. In practical devices the nozzle spacings always exceed 4.5 nozzle diameters.

The emitter pressure gain also depends on the value of the plate decay factor. When the decay factor is large the balance point is less sensitive to emitter pressure changes. As a result, jets with more turbulence and larger decay factors produce emitter modulators with lower pressure gains.

The results of Eq. (7.17) (Fig. 7.15) demonstrate the way in which the important parameters $(L/d, z/d,$ and $C_p)$ influence the emitter modulator pressure gain. However, the exact magnitudes of emitter pressure gain may differ considerably from those presented here. This occurs because of the simplifying assumptions made in the derivation. The most questionable assumption is the

7.5 Impact Modulator Pressure Gain

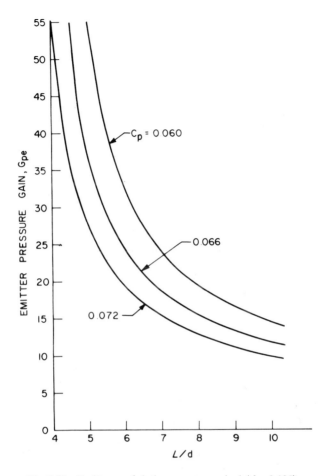

Fig. 7.15 Emitter modulation pressure gain ($z/d = 0.125$).

representation of the dividing streamsurface by a flat plate. A more complete treatment of the modulator [6, 7] gives pressure gain values that are 30–50% of those given here, but the relative dependency of the pertinent factors is about the same.

7.5.2 Annular and Transverse Pressure Gain

For the annular and transverse impact modulators the basic operating mechanism is the control pressure effect on the plate decay factor. In these modulators the emitter pressure is fixed at a constant value that puts the streamsurface either at the cutoff or the beginning of flow modulation positions, depending on whether the control is transverse or annular.

A convenient derivative chain for determining the annular and transverse pressure gains is

$$G_p = \left[\frac{d(p_o/p_s)}{d(z/d)}\right]\left[\frac{d(z/d)}{dC_p}\right]\left[\frac{dC_p}{d(p_c/p_s)}\right] \tag{7.18}$$

The first bracketed term is the one already derived in Eq. (7.15) for the emitter modulator.

To obtain the second bracketed term requires some manipulation of Eq. (7.8). The procedure is to find the emitter pressure that is necessary before control pressure is applied. Then, by maintaining this emitter pressure for the modulating condition, Eq. (7.8) yields

$$(C_p)_{a,t}[L/d - z/d] = C_p[L/d - (z_i/d)] \tag{7.19}$$

where $(C_p)_{a,t}$ is the plate decay factor for annular and transverse control and (z_i/d) is the initial streamsurface position. For annular control $(z_i/d) = 0.250$, and for transverse control $(z_i/d) = 0$. Thus the second bracketed term in Eq. (7.18) is

$$\frac{d(z/d)}{dC_{pa}} = \frac{C_p}{C_{pa}^2}\left[\frac{L}{d} - 0.25\right] \quad \text{(annular)} \tag{7.20a}$$

$$\frac{d(z/d)}{dC_{pt}} = \frac{C_p}{C_{pt}^2}(L/d) \quad \text{(transverse)} \tag{7.20b}$$

The third bracketed term in Eq. (7.18) follows directly from Eqs. (7.10) and (7.11) as

$$\frac{d(C_{pa})}{d(p_{ca}/p_s)} = K_a\left(\frac{A_{ca}}{A_e}\right)\left(\frac{p_s}{p_{em}}\right) \quad \text{(annular)} \tag{7.21a}$$

$$\frac{d(C_{pt})}{d(p_{ct}/p_s)} = K_t\left(\frac{A_{ct}}{A_e}\right)\left(\frac{p_s}{p_{em}}\right) \quad \text{(transverse)} \tag{7.21b}$$

Now the combination of Eqs. (7.8), (7.15), (7.18), (7.20), and (7.21) yields expressions for the annular pressure gain G_{pa} and the transverse pressure gain G_{pt} as

$$G_{pa} = -32(z/d)(K_a)\frac{C_p}{C_{pa}^2}(L/d - 0.25)\frac{A_{ca}}{A_e}(1 - Y_a) \tag{7.22a}$$

$$G_{pt} = -32(z/d)(K_t)\frac{C_p}{C_{pt}^2}(L/d)\frac{A_{ct}}{A_e}(1 - Y_t) \tag{7.22b}$$

where

$$Y_a = \exp[-1/(4C_p^2(L/d - 0.25)^2)]$$
$$Y_t = \exp[-1/(4C_p^2(L/d)^2)]$$

The pressure gains given in Eq. (7.22) are shown plotted versus nozzle spacing in Fig. 7.16 for the case where the control-to-emitter area ratio is unity, $z/d = 0.125$, $K_a = -0.163$, and $K_t = 0.332$. Due to the sign of the control factors

7.5 Impact Modulator Pressure Gain

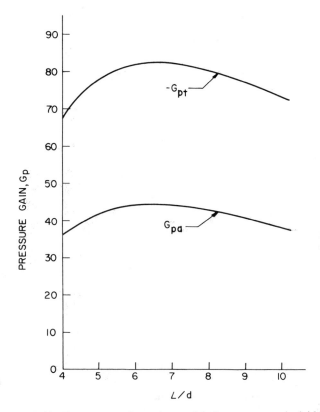

Fig. 7.16 Transverse and annular modulation pressure gain ($z/d = 0.125$).

(K_a and K_t) the annular pressure gain is positive and the transverse pressure gain is negative. Both gains maximize at a nozzle spacing of about 6.5 nozzle diameters. However, the maxima are rather flat. Lower power consumption can be obtained by decreasing the nozzle spacing below the optimum spacing with only a small sacrifice in pressure gain. Usually practical devices have nozzle spacings between 4.5 and 6.5 nozzle diameters.

Here again, though, the results are somewhat qualitative. The flat-plate representation of the streamsurface leads to errors in both the source flow modulation and the balance position. With regards to the source modulation the output pressure due to a flat plate is generally larger than it would be for a streamsurface. The shape of the input–output pressure relation would, therefore, be quite different. However, the pressure gains presented here are made at the midposition ($z/d = 0.125$) where the slopes of the output pressure–distance relations (Fig. 7.14) are about the same for both the flat plate and the streamsurface. Thus the main discrepancy of the flat-plate assumption in describing pressure gain comes from errors in the determination of the balance point.

PROBLEMS

7.1 Two circular nozzles 1.0 mm in diameter are directed towards each other along a common axis. The nozzles are 10.0 mm apart. The pressure on one nozzle is 20 kN/m² and the resulting jet issuing from this nozzle has a jet spread factor (measured on a flat plate) of 0.070.

(a) What is the pressure on the other nozzle if the centerline balance point is 0.25 mm from the nozzle face (beginning of flow modulation)?
(b) What is the pressure on this nozzle if its flow is cut off by the jet?
(c) Repeat parts (a) and (b) if the jet spread factor is 0.060.

7.2 A jet from a 2.0-mm diameter nozzle is stagnated on a flat plate that is perpendicular to the jet and 12.0 mm downstream. The stagnation pressure of the jet on the plate is measured through a 0.5-mm-pressure tap hole in the plate. The tap is aligned with the jet axis.

(a) What would happen to the pressure recovered by the tap if a circular cylinder of 4.0-mm inside diameter and 4.0-mm long were fastened to the front of the plate so that the axis of the cylinder lined up with the jet axis?
(b) What would happen to the pressure recovered if the cylinder length was increased to 6.0 mm.

7.3 A jet stagnates on a perpendicular plate 10 nozzle diameters away. There is a pressure tap in the plate which lines up with the jet axis. The jet is modulated by a transverse jet at pressure p_c, so that the jet plate decay factor may be expressed as

$$C_p = 0.070 + k p_c / p_{em}$$

where p_{em} is the jet pressure and k is a constant. Show that the change in plate pressure is related to a change in control pressure by

$$dp_p/dp_c = -8.75k$$

7.4 Two 1.0-mm diameter nozzles oppose each other along a common axis. The nozzles are 5.0 mm apart. There is a static pressure tap p_o near the exit of one nozzle. When a flat plate is brought close to this nozzle the static pressure is measured approximately as

$$p_o/p_s = 1 - 5z/d \quad (0 \le z/d \le 0.3)$$

where p_s is the pressure supply to the source nozzle and z/d is the distance of the plate from the nozzle. With the plate removed a jet of fluid from the other nozzle is used to modulate the output pressure p_o. If C_p of the emitter nozzle is 0.080, find the emitter modulator pressure gain at $z/d = 0.1$. What is the pressure gain if C_p is changed to 0.065? (Assume that the stagnated emitter jet acts like a flat plate.)

NOMENCLATURE

A_c Control area, m^2
A_e Emitter area, m^2
C_2 Free-jet decay factor
c_d Discharge coefficient
C_p Impinging jet decay factor on flat plate
d Diameter, m
d_c Control diameter, m
G_p Pressure gain
H Impingement distance, m
K_a Annular control constant
K_t Transverse control constant
L Emitter–source spacing, m
p Jet centerline pressure, kN/m^2
p_c Control pressure, kN/m^2
p_{em} Emitter pressure, kN/m^2
p_o Output pressure, kN/m^2
p_p Impinging jet pressure on flat plate, kN/m^2
p_s Source pressure, kN/m^2
x Axial distance from emitter nozzle, m
x_p Axial distance from emitter nozzle to measuring position, m
z Distance from source to balance point, m

REFERENCES

1. B. G. Bjornsen, The impact modulator. *Proc. HDL Fluid Amplification Symp.* **11**, 5–32 (1964).
2. T. J. Lechner and P. H. Sorenson, Some properties and applications of direct and transverse impact modulators. *Proc. HDL Fluid Amplification Symp.* **11**, 33–59 (1964).
3. J. F. Blackburn, G. Reethof, and J. L. Shearer, "Fluid Power Control." Technology Press of M.I.T., Cambridge, Massachusetts and Wiley, New York, 1960.
4. S. Katz, Pressure gain analysis of an impacting jet amplifier. *Advances in Fluidics.* ASME Publ. (May 1967).
5. A. Le Clerc, Deviation d'un jet liquide par une plaque normale a son axe. *La Houille Blanche* **5**, 816 (1950).
6. S. Katz, "A Static Model of Direct and Transverse Impact Modulators." Ph.D. Thesis, Oklahoma State Univ., July 1970.
7. S. Katz and K. N. Reid, A generalized static model for fluid impact modulators. *Trans. ASME J. Basic Eng. Ser. D* **93**, No. 2 (1971).
8. J. M. Kirshner, "Fluidics—1. Basic Principles." Harry Diamond Lab. TR 1418, November 1968.
9. K. N. Reid and S. Katz, "Characterization of Free and Impinging Axisymmetric Jets With and Without Auxiliary Flows." ASME paper 70-WA/Flcs-6, December 1970.
10. J. M. Kirshner, A survey of fluid sensors, Part II. *HDL Fluidics State-of-the-Art Symp.*, Sept. 30–Oct. 3, 1974.

Chapter 8

THE VORTEX TRIODE

8.1 HISTORICAL INTRODUCTION

Sometime in 1927 or 1928, Dieter Thoma conceived of the vortex diode [1] depicted in Fig. 8.1 and suggested it as a thesis topic to Richard Heim. As a result Heim [2] made the earliest known investigations of the vortex diode (published in 1929).

Thoma's concept is illustrated in Fig. 8.1 where flow entering in the reverse direction R (the hard or high resistance direction) spirals in from the outside toward the axis. As the fluid particles spiral inward, their tangential velocity u_θ increases due to the conservation of momentum. Since their radius of curvature r simultaneously decreases, the centrifugal force u_θ^2/r increases rapidly as the particles move toward the axis. This force tends to make flow in this direction difficult and therefore results in a throttling action on the flow. Flow from the forward (easy) direction F, however, sees only conventional resistance so that the device exhibits diodicity.

Diodicity is usually defined in terms of the loss coefficients in the two directions, where the loss coefficient associated with a measured pressure drop Δp and a measured average velocity $\bar{u}$ is given by

$$K_L = \Delta p/(\rho \bar{u}^2/2) \qquad (8.1)$$

The diodicity D_i is defined as

$$D_i = K_{LR}/K_{LF} \qquad (8.2)$$

8.1 Historical Introduction

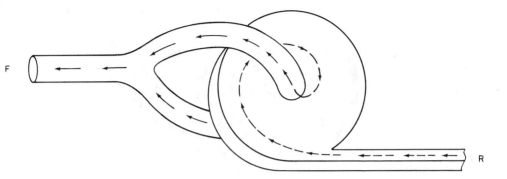

Fig. 8.1 The vortex diode.

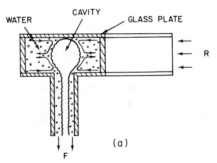

Fig. 8.2 Vortex diode flow visualization: (a) flow pattern, (b) chamber shape, and (c) drain region.

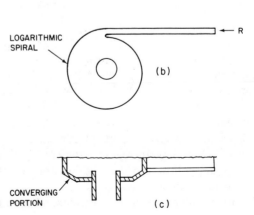

where K_{LR} and K_{LF} are the loss coefficients in the reverse and forward directions, respectively.

Heim's experiments were carried out on vortex chambers having only one axial outlet (Fig. 8.2a). The outer curved surface of the chamber he used was a logarithmic spiral (Fig. 8.2b) and the configuration of that side of the chamber containing the axial duct was varied by contouring part of it so as to converge (Fig. 8.2c) toward the axial tube. The amount of projection of the tube into the chamber (Fig. 8.2c) was also varied. In a number of cases, guide vanes were installed within the axial duct.

The greatest diodicity (43.3) was obtained by Heim for the larger of the two chambers tested (about 100 mm in diameter) containing a contoured face and guide vanes.

Heim also made flow visualization studies using water. He observed that the water moved outward from the center resulting in an air bubble near the axis (Fig. 8.2a) and that the flow moved toward the axis near the end walls but away from the axis in the vicinity of the midplane. He recognized that these secondary eddies were due to the fact that the tangential velocity was relatively low in the boundary layers near the end walls so that the force acting to retard the particles was much less there than in the region of the midplane.

The first use of auxiliary jets to change the circulation within a vortex device seems to have been by Kadosch and LeFoll [3] who describe a vortex device in which a control jet is used to attach a jet to a curved wall leading into the circular housing. They also describe a vortex device with internal jets that can be turned on in the same or opposite direction as the main flow so as to increase or decrease the circulation when they are turned on. This yields a three-state device.

The vortex amplifier (triode) was conceived by Bowles and Horton in 1961 and a patent was issued to them in 1966 [4]; however, Gray and Stern [5] state that a vortex triode had been investigated earlier by J. M. Rhoades and D. E. Cain. Gray and Stern reference an unpublished report by Cain.

8.2 BASIC DESCRIPTION OF VORTEX TRIODE

One form of the vortex triode is illustrated in Fig. 8.3. The main or power flow is supplied by p_s. Control flow from a source p_c modulates the supply flow. In the absence of control flow, the power flow moves in essentially a straight line toward the axial drain. The addition of control flow provides tangential momentum and results in the same sort of throttling action that occurs in the vortex diode.

The triode may be separated into three distinct flow regions for purposes of analysis (Fig. 8.3). In region 1 the control flow imparts tangential momentum to the supply flow and the two flows mix together. The mixed flow then enters the vortex chamber (region 2) where the fluid velocity increases. Region 3 leads from the outlet hole in the vortex chamber to the collector.

8.2 Basic Description of Vortex Triode

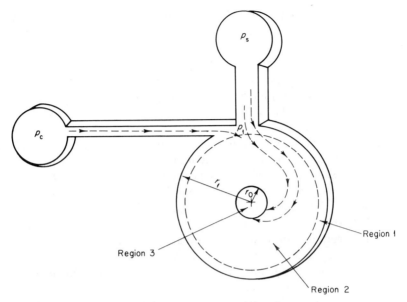

Fig. 8.3 The vortex triode.

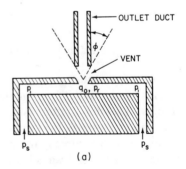

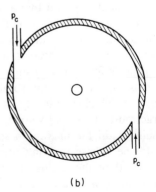

Fig. 8.4 Multiinput vortex triode: (a) supply flow and (b) control flow

Figure 8.4 shows another type of vortex triode. In general the triode may have two (Fig. 8.4b) or more control inputs and the supply flow may be inserted from an inlet parallel to the axis (Fig. 8.4a) or by means of radial inlets.

The completeness of the mixing occurring at the outer wall (region 1) depends on the number of control ports and on the amount of control flow. Flow visualization studies by Mayer [6] show that, for a single port, the mixing is poor at the outer wall and the control flow can be seen moving into the main flow as a distinct entity for an appreciable distance.

When four control inputs are used, the mixing is better, but when the tangential flow is small, sections of tangential flow alternate with sections of radial flow forming a cell-like structure. At higher tangential flows the mixing improves and the vortex fills the chamber. The vortex triode may have a single outlet as in Fig. 8.4, or a double outlet similar to that shown in Fig. 8.1.

Figure 8.4a shows a vented outlet. This serves two purposes. First, it isolates the amplifier from any downstream loading effects. Second, it allows the flow into the outlet duct to become appreciably lower at cutoff ($q_c = q_o$, $q_s = 0$) than the flow q_o out of the amplifier proper. The reason for this is that as the swirl increases the flow out of the axial hole takes on the shape of a hollow cone with the angle ϕ increasing from zero to a maximum as the angular velocity increases. If the geometry is properly adjusted, almost no flow will enter the outlet duct when the control flow is sufficient to cause cutoff.

8.2.1 Flow Field in Vortex Triode

One of the most important geometric parameters in the vortex triode is the radius of the outlet hole in the vortex chamber with respect to the chamber radius. The flow pattern within the chamber depends to a large extent on this radius ratio. As the outlet radius decreases the viscous losses increase.

Figure 8.2 has already indicated that the tendency for the radial flow to move fastest near the end walls is accentuated as the flow approaches the axis. These flow patterns have been confirmed by a number of measurements (e.g., Savino and Keshock [7]) and by the visualization studies of Wormley [8].

The foregoing studies of the profile show that, for relatively small control flow (and low tangential velocity near the input), the radial velocity is zero at the end walls, rising to a maximum at the midplane, and continues to have a similar shape even for relatively small radii (where the tangential velocity is appreciably increased). When the control flow is sufficiently great, however, the radial profile begins to change radically in shape as it approaches the axis (Fig. 8.5c). For large radii, the radial velocity profile shows a maximum at the center, but for smaller radii the maximum splits into two maxima, one on either side of the midplane. These maxima are close to the wall for radii that are sufficiently smaller than the radius of the chamber r_i. In addition, for sufficiently small radii the flow in the vicinity of the midplane moves away from the axis, as indicated in Fig. 8.5b. The flow then splits as illustrated, the portions on each

8.2 Basic Description of Vortex Triode

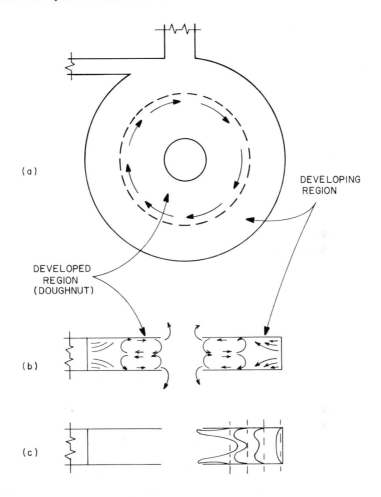

Fig. 8.5 Viscous effects on flow in vortex triode: (a) spinning doughnut, (b) recirculation within doughnut, (c) radial velocity profiles.

side of the midplane moving toward the respective top and bottom plates. This results in a doughnut shape recirculating flow pattern [8] whose axis coincides with the amplifier axis (Figure 8.5a).

The whole doughnut spins about the axis impelled by the tangential velocity. For low swirls the developing region fills the chamber and no doughnut exists. For higher swirls the doughnut forms near the drain region and increases in radius as the swirl increases.

Because of recirculation, the velocity gradient near the top and bottom plates increases as the doughnut radius grows. This effect appreciably increases the viscous losses, which depend on the velocity gradient near the walls and thus shows why a small outlet results in greater losses than a large one.

8.2.2 Performance of Vortex Triode

Figure 8.6 [9] shows the effect of control pressure p_c on the output flow q_o for various supply pressures p_{s1}, p_{s2}, p_{s3} where

$$p_{s3} > p_{s2} > p_{s1}$$

First we note that control flow is not initiated until the control pressure reaches a value large enough to overcome the pressure at the control nozzle exit; increase of the control pressure beyond this point causes the total output flow to decrease until the supply flow is completely cut off.

The curve $q_c = 0$ is obtained if the supply pressure is raised above the output (drain) pressure when the control flow is zero. It is essentially an orifice characteristic for the supply nozzle. Conversely the curve designated as $q_s = 0$ is the orifice characteristic of the control nozzle. The curves in Fig. 8.6 are single valued; the output flow decreases as the control pressure increases. The slope of the curve is a measure of gain. When used as an amplifier, it is desirable that the slope (gain) be reasonably large as well as approximately constant over a significant portion of the curve. The triode may also have bistable characteristics. In this case the curves shown in Fig. 8.6 will have regions of positive slope and the output flow will be a multivalued function.

One of the more useful characteristic curves for the vortex triode is the output flow as a function of control flow q_c. This type of characteristic normalized by dividing by the maximum output flow q (the output flow when the control flow is zero) is illustrated in Fig. 8.7. Figure 8.7a shows the single-valued or proportional amplifier mode of operation. Note that the ordinate and abscissa are plotted on different scales since, at the cutoff point, the control flow equals the output flow.

Figure 8.7b, with its multivalued region, indicates a digital type of operation and a corresponding hysteresis loop. Thus, as the control flow is increased from zero, that portion of the curve from O to B is traced out. A further increase in control flow results in a sudden drop in output flow to point C, and the curve from C to the cutoff point is traced out. At the cutoff point the supply flow is completely cut off so that the output flow consists entirely of control flow. Further increase of control flow beyond this point serves no useful purpose. If now the control flow is decreased toward zero, a new path $CDAO$ will be traced out in the multivalued region. We will discuss in Sections 8.3 and 8.4 the parameters that determine whether the flow characteristic will be single or multivalued.

The vortex triode may serve as either a throttle, an amplifier, or a bistable device.

It may be used as a throttle since the supply flow is decreased rather than diverted by the control flow. A vortex device used in this way is often called a vortex throttle (or vortex valve), and a measure of its performance is the turn-down ratio T_R. The turn-down ratio for incompressible flow is the total flow at cutoff q_{oc} divided into the maximum flow q_{om}.

8.2 Basic Description of Vortex Triode

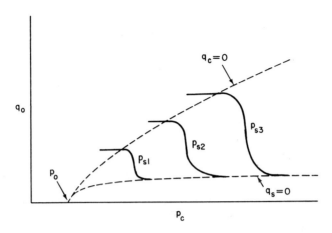

Fig. 8.6 Output flow versus control pressure for vortex triode.

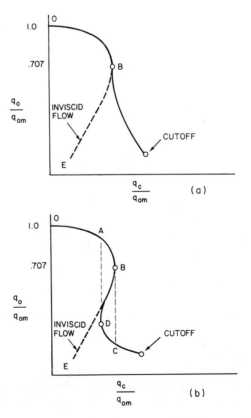

Fig. 8.7 Output flow versus control flow for vortex triode: (a) proportional amplifier mode of operation, (b) digital type of operation.

The control flow at cutoff is designated as q_{cc}. Since at cutoff all of the output flow comes from the control, $q_{cc} = q_{oc}$ and we may express the turn-down ratio as

$$T_R \equiv q_{om}/q_{oc} = q_{om}/q_{cc} \qquad (8.3)$$

The control pressure at which cutoff occurs is designated as p_{cc}.

A second figure of merit for the vortex valve is the ratio of the minimum power output divided by the maximum power output. For incompressible flow this is given by

$$I_g = p_{cc}q_{cc}/p_s q_{om} \qquad (8.4)$$

A similar but slightly differently defined figure of merit for compressible flow is given by Gebhen [10].

The ability of the vortex valve to throttle rather than divert is a distinct advantage over other types of fluidic elements when appreciable power is involved or when power is at a premium.

When used as a proportional amplifier, it is desirable that there be a region of the curve Fig. 8.7a that is reasonably linear and that the slope in this region (the gain) be relatively high. Noise should be low and the bandwidth adequate for its intended use. A disadvantage of the vortex triode, particularly when used as an amplifier, is that the control pressure must be greater than the supply pressure. Furthermore, it is doubtful that the vortex amplifier can ever be improved to obtain the dynamic ranges possible with other fluidic devices.

The geometry of the vortex triode can be designed so as to yield a bistable device. In this mode it can be used either as a bistable throttle or as a logic device.

8.3 ANALYSES OF THE VORTEX TRIODE

The vortex triode has a very complex flow field. The analysis of the triode is made more tractable if the device is separated into three distinct flow regions: (1) the inlet mixing region, (2) the vortex chamber region, and (3) the output region. In this section we will begin with a general development of the analytical forms that represent the three regions. Then to reduce the complexity of the analysis, we will make some specific simplifying assumptions. In each case we will indicate the extent to which the simplified theory agrees with experimental results.

8.3.1 General Analysis

The development presented here deals only with incompressible flow. The compressible flow analysis is more involved algebraically but would follow along the same lines.

Figure 8.8 shows a static equivalent circuit for the vortex triode.

8.3 Analyses of the Vortex Triode

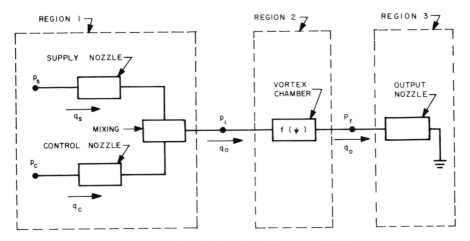

Fig. 8.8 Static equivalent circuit for vortex triode.

Region 1 From the energy equation the incompressible relations between pressure and flow in the supply and control nozzles are

$$q_s = c_{ds} A_s [2(p_s - p_1)/\rho]^{1/2} \tag{8.5a}$$

$$q_c = c_{dc} A_c [2(p_c - p_1)/\rho]^{1/2} \tag{8.5b}$$

where p_1 is the pressure near the outer wall of region 1 (region 1 is of infinitesimal radial thickness); c_{ds} is the discharge coefficient for the supply nozzle, and c_{dc} is the discharge coefficient for the control nozzle. A_s is the area of the supply nozzle. A_c is the area of the control nozzle [the discharge coefficients, c_d, are related to the loss factors K_L used in Chapter 2 by $c_d = (1 + K_L)^{-1/2}$].

It is assumed that the mixing is lossless and that all the mixing occurs in region 1, that is, before entering the main chamber.

The total flow which is also the flow out is then given by

$$q_o = q_s + q_c \tag{8.6a}$$

The initial radial velocity u_{ri} is

$$u_{ri} = q_o/A_i \tag{8.6b}$$

where

$$A_i = 2\pi r_i h \tag{8.6c}$$

where h is the chamber height and r_i is the chamber radius.

The angular momentum flux into the mixing block in region 1 is $\rho r_i q_c^2/(A_c c_{dc})$. The angular momentum flux out of the mixing block is $\rho r_i q_o u_{\theta i}$, where $u_{\theta i}$ is the tangential velocity in region 1.

If we assume a momentum flux loss of the form $u_{\theta i} f_{\tau 1}$, then conservation of angular momentum leads to

$$u_{\theta i} = \frac{q_c^2/(q_o A_c c_{dc})}{(1 + f_{\tau 1})} \tag{8.7}$$

where $f_{\tau 1}$ is a loss factor that tends to reduce the initial tangential velocity.

Region 2 Since vortex triode data are sometimes presented in terms of the flow angle we find it convenient at this time to introduce the flow angle ψ as

$$\tan \psi = u_r(r)/u_\theta(r) \tag{8.8}$$

The flow angle is generally a function of the radius. However, as we shall observe shortly, the angle is independent of radius for inviscid flow in the vortex. From Eqs. (8.6b) and (8.7) the initial flow angle ψ_i is

$$\tan \psi_i = [c_{dc} A_c q_o^2 (1 + f_{\tau 1})]/A_i q_c^2 \tag{8.9}$$

The energy equation from the inlet to outlet in the vortex chamber is

$$p_1 + (\rho/2)(u_{\theta i}^2 + u_{ri}^2) = p_o + (\rho/2)(u_{\theta o}^2 + u_{ro}^2) + f_{\tau 2} \tag{8.10}$$

where p_o is the static pressure at the point in the vortex chamber where $r = r_o$, $u_{\theta o}$ and u_{ro} are the values of the tangential and radial velocity, respectively, at $r = r_o$, and $f_{\tau 2}$ is an energy loss that takes place within the vortex chamber. We may also write Eq. (8.10) in terms of the flow angle so that

$$p_1 + \frac{\rho u_{ri}^2}{2}\left(\frac{1}{\tan^2 \psi_i} + 1\right) = p_o + \frac{\rho u_{ro}^2}{2}\left(\frac{1}{\tan^2 \psi_o} + 1\right) + f_{\tau 2} \tag{8.11}$$

Region 3 At the outlet the tangential component of dynamic pressure does not contribute to the output signal. The output flow depends only on the static pressure and the radial dynamic pressure. If we assume that the distribution of these quantities is uniform and that the flow passes to an ambient pressure, then

$$q_o = N c_{do} A_o (2 p_o/\rho + u_{ro}^2)^{1/2} \tag{8.12}$$

where $A_o = \pi r_o^2$ is the drain area, c_{do} is the discharge coefficient, $N = 1, 2$ is the number of outlets, and it is assumed that the hole area ($N \pi r_o^2$) is less than the drain curtain area $2\pi r_o h$; i.e.,

$$N r_o < 2h$$

We will not discuss the case for which $N r_o > 2h$, in which event $N A_o$ should be set equal to $2\pi r_o h$.

8.3.2 Semi-Inviscid Analysis

For inviscid flow in the vortex chamber both the tangential and radial velocities are inversely proportional to the radial distance from the axis. Thus

$$u_{\theta o}/u_{\theta i} = u_{ro}/u_{ri} = r_i/r_o \tag{8.13}$$

8.3 Analyses of the Vortex Triode

As a consequence, the flow angle ψ remains constant throughout the chamber and is expressed from Eq. (8.9) by

$$\tan \psi = c_{dc} A_c q_o^2 / A_i q_c^2 \tag{8.14}$$

where, for the semi-inviscid analysis, $f_{t1} = f_{t2} = 0$, but the discharge coefficients are retained.

Although the flow angle does not depend on radial position, it does depend on the momentum interaction in the mixing region. For viscous flows, the average (over the chamber height) radial velocity must still increase as the reciprocal of the radial distance from the drain; however, because of frictional losses, the average tangential velocity does not increase as fast as for inviscid flow so that the average angle ψ increases toward 90 degrees as the flow approaches the drain.

The energy equation (8.11) and Eq. (8.13) may be combined to yield

$$p_1 - p_o = \frac{\rho u_{ri}^2}{2} \left(\frac{1}{\tan^2 \psi} + 1 \right) \left[\left(\frac{r_i}{r_o} \right)^2 - 1 \right] \tag{8.15}$$

Our objective now is to obtain a relation between output flow and control flow. We can accomplish this through the substitution of Eqs. (8.5), (8.6), and (8.12)–(8.14) into Eq. (8.15). To make the procedure somewhat easier to follow let us hold off for a moment the use of Eq. (8.14). In this case, Eq. (8.15) reduces to

$$\frac{q_o^2}{2c_N^2 c_{do}^2 A_o^2} = \frac{p_s}{\rho} + \frac{q_c q_o}{c_{ds}^2 A_s^2} - \frac{q_c^2}{2c_{ds}^2 A_s^2} + \frac{q_o^2}{2A_i^2} \left[\frac{1}{\tan^2 \psi} \left(1 - \frac{r_i^2}{r_o^2} \right) \right] \tag{8.16}$$

where

$$c_N^2 = \frac{N^2}{1 + c_{do}^2 N^2 A_o^2 / c_{ds}^2 A_s^2 - c_{do}^2 N^2 A_o^2 / A_i^2}$$

It is useful to use as a normalization factor the maximum value q_{om}, i.e., the output flow when the control flow is zero:

$$q_{om}^2 = 2c_N^2 c_{do}^2 A_o^2 p_s / \rho \tag{8.17}$$

Thus if we let $q_o' = q_o / q_{om}$ and $q_c' = q_c / q_{om}$, Eq. (8.16) may be expressed as

$$q_o'^2 = 1 + \frac{c_N^2 c_{do}^2 A_o^2}{c_{ds}^2 A_s^2} (2 q_o' q_c' - q_c'^2) + \frac{c_N^2 c_{do}^2 A_o^2}{A_i^2} \left[\frac{(1 - r_i^2/r_o^2)}{\tan^2 \psi} \right] q_o'^2 \tag{8.18}$$

Now if A_s/A_o is of the order of 5 or larger, the second term on the right may be neglected in comparison with the third term. Thus, from Eqs. (8.18) and (8.14) we obtain

$$q_o'^2 = 1 - c_G(q_c'^4 / q_o'^2) \tag{8.19}$$

where

$$c_G = -(1 - r_i^2/r_o^2) c_N^2 c_{do}^2 A_o^2 / c_{dc}^2 A_c^2$$

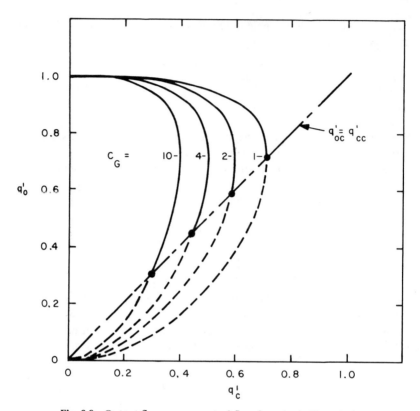

Fig. 8.9 Output flow versus control flow from inviscid analysis.

The solution of Eq. (8.19) by the quadratic formula leads to

$$q_o' = (\sqrt{2}/2)[1 \pm (1 - 4c_G q_c'^4)^{1/2}]^{1/2} \tag{8.20a}$$

Figure 8.9 shows the above relation between control flow and output flow for several values of c_G. The output flow at cutoff q_{oc}' equals the control flow at cutoff q_{cc}'. To determine the extent of the operating region we therefore superimpose the cutoff line ($q_{oc}' = q_{cc}'$) on the figure. The region above the cutoff line (shown solid) is the operating region of the vortex triode. The slope of the curve in the operating region is infinite at a point we shall designate as (q_{of}', q_{cf}') where the values are

$$q_{of}' = q_{cf}'(c_G)^{1/4} = 0.707 \tag{8.20b}$$

Note that the turndown ratio is merely the inverse of q_{oc}'. Thus, from Eq. (8.19) and the cutoff condition we readily find that

$$T_R = 1/q_{oc}' = (1 + c_G)^{1/2} \tag{8.21}$$

8.3 Analyses of the Vortex Triode

This analysis shows why the two types of curves of Fig. 8.7 exist, since for inviscid flow the shape of the characteristic is like that of curve OBE of Fig. 8.7.

Wormley and Richardson's experimental results [11] show that in general the experimental curves (Fig. 8.7) are very close to the inviscid flow analysis for $q_o' \geq 0.707$ under the restrictions that apply to all the amplifier geometries and pressures they examined; that is, for small control flows the inviscid analysis holds (approximately) for all experimental cases considered by them.

The agreement between the inviscid and viscous cases actually depends on the swirl, i.e., the flow angle, and on the ratio r_i/r_o. Most of the data seem to show relatively good agreement for initial flow angles greater than about 20 degrees. Whether the rest of the flow characteristic curve is like that of Fig. 8.7a depends on how great the viscous effects are. If the ratio of exit hole radius r_o to the amplifier radius r_i is greater than about 0.3, the viscous effects are relatively small so that the inviscid flow double-valued characteristic is followed for some values of $q_o' < 0.707$ (Fig. 8.7b). On the other hand, if r_o/r_i is less than 0.15, the viscosity effects are sufficient to cause a curve of the form of Fig. 8.7a to be obtained.

8.3.2.1 Application to the Vortex Diode

The preceding analysis can easily be applied to the vortex diode which is, of course, only a special case of the vortex triode, i.e., one for which the supply flow is zero.

Therefore, if we return to Eqs. (8.6b) and (8.7), in which we let $q_o = q_c$, and then insert the values of u_{ri} and $u_{\theta i}$ thus obtained into Eq. (8.10) together with p_1 from Eq. (8.5b) and p_o from Eq. (8.12), we find

$$q_c^2 = \frac{2p_c}{\rho} \frac{1}{(1/N^2 c_{do}^2 A_o^2) + (r_i^2/r_o^2 A_c^2 c_{dc}^2) - (1/A_i^2)} \tag{8.22}$$

To obtain a good diode it is necessary that the second term in the denominator be appreciably larger than the first term, since in the forward direction the flow will depend on the first term (the outlet orifices).

8.3.3 Semi-Inviscid Analysis in Terms of Flow Angle

Mayer [12] has taken a great deal of vortex triode data in terms of flow angle. Although he has modeled the triode according to compressible flow relations, the correspondence between experiment and the semi-inviscid theory is more readily demonstrated with the incompressible flow relations.

Mayer's method consists of multiplying the orifice equation for the outlet by a coefficient c_M that is assumed to be primarily a function of the initial flow angle ψ_i. Thus

$$q_o^2 = 2c_M^2 c_{do}^2 A_o^2 (p_s/\rho) \tag{8.23a}$$

If Eq. (8.23a) is normalized by Eq. (8.17) the result is

$$q_o' = c_M/c_N \equiv c_M' \tag{8.23b}$$

where c_M' is the normalized parameter used by Mayer and is equivalent to the normalized output flow.

Mayer's experiments were done on amplifiers having the configuration of Fig. 8.4a, for which $A_s \cong A_i$; hence, for r_i/r_o greater than about 4, Eq. (8.18) with Eq. (8.6c) becomes

$$q_o' = \left[\frac{1}{1 + c_N^2 c_{do}^2 r_o^2/4(\tan^2 \psi)h^2} \right]^{1/2} \tag{8.24}$$

Mayer's [12] data show the relation between c_M' ($=q_o'$) and ψ_i. These data indicate very little dependence on control pressure, temperature, or control port size. In this connection, then, the data agree well with the semi-inviscid theory of Eq. (8.24). On a quantitative basis, however, the agreement is not as good. Figure 8.10 compares Mayer's experimental results ($r_o/h = 0.25$) with the theory of Eq. (8.24) ($c_N = 1$, $c_{do} = 1.0$). When $q_o' > 0.707$, the inviscid theory is always within 3.0% of the experimental data. For values of q_o' less than 0.707, the discrepancy grows progressively larger.

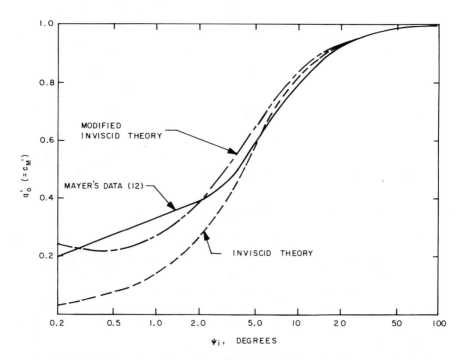

Fig. 8.10 Comparison of experimental and analytical results for vortex triode.

8.3 Analyses of the Vortex Triode

Mayer's data also show that the c_M' versus ψ_i relation depends upon the chamber height h, the number of outlet ports N, and the outlet radius r_o. The same trends are demonstrated in Eq. (8.24).

Although Mayer indicates that other types of characteristics can be calculated once $q_o'(\psi)$ is known, our efforts to calculate the q_o' vs q_c' curve from Mayer's curves on the assumption that Eq. (8.14) could be used to relate q_c' to ψ_i gave poor results.

8.3.4 Modified Semi-Inviscid Analysis

We present here a modification of Eq. (8.24) to account somewhat for viscosity in the vortex chamber (region 2).

Since the viscosity affects the tangential velocity more than any other variable, we will attempt to improve the inviscid equation by modifying the tangential velocity.

First we note that Eq. (8.24) is actually in terms of the flow angle at output ψ_o; i.e.,

$$q_o' = \left[\frac{1}{1 + c_N^2 c_{do}^2 r_o^2 / 4h^2 \tan^2 \psi_o}\right]^{1/2} \qquad (8.25)$$

Since we are primarily interested in the flow angle at the input, we seek a relation between ψ_i and ψ_o to replace the inviscid relation ($\psi_i = \psi_o$).

Bichara and Orner [13] derive a relation between $u_{\theta i}$ (the tangential velocity at the input) and $u_{\theta o}$ (the tangential velocity at the output). From this we may obtain the link between ψ_o and ψ_i. Bichara and Orner assume that in the vortex chamber the radial losses are negligible and that the tangential velocity profile is not a function of the axial direction z. Following them we thus write

$$u_r \, du_\theta/dr + u_r u_\theta/r = (1/\rho) \, \partial \tau_{\theta 2}/\partial z \qquad (8.26)$$

where $\tau_{\theta 2}$ is the tangential shear stress in region 2.

Now assume that u_θ is not a function of z, except for the discontinuity at $z = \pm h/2$ where $u_\theta = 0$, and that the wall shear stress $\tau_{\theta 2w} = f_2 \rho u_\theta^2$ (where f_2 is a friction factor). If Eq. (8.26) is integrated with respect to z from $-h/2$ to $+h/2$ and we make use of the continuity equation ($u_r r = u_{ri} r_i$) the result is

$$d(ru_\theta)/dr = (2f_2/u_{ri} r_i h)(ru_\theta)^2 \qquad (8.27)$$

If the friction factor f_2 does not depend on r, then we may integrate Eq. (8.27) to obtain

$$u_{\theta o} = \frac{u_{\theta i} r_i / r_o}{1 + \alpha_B(1 - r_o/r_i)} \qquad (8.28)$$

where

$$\alpha_B = 2f_2 r_i u_{\theta i}/hu_{ri}$$

The continuity equation can now be used again to put Eq. (8.28) in terms of the flow angles. Thus

$$\tan \psi_o = \tan \psi_i + (f_2 r_i/h)(1 - r_o/r_i) \tag{8.29}$$

For small flow angles the output flow angle is always greater than the input flow angle by the fixed amount of the friction term.

Bichara and Orner [13] give values of f_2 lying between about 0.005 and 0.025 depending on tangential Reynolds number at cutoff.

Inserting Eq. (8.29) into Eq. (8.25)

$$q_o' = \left[\frac{1}{1 + c_N^2 c_{do}^2 r_o^2/4h^2 [\tan \psi_i + (f_2 r_i/h)(1 - r_o/r_i)]^2} \right]^{1/2} \tag{8.30}$$

Figure 8.10 also shows the modified theory for Eq. (8.30) for the case of $r_i/h = 2$, $f_2 = 0.01$, and $c_{do} = 1$. The modified theory shows better agreement with the data at the small flow angles; however, using Eq. (8.14) it is still not possible to relate this equation to the output flow versus control flow with any accuracy.

8.3.5 Analytical Methods Including Viscous Effects

Wormley's visualization studies led him to model the vortex chamber as consisting of a developing region and a developed region (Fig. 8.5). His analysis gives the pressure distribution within the chamber and shows how the circulation decays [8]. He did not, however, relate his results to the amplifier characteristics.

Building partially on Wormley's results but using some simplifying assumptions, Bichara and Orner [13] were able to obtain characteristic curves primarily on the basis of theory, although empirical relations are necessary for their friction factor f_2 and for the various discharge coefficients. In their analysis the amplifier is also divided into the three regions we have described in Section 8.3.1. We will develop the viscous effects in each region separately.

Region 1 Region 1 is modeled in a manner similar to that for the inviscid case except that the momentum dissipated by the circular cylindrical wall is taken into account. The angular momentum equation is

$$\rho r_i q_c^2 / A_c c_{dc} = \rho r_i q_o u_{\theta i} + 2\pi r_i^2 h \tau_{\theta 1} \tag{8.31}$$

where $2\pi r_i^2 h \tau_{\theta 1}$ is the rate of dissipation of angular momentum by the circular wall and $\tau_{\theta 1}$ is the circular wall shear stress.

By analogy with turbulent flow over a flat plate, it is assumed that

$$\tau_{\theta 1} = \frac{0.0225 \rho u_{\theta i}^2}{(u_{\theta i} \delta_1 \rho/\mu)^{1/4}}$$

and that

$$\delta_1 = \frac{r_i}{(u_{\theta i} \rho r_i/\mu)^{1/5}}$$

8.3 Analyses of the Vortex Triode

so that

$$\tau_{\theta 1} = \frac{0.0225 \rho u_{\theta i}^2}{(u_{\theta i} \rho r_i / \mu)^{1/5}} \tag{8.32}$$

Then, from Eqs. (8.31) and (8.32) and using $q_o = 2\pi u_{ro} h$,

$$u_{\theta i} = \frac{q_c^2 / q_o A_c c_{dc}}{1 + f_{\tau 1}} \tag{8.33a}$$

where

$$f_{\tau 1} = 0.0225 N_{Rr}^{-1/5} (\cot \psi_i)^{4/5} \tag{8.33b}$$

and $N_{Rr} = \rho u_{ri} r_i / \mu$ = radial Reynolds number.

The maximum radial Reynolds number N_{Rrm} is the radial Reynolds number at zero control flow.

If we designate $u_{\theta i}^*$ as the inlet tangential velocity in inviscid flow, Eq. (8.33a) can be normalized to yield a recovery factor $u_{\theta i}/u_{\theta i}^*$:

$$u_{\theta i}/u_{\theta i}^* = 1/(1 + f_{\tau 1}) \tag{8.34a}$$

and $f_{\tau 1}$ (8.33b) can also be written as

$$f_{\tau 1} = 0.0225[u_{\theta i}/u_{\theta i}^*)(u_{\theta i}^*/u_{ri})]^{4/5} \tag{8.34b}$$

Figure 8.11 is a graph of $u_{\theta i}/u_{\theta i}^*$ as a function of $u_{\theta i}^*/u_{ri}$ for two values of the radial Reynolds number as given by Bichara and Orner [13].

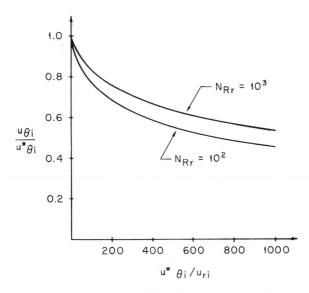

Fig. 8.11 Jet recovery factor as a function of ideal tangential to radial velocities.

The importance of the dissipation in region 1 was noted earlier by Wormley [8], who obtained an empirical curve for the jet recovery factor $u_{\theta i}/u_{\theta i}^*$ as a function of $u_{\theta i}^*/u_{ri}$ that agrees fairly well at low swirls ($u_{\theta i}^*/u_{ri} < 100$) with the curves of Fig. 8.11.

For $u_i^*/u_{ri} < 400$, Wormley's results are given approximately by

$$u_{\theta i}/u_{\theta i}^* = 0.30 + 1.33 \times 10^{-6}[(u_{\theta i}^*/u_{ri}) - 725]^2 \quad (8.34c)$$

Equation (8.34c), which indicates no dependence on Reynolds number, was determined from data for which N_{Rrm} lay between 750 and 3300, yet Wormley's recovery factor is in general less than that obtained by Orner and Bichara, even at $N_{Rrm} = 100$. The difference between the two curves increases with increasing swirl. Thus, at $u_{\theta i}^*/u_{ri} = 300$, Eq. (8.34c) gives a value for the recovery factor of 0.54 whereas Fig. 8.11 gives a value of 0.64, even for $N_{Rrm} = 100$.

It is worth noting that the use of either Eq. (8.34a) or (8.34c) allows one to convert Mayer's curves into flow characteristics more accurately than if Eq. (8.14) is used. The appropriate modified equation from Eqs. (8.9) and (8.34a) is

$$\tan \psi_i = \frac{c_{dc} A_c q_o^2}{A_i q_c^2} \left(\frac{1}{u_{\theta i}/u_{\theta i}^*}\right) \quad (8.35)$$

Note that the input angle is always larger when viscous effects are included.

Region 2 Wormley [8] investigated the flow in the main chamber, taking into account both the effects of radial losses and of an assumed tangential velocity profile within the chamber. In doing so, he considered that the main chamber consisted of a developing region and a developed region (Fig. 8.5). In the developing region the radial flow moves toward the drain. In the developed region all of the flow moving toward the drain is near the top and bottom plates.

Wormley was able to show the relationship between pressure at any point in the chamber, swirl, and a parameter he defined as the boundary layer coefficient. Although he obtained very good agreement between his theory and his experimental results, he did not try to treat regions 1 and 3 analytically. He did, however, obtain an empirical relation for region 1, as previously noted.

We obtained a coefficient α_B in the vortex chamber by integrating the tangential momentum equation in Section 8.3.4. The results from Eq. (8.28) were that

$$\alpha_B = 2f_2 r_i u_{\theta i}/hu_{ri} = 2f_2 r_i/(\tan \psi_i)h \quad (8.36a)$$

α_B plays the same role as Wormley's modified boundary layer coefficient; that is, it relates the circulation ru_θ to the chamber radius and thus indicates the decrease in circulation due to viscous losses. Wormley's modified boundary layer coefficient α_w is given by

$$\alpha_w = \frac{2r_i u_{\theta i}(0.0225)}{hu_{ri}(N_{Rr}/2)^{1/4}} \quad (8.36b)$$

8.3 Analyses of the Vortex Triode

We see that $\alpha_B = \alpha_w$ if f_2 is given by

$$f_2 = \frac{0.0225}{(N_{Rr}/2)^{1/4}} \tag{8.36c}$$

Bichara and Orner, however (as will be discussed later), have chosen to relate f_2 to the maximum tangential Reynolds number $N_{R\theta m}$. An important difference between the two is that $N_{R\theta m}$ is a constant whereas N_{Rr} depends on q_o.

The relationship between circulation and radial position given by Wormley indicates no loss of circulation in the outer regions $r \sim r_i$ (which he defines as the developing region). The loss begins in the vicinity of the doughnut (the developed region) and increases as the flow moves toward the drain. As the swirl increases, the developed region (the doughnut) becomes larger and the circulation begins to decay closer to the outer radius; however, in actual fact at high swirls the decay of circulation begins at $r = r_i$, as is evidenced by the initial decrease of tangential velocity inferred by Wormley [8] and measured by Syred and Royle [14].

In Wormley's discussion of his measurements of the pressure distribution as a function of radius, he notes that at high swirls the measurements indicate that in the outer regions of the chamber (vicinity of region 1) the tangential velocity decreases as the radius decreases, an inference verified by Syred and Royle. This effect is due to the large losses of tangential momentum in the vicinity of the outer wall.

Except for the losses due to the outer wall (region 1), Wormley's analysis, however, allows for no decay of the circulation in the outer (the developing) region. Thus, in this respect, the results of Bichara and Orner agree better with the facts since their model allows the circulation to decay beginning at $r = r_i$. Bichara and Orner's results indicate a greater circulation decay within the chamber (region 2) than do Wormley's. This tends to compensate for the smaller loss (than that of Wormley) that they obtain in region 1.

The studies of Syred and Royle [14] included attempts to determine the effect of the top and bottom plates on the viscous dissipation near the outer wall. They hypothesized that, because of friction with the top and bottom plates, control flow at the midplane should be more effective than that near the top and bottom plates. Syred and Royle tested this hypothesis by using horizontal slits at the midplane, vertical slits in the conventional manner, and small circular inlets at the midplane. The reduced friction associated with the horizontal slits and the circular inlets decreased the turn-down ratio slightly (and increased the noise), but the greatest change occurred in the general form of the characteristic curves so that they appear to follow the inviscid equations over a larger portion of the curves. In particular, maximum slope occurs at $q_o' = 0.707$ for the horizontal slits and the circular inlets, but not for the vertical slits.

Unfortunately the reduced losses cannot be assigned with certainty to the effect of the top and bottom plates because the amount of fluid contact with the outer wall is also reduced by horizontal slits and circular inlets. This question could be resolved if the experiment were to be repeated with the horizontal slits near the top and bottom plates and the results compared with the effect of horizontal slits in the vicinity of the midplane.

The results of Syred and Royle on insertion of flow in the midplane may be of particular significance in vortex diodes.

We have previously used α_B to modify the inviscid equations and have seen that this alone is not sufficient. However, we can use Eq. (8.28) to integrate the radial momentum equation, namely

$$u_r \, du_r/dr - u_\theta^2/r = (1/\rho) \, dp/dr \tag{8.37}$$

and obtain the result that

$$p_1 - p_o = (\rho u_{ri}^2/2)[(r_i^2/r_o^2) - 1] - (\rho/2)(u_{\theta i}^2 - u_{\theta o}^2) + f_{\tau 2} \tag{8.38a}$$

where

$$f_{\tau 2} = (\rho/2)(u_{\theta i}^2 - u_{\theta o}^2) + B_3 \rho u_{\theta i}^2/(1 + \alpha_B)^4 \tag{8.38b}$$

and

$$B_3 = \tfrac{1}{2}(1 + \alpha_B)^2 \left[\left(\frac{r_i}{r_o}\right)^2 - 1\right] + 2\alpha_B(1 + \alpha_B)\left(\frac{r_i}{r_o} - 1\right)$$

$$+ 3\alpha_B^2 \ln\left[\alpha_B\left(\frac{r_i}{r_o} - 1\right) + \frac{r_i}{r_o}\right] + \alpha_B^3 \left[1 - \frac{1}{\alpha_B[(r_i/r_o) - 1] + r_i/r_o}\right] \tag{8.38c}$$

We see that, for $\alpha_B = 0$, this equation reduces to Eq. (8.15).

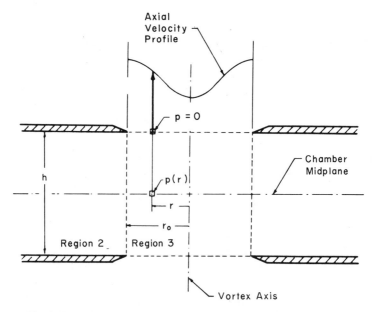

Fig. 8.12 Relation between axial velocity and midplane pressure.

8.3 Analyses of the Vortex Triode

Region 3 Bichara and Orner [13] take into account the fact that, because a forced vortex exists in region 3, the static pressure is not simply p_o throughout region 3. Therefore, they write for the pressure $p(r)$ at the chamber midplane (Fig. 8.12) that causes the output flow

$$p(r) = p_o + \tfrac{1}{2}\rho u_{ro}^2 - \rho \int_r^{r_o} (u_\theta^2/r)\, dr \tag{8.39}$$

For a forced vortex

$$u_\theta = u_{\theta o} r/r_o \tag{8.40}$$

so that Eq. (8.39) becomes

$$p(r) = p_o + \tfrac{1}{2}\rho u_{ro}^2 - \tfrac{1}{2}\rho u_{\theta o}^2 [1 - (r/r_o)^2] \tag{8.41}$$

It is apparent that for $u_{\theta o}$ sufficiently large, $p(r)$ can be negative (Fig. 8.13). The radius r_N at which the $p(r) = 0$ is given from Eq. (8.41) is

$$r_N = r_o[1 - (2p_o/\rho u_{\theta o}^2) - (u_{ro}^2/u_{\theta o}^2)]^{1/2} \tag{8.42}$$

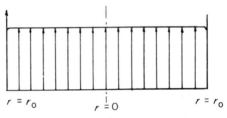
Low tangential velocities

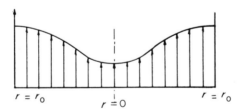
Intermediate tangential velocities

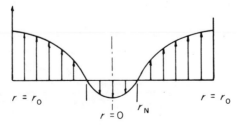
High tangential velocities

Fig. 8.13 Axial velocity profile for various tangential velocities.

Assuming that the external pressure is zero, the axial velocity distribution $u_z(r)$ (Fig. 8.13) is given by

$$u_z(r) = [2p(r)/\rho]^{1/2}, \qquad p(r) \geq 0 \qquad (8.43a)$$

$$u_z(r) = -[-2p(r)/\rho]^{1/2}, \qquad p(r) \leq 0 \qquad (8.43b)$$

the output flow is then given by

$$q_o = N 2\pi \int_0^{r_o} u_z(r)\, r\, dr \qquad (8.44a)$$

where $N = 1, 2$ is the number of outlets, or, using Eq. (8.43),

$$q_o = 2\pi N \left\{ \int_{r_N}^{r_o} \left[\frac{2p(r)}{\rho}\right]^{1/2} r\, dr - \int_0^{r_N} \left[\frac{-2p(r)}{\rho}\right]^{1/2} r\, dr \right\} \qquad (8.44b)$$

where r_N is given by Eq. (8.42) if the quantity under the radical in Eq. (8.42) is positive and $r_N = 0$ otherwise.

The result of the integration is

$$q_o = \frac{2\pi r_o^2 N c_{do}}{3 u_{\theta o}^2} \left(\frac{2p_o}{\rho} + u_{ro}^2\right)^{3/2} \left(1 - \left|1 - \frac{u_{\theta o}^2}{2p_o/\rho + u_{ro}^2}\right|\right)^{3/2} \qquad (8.45)$$

where the parallel lines designate absolute value.

Normalized Equations The flows and pressures can now be normalized by dividing by the maximum flow. Since this model attributes all viscous losses to tangential flow, the maximum flow which is purely radial is identical to that for inviscid flow given in Eq. (8.17).

We define

$$p_1' = p_1/p_s, \qquad q_s' = q_s/q_{om}$$
$$p_c' = p_c/p_s, \qquad q_c' = q_c/q_{om}$$

and

$$u_{ri}' = u_{ri}/u_{rim}$$

where u_{rim} is the maximum value of u_{ri}, i.e., its value when the tangential velocity is zero.

Now there are nine unknowns: (a) the flows q_o', q_c', and q_s'; (b) the pressures p_o' and p_1'; and (c) the velocities u_{ri}', $u_{\theta i}'$, u_{ro}', and $u_{\theta o}'$. To find the triode characteristics we, therefore, need nine equations which are:

From Eq. (8.5a)

$$p_1' = 1 - q_s'^2 c_N^2 c_{do}^2 A_o^2 / A_s^2 c_{ds}^2 \qquad (8.46a)$$

and from Eq. (8.5b)

$$p_c' - p_1' = A_o^2 c_{do}^2 c_N^2 q_c'^2 / A_c^2 c_{ds}^2 \qquad (8.46b)$$

8.3 Analyses of the Vortex Triode

From Eq. (8.6a)
$$q_o' = q_c' + q_s' \tag{8.46c}$$
and from Eq. (8.6b)
$$u'_{ri} = q_o' \tag{8.46d}$$
From Eq. (8.33)
$$u'_{\theta i} = \frac{A_i q_c'^2}{A_c c_{dc}[q_o' + 0.0225 N_{Rrm}^{-1/5}(u'_{\theta i})^{4/5}]} \tag{8.46e}$$
where
$$N_{Rrm} = u_{rim}\rho r_i/\mu$$
From Eq. (8.13), the continuity equation becomes
$$u'_{ro} = u'_{ri}(r_i/r_o) \tag{8.46f}$$
and from Eq. (8.28)
$$u'_{\theta o} = \frac{u'_{\theta i}(r_i/r_o)}{1 + \alpha_B(1 - r_o/r_i)} \tag{8.46g}$$
where
$$\alpha_B = 2f_2 r_i u'_{\theta i}/hu'_{ri}$$
In normalized form Eq. (8.38) is
$$p_o' = p_1' - \frac{\rho u_{ri}'^2}{2}c_N^2 c_{do}^2 A_o^2\left[\left(\frac{r_i}{r_o}\right)^2 - 1\right] - \frac{\rho u_{\theta i}'^2 c_N^2 c_{do}^2 A_o^2 B_3}{(1 + \alpha_B)^4} \tag{8.46h}$$
and Eq. (8.45) normalizes to
$$q_o' = \frac{2\pi r_o^2 N c_{do}}{3 u_{\theta o}'^2}\left(\frac{p_o'}{c_N^2 c_{do}^2 A_o^2} + u_{ro}'^2\right)^{3/2}\left(1 - \left|1 - \frac{u_{\theta o}'^2}{(p_o'/c_N^2 c_{do}^2 A_o^2) + u_{ro}'^2}\right|\right)^{3/2} \tag{8.46i}$$

Bichara and Orner set up a computer program using the normalized Eq. (8.46) to obtain amplifier characteristics. By selecting discharge coefficients and the friction factor f_2 to give a best fit with experimental data, it was found that good agreement could be obtained. It was necessary to assign values to c_{ds} from 0.60 to 0.96, c_{do} required values from 0.70 to 0.94, c_{dc} varied from 0.60 to 1.00, and f_2 varied from 0.0055 to 0.0228.

The friction factor f_2 was correlated with the tangential Reynolds number at cutoff (Fig. 8.14) and the following empirical relation was obtained
$$f_2 = 1.462/N_{R\theta m}^{0.405} \tag{8.47}$$
where
$$N_{R\theta m} = u_{\theta i} r_i \rho/\mu|_{q_s = 0}$$

The points on Fig. 8.14 were obtained by finding the value of f_2 (and the discharge coefficients) that fit an experimental curve best. It is seen that the points are rather scattered so that some of the points necessary to obtain a best fit for the experiment deviate appreciably from the curve drawn through the

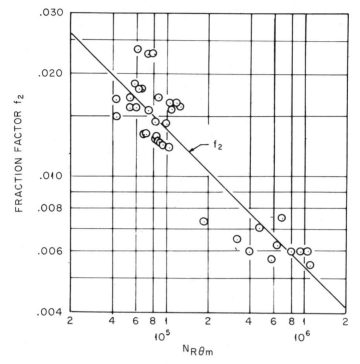

Fig. 8.14 Friction factor as a function of cutoff tangential Reynolds number.

points in Fig. 8.14. Since the discharge coefficients used for a best fit also deviate over a wide range of values, the assumption that each of the coefficients is approximately equal to 0.8 and that the friction coefficient is given by Eq. (8.47) can result in an analytical representation of the characteristic curves that deviates at some points as much as 30% from the experimental curves.

As pointed out in the discussion of Eq. (8.36), it is quite possible that better agreement can be obtained if Eq. (8.36c) is used rather than Eq. (8.47).

It should be noted that the results are easily applied to the vortex diode as was shown for the inviscid case.

8.4 VORTEX TRIODE DESIGN CHART

The simultaneous solution of Eqs. (8.46) is both cumbersome and time consuming. When, as is often the case, we are only interested in the q_o' versus q_c' relation, a graphical procedure introduced by Wormley and Richardson [11] provides a significant simplification. The procedure is based on their experimental results wherein the q_o' versus q_c' relation is independent of Reynolds number, ratio of supply area to output area, and height if the following restrictions are placed on those three parameters:

8.4 Vortex Triode Design Chart

Condition 1 $N_{Rrm} > 750$, where the maximum radial Reynolds number is defined as

$$N_{Rrm} = q_{om}/2\pi r_i \nu$$

The largest value of Reynolds number used in the testing was 3300, so that strictly speaking they have shown only that the dependence on Reynolds number is weak for

$$750 \leq N_{Rrm} < 3300$$

Condition 2 $A_s/A_o > 3.0$. This condition also leads to a large turn-down ratio, which is in general desirable but is particularly important if throttling of power is necessary.

Condition 3 $0.144 \leq h/r_i \leq 0.64$ and $2.1 \leq h/r_o \leq 8.65$. Wormley and Richardson assume that if $u_{ri} = q_o/A_i$ [Eq. (8.6b)] and the above conditions are satisfied, then the flow characteristic depends only on the area ratio A_c/A_o and the radius ratio r_o/r_i. In particular, the cutoff point depends only on these ratios. On the basis of their experimental data they were thus able to draw up a chart of cutoff pressures and flows (Fig. 8.15) as a function of these ratios.

Wormley and Richardson's experiments were done on an amplifier having two symmetrically placed tangential inputs (as in Fig. 8.4b); however, Lawley and Price [15] have shown that the chart also applies to a single input device and have taken some additional data extending the original curves. Figure 8.15 also includes some curves Lawley and Price postulate but for which data were not taken.

We will now briefly outline the way that the chart may be used to design a vortex triode and to obtain an approximation of its q_o' versus q_c' characteristics. The characteristic approximation is a faired curve through three known points: (1) no control flow (0, 1); (2) $(q_{cf}', 0.707)$; and (3) cutoff (q_{cc}', q_{oc}').

The design procedure [11] is as follows:

(a) The operating fluid and tentative values of supply pressure, maximum supply flow, maximum control pressure, and maximum control flow are specified. (These values are tentative since the set chosen may not satisfy the conditions on Reynolds number after the calculation has been completed.)

(b) The specified quantities are normalized to obtain $p_{cc}' = p_{cc}/p_s$ and $q_{cc}' = q_{cc}/q_{om}$. (Note that the pressures are actually the pressures across the device; i.e., it is assumed that the exit pressure is ambient.)

(c) Determine whether the value for p_{cc}', q_{cc}' lies in the desired region of the chart (i.e., does it lead to a bistable or a single-valued characteristic curve). This information is obtained from the $q_{cf}' = q_{cc}'$ curve on the chart. If q_{cc}' is less than q_{cf}' (i.e., for points on the chart lying to the right and above the multivalued curve), the device is bistable. This is apparent from Fig. 8.7b, where it can be seen that if the cutoff is in the region between D and C, i.e., if the cutoff flow q_{cc}' is

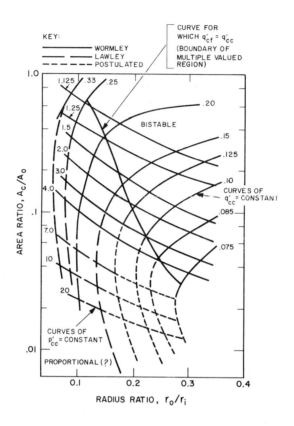

Fig. 8.15 Vortex triode design chart.

less than q'_{cf}, then the device must be multivalued at cutoff. On the other hand, if the cutoff is to the right of B, the cutoff is single valued, although the device may have a bistable region of operation. Thus, to repeat, the region above the $q'_{cf} = q'_{cc}$ curve gives a definitely bistable device, but the region below $q'_{cf} = q'_{cc}$ does not necessarily yield a single-valued device. If in addition to lying below the $q'_{cf} = q'_{cc}$ curve the point also lies in a region for which $r_o/r_i < 0.15$, then, by reason of the previous arguments on viscous losses, we would expect a single-valued device to be obtained.

(d) If the point (p'_{cc}, q'_{cc}) does not lie in the desired portion of the chart, another set of values is necessary and the process begins again with step a. If (p'_{cc}, q'_{cc}) does lie in the desired region, then the calculation continues.

(e) Values of A_c/A_o and r_o/r_i are found from Fig. 8.15 for the values of p'_{cc} and q'_{cc} obtained in step b.

(f) Values of A_s/A_o, h/r_i, and h/r_o are chosen to satisfy the three conditions and the value of r_o/r_i found in step e.

(g) The Reynolds number is calculated. If condition 1 is not satisfied, the results must be discarded. If condition 1 is satisfied, the design of the unit is complete and we may now determine the q_o' vs q_c' characteristic.

(h) q_{cf}' can be calculated from Eq. (8.20b). However, a good approximation of q_{cf}' can be obtained directly from the chart; q_{cf}' is the q_{cc}' value that occurs at the intersection of the line of constant p_{cc}' and the boundary line ($q_{cf}' = q_{cc}'$).

(i) Plot the three points (0, 1), (q_{cf}', 0.707), and (q_{cc}', q_{oc}') where we remember that $q_{oc}' = q_{cc}'$.

(j) Draw a curve through the three points.

(k) A similar procedure is used for the flow versus control pressure curve.

8.5 THE VORTEX TRIODE AS A PROPORTIONAL AMPLIFIER

Perhaps an undue stress has been placed in the previous discussion on the value of a high turn-down ratio. Although this is indeed important for a throttling device, we repeat that if the device is to be used as an amplifier the important considerations are gain, bandwidth, and dynamic range. Since the dynamic range is the ratio of largest to smallest acceptable signals, it depends on how large a portion of the curve is linear and on the noise.

Very little on dynamic range of vortex amplifiers has appeared in the literature, although some studies of noise have been carried out by Syred et al. [16] and by Syred and Royle [14].

It was found that in general the factors that increase the turn-down ratio (TDR) also increase noise; however, significant reductions in noise could be obtained by the use of small angle diffusers at the vortex outlets so that low (5%) noise was obtained at TDR values of 10 and medium (40%) noise levels were obtained at TDR of 20. The noise-to-signal percentages were defined as 100 times the wall static pressure fluctuation divided by the mean static pressure across the device.

8.5.1 Small Signal AC Analysis

The first ac analyses of the vortex amplifier were given by Taplin at a Pennsylvania State University seminar in 1965 [17] and in an AGARD lecture [18] in 1966 (also presented as part of a summer course at MIT in 1966). A further extension including the effects of backloading and with some modifications in notation was given in 1970 [19].

The method used is to set up an equivalent circuit based on physical reasoning and then to determine the coefficients experimentally.

Taplin pointed out the significance of the chamber fill time (the chamber volume divided by the total volume flow rate) in his original discussions and indicated [17] that there were two time constants of significance: a pure time delay and a lag time, each of which he set equal to one quarter of the fill time. In a subsequent paper [19], however, only a first-order lag is indicated in the equations, although the text indicates that two time constants were obtained by varying the parameters of one of two parallel paths in the equivalent circuit so as to obtain the correct (corresponding to experimental data) frequency response.

In an attempt to obtain a dynamic model useful for design purposes, Anderson [20] has used the method of Wormley's [8] analysis as a basis from which to obtain the dynamic response of the chamber. From this response he approximates a lumped circuit pure delay and a first-order lag as a function of the vortex amplifier parameters. He is then able to obtain an equivalent circuit for a small-amplitude signal about a given level from the steady state characteristics and the time constants.

8.5.2 Inviscid Time Response

To get some idea of how the time constants arise, we first consider inviscid incompressible flow for which the tangential momentum equation is

$$\frac{\partial}{\partial t}(ru_\theta) + u_r \frac{\partial}{\partial r}(ru_\theta) = 0 \tag{8.48}$$

If we take the Laplace transform of Eq. (8.48) with respect to time and apply the continuity equation $u_r r = u_{r_i} r_i$ the result is

$$d(rU_\theta)/rU_\theta = -sr\, dr/u_{r_i} r_i \tag{8.49}$$

where $U_\theta(r, s) = \mathscr{L}\{u_\theta(r, t)\}$, s is the Laplace variable, and $u_\theta(r, 0) = 0$.

If Eq. (8.49) is integrated between the limits r_i and r, and then transformed back into the time domain, we obtain

$$u_\theta(r, t) = \frac{r_i}{r} u_{\theta i}\left(r_i, t + \frac{r_i^2 - r^2}{2u_{r_i} r_i}\right) \tag{8.50}$$

Equation (8.50) expresses the relation between the input tangential velocity and the tangential velocity at any other radial position r. To determine the tangential velocity step response we apply a step function at the input; that is,

$$u_{\theta i} = u_\theta(r_i, t) = u_{\theta 1} H(t) \tag{8.51}$$

where $u_{\theta 1}$ is constant and $H(t)$ is the unit step function. For a step function input, Eq. (8.50) becomes

$$u_\theta(r, t) = \frac{r_i u_{\theta 1}}{r} H\left(t + \frac{r_i^2 - r^2}{2u_{r_i} r_i}\right) \tag{8.52}$$

8.5 The Vortex Triode as a Proportional Amplifier

Now since the radial velocity increases in the opposite direction to the radial coordinate, $u_{ri} = -|u_{ri}|$ and Eq. (8.52) may also be written as

$$u_\theta(r, t) = \frac{r_i u_{\theta 1}}{r} H\left(t - \frac{r_i^2 - r^2}{2|u_{ri}|r_i}\right) \qquad (8.53)$$

Equation (8.53) indicates that a tangential velocity step initiated at r_i at time $t = 0$ occurs at r as a velocity step increased in magnitude by the factor r_i/r at time t_{dr} where

$$t_{dr} = \frac{r_i^2 - r^2}{2|u_{ri}|r_i} \qquad (8.54)$$

In particular, the time for a tangential velocity change at r_i to reach the outlet r_o is given by

$$t_{dro} = \frac{r_i^2 - r_o^2}{2|u_{ri}|r_i} = \frac{\pi h(r_i^2 - r_o^2)}{2\pi h|u_{ri}|r_i} \qquad (8.55a)$$

Since $r_o^2 \ll r_i^2$,

$$t_{dro} \cong \frac{r_i^2 \pi h}{2|u_{ri}|\pi h r_i} \qquad (8.55b)$$

Thus the transport time for the inviscid flow case is given by the fill time, i.e., the volume of the amplifier divided by the volume flow rate.

8.5.3 Pressure Effects

We now determine the pressure distribution within the vortex chamber as a function of both time and radial position. The radial momentum equation for inviscid flow can be written as

$$\partial u_r/\partial t + u_r \, \partial u_r/\partial r - u_\theta^2/r = -(1/\rho) \, \partial p/\partial r \qquad (8.56a)$$

Using the continuity equation this can be rewritten as

$$\frac{r_i}{r} \frac{\partial u_{ri}}{\partial t} - \frac{r_i^2 u_{ri}^2}{r^3} - \frac{u_\theta^2}{r} = -\frac{1}{\rho} \frac{\partial p}{\partial r} \qquad (8.56b)$$

The tangential velocity response to a step function has already been given in Eq. (8.53). We may also write that solution as

$$u_\theta(r, t) = 0, \qquad t < \frac{r_i^2 - r^2}{2|u_{ri}|r_i} \quad \text{or} \quad r < [r_i^2 - 2|u_{ri}|r_i t]^{1/2} \qquad (8.57a)$$

$$u_\theta(r, t) = \frac{r_i u_{\theta 1}}{r}, \qquad t > \frac{r_i^2 - r^2}{2|u_{ri}|r_i} \quad \text{or} \quad r > [r_i^2 - 2|u_{ri}|r_i t]^{1/2} \qquad (8.57b)$$

If Eq. (8.57) is substituted into Eq. (8.56b) and the integration performed the pressure distribution is

$$p(r, t) = p_o - \rho r_i \frac{\partial u_{ri}}{\partial t} \ln \frac{r}{r_o} - \frac{\rho r_i^2 u_{ri}^2}{2}\left[\frac{1}{r^2} - \frac{1}{r_o^2}\right]$$

$$\text{for} \quad r_o \leq r < (r_i^2 - 2|u_{ri}|r_i t)^{1/2} \tag{8.58a}$$

$$p(r, t) = p_o - \rho r_i \frac{\partial u_{ri}}{\partial t} \ln \frac{r}{r_o} - \frac{\rho r_i^2 u_{ri}^2}{2}\left[\frac{1}{r^2} - \frac{1}{r_o^2}\right]$$

$$- \frac{\rho r_i^2 u_{\theta 1}^2}{2}\left(\frac{1}{r^2} - \frac{1}{r_i^2 - 2|u_{ri}|r_i t}\right)$$

$$\text{for} \quad r_i \geq r > (r_i^2 - 2|u_{ri}|r_i t)^{1/2} \tag{8.58b}$$

As written above, the pressure distribution at a given time t is given as a function of r. It is seen that the fluid is separated into two regions, an outer region within which the tangential velocity has changed at the time t and an inner region within which the tangential velocity is still zero. The same equations can also be used to give the time response at a particular radius r by putting the conditions in terms of time as indicated in Eq. (8.57). Thus the conditions for the pressure distribution in Eq. (8.58) in terms of time are

$$0 < t < \frac{r_i^2 - r^2}{2|u_{ri}|r_i} \equiv t_{dr} \tag{8.59a}$$

and

$$t_{dro} \equiv \frac{r_i^2 - r_o^2}{2|u_{ri}|r_i} > t > \frac{r_i^2 - r^2}{2|u_{ri}|r_i} \equiv t_{dr} \tag{8.59b}$$

For $t > t_{dro}$, the pressure level remains constant at the same pressure as that for which $t = t_{dro}$.

Besides the limitations on Eq. (8.58b) indicated by the inequalities (8.59b), there is an additional limitation, namely the sum of the terms on the right must be nonnegative. If the conditions are such that $p(r, t) < 0$, the equation is not valid.

According to Eqs. (8.59a) and (8.59b), the pressure remains constant at any radius r for a time t_{dr} before beginning to decrease as a result of the tangential velocity step; however, a change in radial velocity is transported with no time delay. The zero delay time for the change in radial velocity is due, of course, to ignoring the term $\partial p/\partial t$, that is, the assumption of incompressibility. If compressibility were considered, the response time for the radial velocity would be determined by the speed of sound.

8.5.4 Effect of Viscosity

It has been pointed out that, as a result of large swirls, the radial velocity is greater near the top and bottom plates (in the boundary layers) than in the core region (the region between the boundary layers).

This affects the time delay in two ways. Since most of the flow occurs near the walls rather than throughout the chamber (as it does for inviscid flow) the effective volume is reduced. This results in a smaller transport time for a given volume flow rate.

In addition to the transport delay there is now, however, a second time constant; namely, the time required to impart the tangential velocity to the core. In the developing portion the core receives its tangential velocity information both by means of the radial flow transport and by viscous shear from the boundary layers. In the developed portion, however, the flow may recirculate for many fill times and the tangential velocity information comes primarily as a result of viscous shear.

Thus, for swirls such that very little or no developed region exists, the time required to bring the entire fluid up to a major fraction $(1 - e^{-1})$ of its steady state value as a result of a small amplitude step function input is still approximately equal to the fill time, but this time is now broken into a transport delay T_1 plus a lag T_2.

The fill time is not constant as the swirl is increased because the flow rate decreases. It follows that, as the swirl is increased, $T_1 + T_2$ increases and T_1/T_2 decreases.

Even if the time constants are normalized by the fill times, e.g., $T_1' = T_1/t_{\text{dro}}$, etc., $T_1' + T_2'$ does not remain constant when the swirl becomes large enough to produce a fully developed region. Because of the recirculation, $T_1' + T_2'$ can become greater than unity.

In order to account for both time constants, Anderson [20] assumes as Taplin [17] did earlier that

$$\Delta(p_1 - p_r) = [k_6 \exp(-T_1 s)/(1 + T_2 s)] \Delta u_{\theta i} + k_7 \Delta u_{ri} \qquad (8.60)$$

where $p_r = p_o + \frac{1}{2}\rho u_{ro}^2$.

Anderson obtained a computer solution for the time response in the chamber for viscous flow using a quasi-steady analysis in which the chamber on each side of the midplane was broken into two sets of forty concentric rings: one set in the boundary layer and the other in the core (the region between the boundary layer and the midplane). This analysis makes it possible to compute the response to a small amplitude step function.

T_1 and T_2 are obtained from this response by choosing as $T_1 + T_2$ the time for which the amplitude becomes equal to $(1 - e^{-1})$ of the steady state response. T_1 is determined by the intersection with the axis of the tangent through the point of inflection of the response curve.

Although it would seem that the time constants should be related to the modified boundary layer coefficient α_w, or to the boundary layer coefficient $\alpha_w u_{ri}/u_{\theta i}$, the relationship is also apparently affected by the height. Anderson does not discuss the relationship between the time constants and α_w, but does show that different curves are obtained for normalized response times as a function of swirl even at the same values of $\alpha_w u_{ri}/u_{\theta i}$ for different values of h/r_i.

Thus the present status requires use of a computer program to obtain T_1 and T_2.

To simplify the analysis and to obtain a model of the amplifier Anderson assumes that the supply port resistance is small so that $p_1 \cong p_s = $ const; hence, Eq. (8.60) becomes

$$-\Delta p_r = \frac{k_6 \exp(-T_1 s)}{1 + T_2 s} \Delta u_{\theta i} + k_7 \Delta u_{ri} \qquad (8.61)$$

Our objective now is to express $\Delta u_{\theta i}$ and Δu_{ri} in terms of the pressure and volume flow signal variables. From Eq. (8.6b),

$$\Delta u_{ri} = k_u \Delta q_o \qquad (8.62)$$

where

$$k_u = 1/2\pi r_i h$$

Since the tangential input velocity in inviscid flow from Eq. (8.7) is

$$u_{\theta i}^* = q_c^2/q_o A_c c_{dc} \qquad (8.63a)$$

or

$$u_{\theta i} = (q_c^2/q_o A_c c_{dc}) u_{\theta i}/u_{\theta i}^* \qquad (8.63b)$$

we may express the differential input tangential velocity in terms of the differential control and output flows as

$$\Delta u_{\theta i} = k_c \Delta q_c - k_o \Delta q_o \qquad (8.64a)$$

where from Eq. (8.63b)

$$k_c = (2q_{c1}/q_{o1} A_c c_{dc}) u_{\theta i}/u_{\theta i}^* \qquad (8.64b)$$

$$k_o = (q_{c1}^2/q_{o1}^2 A_c c_{dc}) u_{\theta i}/u_{\theta i}^* \qquad (8.64c)$$

and where $u_{\theta i}/u_{\theta i}^*$ is given by Eq. (8.34a) or (8.34c), q_{c1} is the control flow at the selected operating point, and q_{o1} is the total output flow at the operating point.

The substitution of Eqs. (8.62) and (8.64a) into Eq. (8.61) yields

$$-\Delta p_r = \frac{k_6 \exp(-T_1 s)}{1 + T_2 s}(k_c \Delta q_c - k_o \Delta q_o) + k_7 k_u \Delta q_o \qquad (8.65)$$

8.5 The Vortex Triode as a Proportional Amplifier

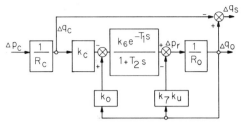

Fig. 8.16 Vortex amplifier dynamics block diagram.

To complete the vortex amplifier model that is shown in Fig. 8.16 it is only necessary to find expressions for the inlet and outlet impedance of the amplifier. If these impedances are purely resistive we may represent them by

$$\Delta q_c = (1/R_c)\,\Delta p_c \tag{8.66a}$$

$$\Delta q_o = (1/R_o)\,\Delta p_r \tag{8.66b}$$

where, from Eq. (8.5b) (with $p_1 = p_s =$ const),

$$R_c = \rho q_{c1}/c_{dc}^2\, A_c^{\,2} \tag{8.66c}$$

and, from Eq. (8.12) (with $p_r = p_o + \rho u_{ro}^2/2$),

$$R_o = \rho q_{o1}/c_{do}^2\, N^2 A_o^{\,2} \tag{8.66d}$$

The complete block diagram model (Fig. 8.16) also makes use of the continuity equation (8.6a).

It remains to find k_6 and k_7. This is done by Anderson using the steady state flow characteristic ($q_o{}'$ vs $q_c{}'$) and the relation between the pressure p_r and the modified boundary layer coefficient α_w for a given radius ratio r_i/r_o.

We recall that

$$\alpha_w = \frac{2r_i\,u_{\theta i}(0.0225)}{hu_{ri}(N_{Rr}/2)^{1/4}} \tag{8.36b}$$

Thus a particular operating point on the flow characteristic specifies a value for α_w and consequently a particular p_r. A small change $u_{\theta i}$ (in terms of q_c) with u_{ri} held fixed (to a point that is not on the flow characteristic) gives a different value of α_w and therefore of p_r. This gives the change in p_r resulting from a change in $u_{\theta i}$. Now with $u_{\theta i}$ held fixed, u_{ri} is changed to its value on the flow characteristic. The new value of α_w gives a new value of p_r which allows the change in p_r resulting from a change in u_{ri} to be found.

In order to use this method the steady state flow characteristic must be known.

To show that the method can be used in designing an amplifier, Anderson uses the three-point design method of Richardson and Wormley to draw three possible curves passing through the three points for a given set of parameters and calculates the equivalent circuit for each case to show that the differences in the three cases are not too large.

If general, whether the measured flow characteristic is used or one is drawn through three points the agreement of the measured frequency response with the calculated response varies from good to fair.

PROBLEMS

8.1

(a) Show from the inviscid analysis that the flow angle at cutoff ψ_c is a function of only the control area (and discharge coefficient), the outer radius of the chamber, and the height.

(b) Show that when all the discharge coefficients are assumed to be unity the flow angle at cutoff may be approximated by

$$\psi_c \cong \tan^{-1}[r_o/2h(c_G)^{1/2}]$$

(c) As the turn-down ratio increases, what happens to the flow angle at cutoff?

8.2 When the supply flow to the vortex triode is zero, the vortex triode should reduce to the vortex diode. From the inviscid triode analysis [Eq. (8.15)] calculate the reverse flow coefficient K_{LR}. Compare this with the value given in Eq. (2.37) for the vortex diode. Explain the discrepancy. What happens when the flow angle is large?

8.3

(a) Draw the characteristic q_o' vs q_c' for the inviscid case when $N = 1$, $r_i = 2.50$ cm, $r_o = 0.41$ cm, $r_c = 0.26$ cm, $h = 0.80$ cm, $A_s = 2\pi r_i h$, and all $c_d = 0.8$.

(b) What is the turn down ratio of this triode?

(c) Repeat parts (a) and (b) for the case where $N = 2$ and all the other dimensions remain the same.

8.4

(a) Plot the relation between output flow angle ψ_o and input flow angle ψ_i for the case where $f_2 = 0.01$, $r_i/r_o = 5.0$, and $h/r_o = 2.0$.

(b) For these same conditions also plot $(\psi_o - \psi_i)$ vs ψ_i.

(c) What effect does an increase in friction have on the curves in part (a) and (b).

(d) How does the change in flow angle $(\psi_o - \psi_i)$ affect the output flow of the triode at various values of the control flow (i.e., at small flows and at large control flows)?

8.5 Use the vortex triode design chart to determine:

(a) If an amplifier with the design parameters $A_c/A_o = 0.1$, $T_R = 10$ and $r_o/r_i = 0.1$ is possible. If so, is it proportional or bistable? If not can A_c/A_o be adjusted to make the amplifier proportional?

(b) If $A_c/A_o = 0.1$ and $r_o/r_i = 0.1$, what is the turn-down ratio T_R? Is the amplifier proportional or bistable?

(c) If $r_o/r_i = 0.3$, what value of A_c/A_o is necessary to achieve a turn-down ratio of 8.0?

8.6

(a) Use the vortex triode design charts to find the dimensions r_i, r_o, h, A_c, and A_s that will meet the following specifications for standard air operation and have proportional operation throughout: $p_{cc} = 62$ kN/m², $p_s = 31$ kN/m², $q_{om} = 0.264 \times 10^{-2}$ m³/sec, and $T_R = 6.36$.

(b) Calculate N_{Rrm} for your design. Does it fall within the acceptable range?

(c) Sketch the q_o' vs q_c' characteristic for your design.

8.7 A step input of tangential velocity $u_{\theta 1} = 10$ m/sec occurs at the outer radius of a vortex chamber $r_o = 2.0$ cm.

(a) How long does it take the step to reach a radius of 1.0 cm.

(b) Show on a graph the time required to reach all radial positions less than 2.0 cm.

8.8 A step input of 25 m/sec is applied to the tangential velocity at the outer radius of a vortex chamber in which $r_i = 5.0$ cm. The radial velocity at the outer radius is 5 m/sec and does not change appreciably with time. Plot the pressure at $r = 2.5$ cm as a function of time.

SUGGESTED TERM PAPERS

8.1 Incorporate Wormley's results [8] into the model of Bichara and Orner [13].

8.2a Write a program to solve the set of equations given in Eq. (8.46). Use the program to predict vortex triode characteristics. Find the geometric parameters which are the most sensitive in determining the turn-down ratio and the power ratio.

8.2b Show how a vortex diode should be optimized by adapting the program for the case in which the supply flow is zero and calculating the forward resistance as in Chapter 2.

NOMENCLATURE*

A_c	Area of control nozzle
A_i	Area of the chamber outer (circular) wall $= 2\pi r_i h$
A_o	Area of outlet
A_s	Area of supply nozzle
c_d	Discharge coefficient

---

* Normalized quantities are designated by means of a prime and are for the most part defined in and preceding Eq. (8.46).

c_{dc}	Discharge coefficient for control nozzle
c_{do}	Discharge coefficient for outlet
c_{ds}	Discharge coefficient for supply nozzle
c_G	Defined in Eq. (8.19)
c_M	Mayer coefficient [defined in Eq. (8.23)]
c_N	Defined in Eq. (8.16)
D_i	Diodicity
f_2	Friction factor [defined preceding Eq. (8.27)]
$f_{\tau 1}, f_{\tau 2}$	Friction factors
h	Distance between top and bottom plates
I_g	Minimum power output divided by maximum power output
k_6, k_7	Coefficients [Eq. (8.61)]
k_c	Coefficient [Eq. (8.64b)]
k_o	Coefficient [Eq. (8.64c)]
k_u	Coefficient [Eq. (8.62)]
K_L	Loss coefficient
N	Number of outlets
N_{Rr}	Radial Reynolds number [Eq. (8.33b)]
N_{Rrm}	Maximum radial Reynolds number [Eq. (8.46e)]
$N_{R\theta m}$	Tangential Reynolds number [Eq. (8.47)]
$p(r)$	Chamber midplane pressure [Eq. (8.39)]
p_1	Pressure at $r = r_i$
p_c	Control pressure
p_{cc}	Control pressure at cutoff
p_o	Pressure at $r = r_o$
q_c	Control flow
q_{cc}	Control flow at cutoff
q_o	Total flow (flow from outlet)
q_{oc}	Total flow at cutoff
q_{om}	Maximum total flow (total flow when control flow is zero)
q_s	Supply flow
q_c'	q_c/q_{om}
q_{cf}'	Normalized control flow [defined preceding Eq. (8.20b)]
q_o'	q_o/q_{om}
q_{of}'	Normalized output flow [defined preceding Eq. (8.20b)]
r_i	Chamber radius
r_o	Outlet radius
r_N	Radius at which $p(r) = 0$ [Eq. (8.42)]
R_c	Resistance of input control [Eq. (8.66c)]
R_o	Resistance of the outlet [Eq. (8.66d)]
s	Laplace variable
t_{dr}	Transport time [Eq. (8.54)]
t_{dro}	Transport time or fill time [defined in Eq. (8.55)]
T_1, T_2	Transport and delay times [Eq. (8.60)]
T_R	Turn-down ratio [Eq. (8.3)]
$\bar{u}$	Average velocity
u_r	Radial velocity
u_{ri}	Radial velocity at $r = r_i$
u_{rim}	Maximum value of u_{ri}
u_{ro}	Radial velocity at $r = r_o$
$u_z(r)$	Velocity in drain
u_θ	Tangential velocity
$u_{\theta i}$	Tangential velocity at $r = r_i$

u_{θ_o} Tangential velocity at $r = r_o$
$U_\theta(r,s) = \mathscr{L}\{u_\theta(r,t)\}$
α_B Bichara and Orner coefficient [Eq. (8.28)]
α_w Modified boundary layer coefficient (Wormley coefficient) [Eq. (8.36b)]
$\tau_{\theta 1}$ Circular wall shear stress
$\tau_{\theta 2}$ Tangential shear stress in region 2
ψ Flow angle $=\tan^{-1} u_r/u_\theta$
ψ_i Flow angle at $r = r_i$
ψ_o Flow angle at $r = r_o$

REFERENCES

1. D. Thoma, "Fluid Lines." U.S. Patent #1,839,618, Patented 5 Jan. 1932.
2. R. Heim, "An Investigation of the Thoma Counterflow Brake," Trans. of the Munich Hydraulic Inst., 1929, ASME Transl., 1935 by M. P. O'Brien.
3. M. Kadosh and J. LeFoll, "Vorrichtung zum Erzeugen einer Kreisenden Stromung in einer runden Raum." Bundesrepublik Deutchland patent number 971,622 issued 26 Feb. 1959. Priority of application in France claimed 27 Sept. 1951.
4. R. E. Bowles and B. M. Horton, "Fluid Amplifier." U.S. Patent 3,276,259, issued 4 Oct. 1966, Patent filed 11 Aug. 1961.
5. W. E. Gray and H. Stern, "Fluid Amplifiers—Capabilities and Applications." Control Engineering, Feb. 1964.
6. E. A. Mayer, Photoviscous flow visualization in fluid state devices. *Proc. Fluid Amplification Symp.* **11** (October 1965), HDL.
7. J. M. Savino and E. G. Keshock, Experimental profiles of velocity components and radial pressure distributions in a vortex contained in a short cylindrical chamber. *Proc. Fluid Amplification Symp.* **11** (October 1965), HDL.
8. D. N. Wormley, "An Analytical and Experimental Investigation of Vortex-Type Fluid Modulators." Ph.D. Thesis. Dept. of Mech. Eng., MIT, Oct. 67; An analytical model for the incompressible flow in short vortex chambers. *J. Basic Eng. Trans. ASME* 264 ff (1969).
9. E. A. Mayer, Other fluidic devices and basic circuits. *Fluidics Quart.* **1**, 76ff (1968).
10. V. D. Gebben, Vortex valve performance power index. *Advan. Fluidics* (May 1967).
11. D. N. Wormley and H. H. Richardson, "Experimental Investigations and Design Basis for Vortex Amplifiers Operating in the Incompressible Flow Regime," 15 October 1967. Prepared for HDL by MIT under contract number DAAG39-67-C-0019; "Experimental Investigation and Design Basis for Vortex Amplifiers Operating in the Incompressible Flow Regime." Tech. Rep. #DSR 70167-1, 25 Dec. 1968. Prepared for HDL by MIT under contract number DAAG39-67-C-0019; A design basis for vortex-type fluid amplifiers operating in the incompressible flow regime. *J. Basic Eng., Trans. ASME, Ser. D* **92**, 568 (1970).
12. E. A. Mayer, Parametric analysis of vortex amplifiers. *Bendix Tech. J.*, Winter 1969.
13. R. T. Bichara and P. A. Orner, Analysis and modeling of the vortex amplifier. *Trans. ASME, J. Basic Eng.* paper #69-Flcs-20 (1969).
14. N. Syred and J. K. Royle, Operating characteristics of high performance vortex amplifiers (*IFAC Symp.*, 2nd). *Fluidics Quart.* **4**, No. 1 (1972).
15. T. J. Lawley and D. C. Price, Design of vortex fluid amplifiers with asymmetrical flow fields. *J. Dynam. Syst. Measurements, Contr. Trans. ASME* 82ff (March 1972).
16. N. Syred, J. K. Royle, and J. R. Tippetts, Optimization of high gain vortex devices. *Cranfield Fluidics Conf. 3rd, 8–10 May 1968.*
17. L. B. Taplin, Phenomenology of vortex flow and its application to signal amplification. *Fluidics Quart.* **1**, No. 2 (January 1968).

18. L. B. Taplin, "Small Signal Analysis of Vortex Amplifiers," Chapter 7, AGARDograph 118, pp. 235–295, December 1968.
19. L. B. Taplin, Dynamic equivalent circuit for a vortex amplifier. *Proc. Cranfield Fluidics Conf.*, *4th* paper B1 (March 1970).
20. W. W. Anderson, "A Dynamic Model of Vortex-Type Fluid Amplifiers." Ph.D. Thesis, MIT, Cambridge, Massachusetts, October 1971.

Chapter 9

THE BEAM DEFLECTION AMPLIFIER

9.1 HISTORICAL INTRODUCTION*

B. M. Horton conceived the beam deflection amplifier partially as a result of the realization that a water hose could be deflected by less power than that issuing from the hose. Consequently, he thought in terms of round nozzles. This concept (shown in Fig. 9.1) included two outputs, two control inputs, and a power jet nozzle, all of which were round. However, after a few months of experimentation it became obvious that round jets could not deflect each other efficiently. Consequently, top and bottom plates were provided, and the nozzles became rectangular rather than round.

This procedure increased the efficiency considerably because the pressure effects (which may be more important than the momentum effects in certain circumstances) came into play. Indeed the pressure effects became so important that, after the addition of top and bottom plates and the associated walls, it was found that a bistable rather than a proportional device had been obtained because the pressure difference across the jet resulting from entrainment in the vicinity of a wall caused the jet to attach.

To prevent attachment, the walls were moved farther apart and openings were made from the right and left cavities to the atmosphere as in Fig. 9.2, or, alternatively, the right and left cavities were connected via an external duct. This duct caused oscillations to occur. After some experimentation, the ducts were shortened, eventually consisting of passages through the cover plate (Fig. 9.3) and resistive material was placed within the passages to damp oscillations.

---

* Reference [1a].

316   9 The Beam Deflection Amplifier

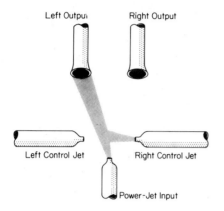

Fig. 9.1 Original beam deflection amplifier concept.

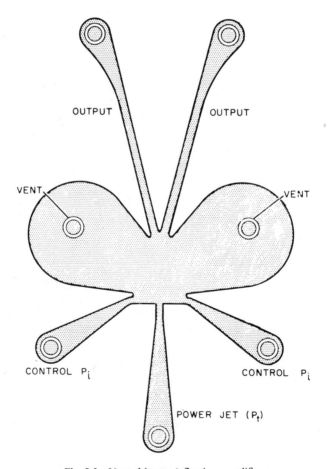

Fig. 9.2 Vented beam deflection amplifier.

9.1 Historical Introduction

When staging of units was attempted, it was necessary to recognize that the second stage had to accommodate the flow from the first stage (which entered it through its controls) plus the flow from its power jet. Obviously, either the second stage had to be larger than the first stage, or some fluid had to be dumped, or both. Figure 9.4 shows one of the originally devised cascaded pairs with interstage bleeds and Figs. 9.3 and 9.5 show staging using progressively larger units

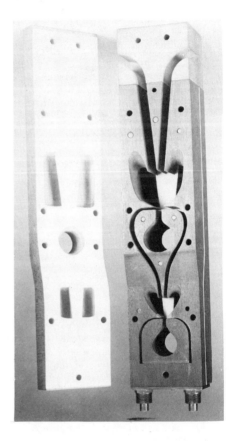

Fig. 9.3 Staged amplifier with cover plate.

without bleeds. (The nozzles for the second and third stages of the amplifier of Fig. 9.4 were removable and one of these nozzles has been removed in the photograph shown.)

Staging closed (without bleeds) devices is appreciably more difficult than staging open (i.e., vented) devices because excess flow cannot be dumped, nor, if needed, can additional flow be obtained from the atmosphere.

The amplifier of Figs. 9.3 and 9.5 was designed by Horton and built in 1961 by Salvadore Peperone as part of the five-stage, closed, proportional amplifier shown in Fig. 9.6. A power gain of over 200,000 was obtained.

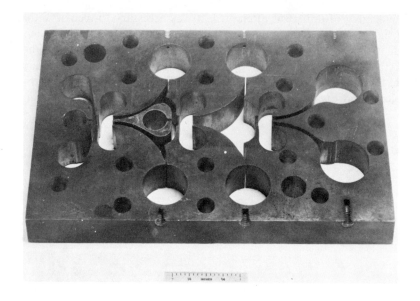

Fig. 9.4 Amplifier with interstage bleeds.

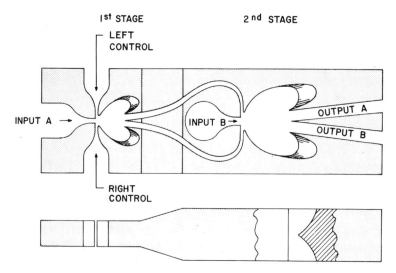

Fig. 9.5 Staged units without vents.

9.1 Historical Introduction

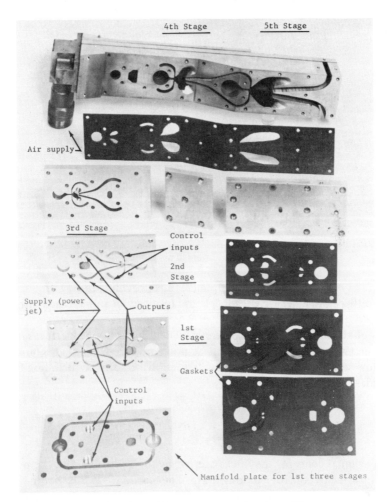

Fig. 9.6 Closed five-stage proportional amplifier.

One of the problems solved in this design was caused by negative feedback, which cuts down the gain. This can best be explained using Fig. 9.7. As the jet is deflected to the right, for example, not all of it will pass into the receivers; some will go into the right cavity where it will act to force the power jet to the left.

In an open device this cavity is vented to the atmosphere and feedback is decreased. Horton [1b] solved this problem for closed devices by building them as shown in Fig. 9.5. The walls of the cavity are shaped to cause the fluid to move up or down into the third dimension. The fluid passes through one of a pair of holes in the gasket, as shown in Fig. 9.6, and then through the cavity in the top plate (Fig. 9.3), back through the companion hole in the gasket, and into the

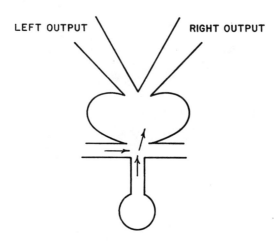

Fig. 9.7 Unvented amplifier.

cavity on the opposite side of the amplifier, thus providing (at low frequencies) a positive rather than a negative feedback and resulting in an increased rather than a decreased gain at low frequencies.

Although the use of the amplifier to steer a demonstration vehicle was reported in October 1962 by Palmisano [2], it was not until May 1964 that Roffman [3], after a number of discussions with Horton, published a paper detailing the concepts underlying this amplifier.

9.2 BASIC OPERATING PRINCIPLES

In the original concepts the forces exerted by the control jets on the power jet were mainly due to control jet momentum. The reason for this is that in the early amplifiers the control nozzles were set back a considerable distance from the edges of the power jet. However, when the setback is reduced sufficiently the control flow no longer depends on the control nozzle width alone. Instead, the control flow becomes a function of the distance from the control edge to the side of the power jet. As a result pressure builds up on the power jet sides and pressure forces act to deflect the power jet. Although the controls may exert both momentum and pressure forces in the same device, we will discuss these separately.

9.2.1 Pressure-Controlled Amplifiers

The first discussion of the control of jets by pressure rather than momentum as well as the first major technical publication on fluid amplifiers was the Sc.D. thesis of Brown [4]. Among the number of interesting concepts discussed in that thesis were pressure-controlled amplifiers and the conditions that determine

9.2 Basic Operating Principles

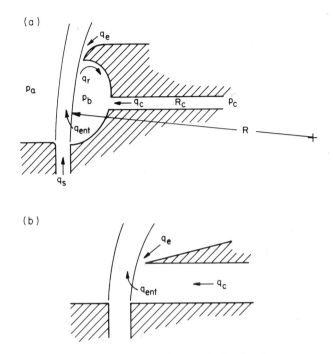

Fig. 9.8 Pressure-controlled one-sided devices: (a) knife edge downstream and (b) control edge setback.

whether a device is proportional or bistable. Both single-sided and double-sided devices were considered. Figure 9.8 shows single-sided models, and Fig. 9.9 illustrates how source and load characteristics may be used to predict the conditions for bistability. The load characteristics are obtained as in Fig. 9.9a, which gives the various flows as functions of the bubble pressure p_b. It is assumed that the jet entrainment flow q_{ent} does not change with pressure, thus giving a straight line as shown. The flow entering from downstream through the region between the power jet and the knife edge is, as shown in Fig. 9.8a, designated as q_e. When the pressure p_b in the bubble is low, the jet seals off the bubble from the downstream region so that q_e is zero; as p_b increases, the space between the power jet and the knife edge allows flow to pass; obviously when the pressure p_b becomes equal to p_a (the ambient pressure), there will no longer be flow from downstream into the bubble. That the return flow q_r varies with bubble pressure, as shown in 9.9a, is also obvious. Since the control flow q_c must be related to the other flows by

$$q_c = q_{ent} - q_e - q_r$$

the load characteristic has the form given in Fig. 9.9b by the solid-line curve.

The source characteristic is obtained by considering the way the control flow q_c changes with bubble pressure p_b for a fixed control pressure p_c, where the resistance R_c of Fig. 9.8a is assumed linear. Three typical control source

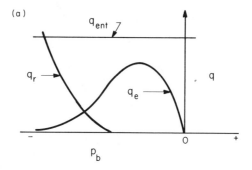

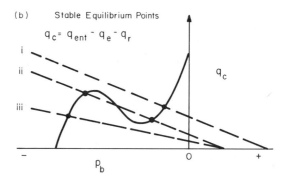

Fig. 9.9 Source and load characteristics: (a) flows into and out of bubble, (b) matching control flow source–load characteristics.

characteristics (i, ii, iii) are shown as dashed lines on Fig. 9.9b. Curve i is obtained for some p_c and some R_c. If p_c is decreased as R remains fixed, curve ii, which has the same slope but a different intercept, is obtained. If we use the same p_c as for curve ii but increase the resistance R_c, source characteristic iii results.

The intersections of the source and load curves that are marked by solid dots are the stable operating points. Examination of these intersections shows that the source–load combination is monostable for sources i and iii. The intersection of the load characteristic with source characteristic ii occurs at three points. The middle point of intersection, however, is unstable. To understand this, suppose that the operating point is at the middle intersection and that there is a small increase in bubble pressure (an infinitesimal disturbance). The source will now supply more flow than the load demands. The additional flow acts to increase the bubble pressure. In effect a regenerative cycle is set up. The cycle terminates when the operating point reaches the rightmost point of intersection. At this point a small increase of bubble pressure causes the load to demand more flow than the source can supply. As a result the bubble pressure decreases. Both the rightmost and leftmost points of intersection of source ii and the load are

9.2 Basic Operating Principles

therefore stable operating points so that the device is bistable under these conditions.

Brown points out that the device can be made monostable for the R_c and p_c of case ii by altering the load characteristics so that the right-hand peak is lowered enough that it does not intersect the ii source curve. This can be done by extending the knife edge so that there is return flow at higher bubble pressures.

The two-sided device is appreciably more complicated because the jet curvature is not uniquely related to the bubble pressure on the same side but depends on the difference in pressure between the two bubbles. Thus, there is a family of load characteristics instead of only one. Brown outlines a technique for using the source and load characteristics of the single-sided device to predict the characteristics of the double-sided device.

9.2.2 Momentum-Controlled Amplifiers

The pressure-controlled amplifier depends primarily on pressure to deflect the jet; the momentum-controlled amplifier may also be somewhat dependent on pressure. The major difference between the two types is that amplifiers depending largely on momentum control have large clearances between the control edge and the power jet edge (the spacing D of Fig. 9.10 is large compared to the jet width).

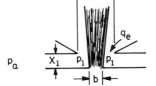

Fig. 9.10 Control region.

In October 1962, the first gain analysis of the jet-deflection proportional fluid amplifier was published. This paper [5] by Peperone et al. examined an amplifier with the configuration shown in Fig. 9.11 to show how the gain is affected by the geometry of the amplifier and took into account such parameters as the distance from the nozzle to the splitter and the width of the receiving apertures. To obtain numerical results, Albertson's profile was used for the velocity distribution. The vents to atmosphere were made very large with respect to the receiver apertures and, in the theoretical development, it was assumed that the velocity profile was unaffected by the presence of the receivers. Both pressure and momentum were assumed to be effective in deflecting the power jet.

Theory and experiment were in reasonably good agreement for small differences in control pressure and flow but differed by about 20% in the vicinity of maximum output difference, both the flow gain and the pressure gain being higher than that predicted by theory. A more recent analysis by Douglas and Neve [6], using similar but somewhat different calculations, shows better agreement with experimental data.

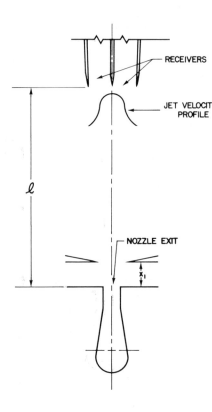

Fig. 9.11 Open amplifier configuration.

9.3 INTRODUCTION TO AMPLIFIER STATIC ANALYSIS

To use a proportional amplifier, the fluid systems designer must know the input impedance, the output impedance, and the transfer characteristic. Under static or slowly varying conditions this is precisely the information provided by the static characteristic curves that were described generally in Chapter 6.

In the following sections of this chapter we will perform a static analysis of the proportional amplifier. This analysis will determine the relation between the amplifier geometry and the signal variables at the input and output terminals. Thus, for any specific geometry the analysis can provide approximately the same information that could be obtained from experimental characteristic curves.

The static analysis is divided into four sections. The first two, Sections 9.4 and 9.5, consider the forces along the edges of the power jet that act to deflect it. Specifically, in Section 9.4 we relate the control signal variables to the power jet deflection and derive an expression for the input resistance characteristic. In Section 9.5 we discuss briefly the ways in which the jet deflection may be enhanced or reduced as the jet passes through the vent region. Section 9.6 then relates the

9.3 Introduction to Amplifier Static Analysis

jet deflection to the maximum pressure that can be recovered at the output terminals. This and the control-deflection relation of Section 9.4 combine to form the transfer characteristic. We are also able to use this information to find the maximum pressure gain of a proportional amplifier stage. Section 9.7 deals with the output resistance in a highly simplified manner and completes the static analysis.

A dynamic analysis of the proportional amplifier begins in Section 9.8 and continues for the remainder of the chapter.

9.3.1 Preliminary Analytical Considerations

We will emphasize pressure-controlled amplifiers in our analysis because pressure control is used for laminar amplifiers, and laminar amplifiers have significant advantages over turbulent amplifiers.

The shape of the nozzle and the power jet pressure have an effect on the jet profile and on the noise present. In particular, when the conditions are such that laminar flow is obtained, there are significant differences from the cases where turbulent flow is used. Powell [7] has discussed the possibility of using laminar jets in proportional devices and described the results of several experiments on the deflection of laminar jets. One of the primary problems in connection with using laminar jets is their instability; however, it is possible to direct a stream against a laminar jet without making it go turbulent, as was shown by Powell in his 1962 studies.

Within the confines of top and bottom plates, moreover, experiments by Manion and Mon [8] have shown that laminar flow is possible at Reynolds numbers as high as 1600 where the critical dimension is the distance between the top and bottom plates.

Manion and Mon [8] also found that a long power nozzle caused their laminar jet to become fully developed (in the height dimension) and that this therefore resulted in greater jet spreading and hence a lower gain and a lower pressure recovery. However, studies by Van Tilburg *et al.* [9] on the effect of power nozzle throat length on the gain for the case of *turbulent* jets show only relatively small dependence.

9.3.1.1 Jet Profile

If laminar jets are to be considered, and particularly, if both laminar and turbulent jets are to be considered, the logical choice for the jet profile is the hyperbolic secant squared distribution which, with slight changes, can be used both for turbulent flow (modified Goertler profile) and for laminar flow (modified Schlichting–Bickley profile). The modified distribution [10], expressed in Eq. (4.144), is repeated here for convenience,

$$u = k_3(x) \operatorname{sech}^2 k_4(x) y \qquad (9.1)$$

where for turbulent flow (Goertler)

$$k_3(x) = \tfrac{1}{2}[3J\sigma_e/\rho(x+x_o)]^{1/2} \tag{9.2a}$$

$$k_4(x) = \sigma_e/(x+x_o) \tag{9.2b}$$

$$x_o = b\sigma_e/3 \tag{9.2c}$$

and for laminar flow (Schlichting–Bickley)

$$k_3(x) = \frac{1.5(J/6\rho)^{2/3}}{v^{1/3}(x+x_o)^{1/3}} \tag{9.3a}$$

$$k_4(x) = \frac{\tfrac{1}{2}(J/6\rho)^{1/3}}{v^{2/3}(x+x_o)^{2/3}} \tag{9.3b}$$

$$x_o = bN_R/36 \tag{9.3c}$$

and $N_R = bu_0/v$, $J = \rho u_0 b^2$ and u_0 is the velocity at the power nozzle.

9.4 ANALYSIS OF INPUT REGION

Forbes Brown's illustrations (Figs. 9.8 and 9.9) show that the pressure in the interaction region is affected by entrainment and by vent flow as well as by the control flow. The power jet acts as a barrier to the control flow because the control flow must pass between the power jet and the control edge. The closer the control edge is to the power jet, the greater the restriction. This fact is illustrated in Fig. 9.12 by data of Van Tilburg et al. [9]. The figure shows that the input impedance decreases as the control edge width (the distance between the edges of the two controls) is increased with a power jet pressure of 5 psig but is independent of spacing when there is no jet to act as a barrier.

Because of the entrainment caused by jets, bringing the control edges closer together (decreasing the setback) appreciably affects the pressure in the vicinity of each edge in a way that is dependent on the distance between the power jet and adjacent edge. Because of this dependence, the pressure tends to decrease as the jet moves toward an edge resulting in easier deflection of the jet. However, when the edge gets too close to the jet, some jet flow spills into the control region and raises the pressure.

If the edge is sufficiently long in the direction of flow of the main jet, the jet will attach so that the device will become a binary switch. However, if the wall is kept relatively short in the direction of the jet (Fig. 9.13), the pressure will be lowered without necessarily obtaining attachment, so that in principle this is a means of increasing the gain.

In the type of geometry originally examined by Manion [11], later by Manion and Goto (unpublished), and then in more detail by Foss [12] (Fig. 9.14), it was found that the effect of the pocket shown was to increase the deflection over that expected from the momentum flux alone. Although the increased gain was

9.4 Analysis of Input Region

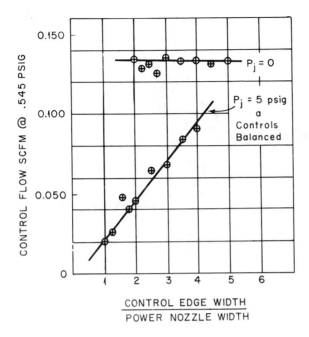

Fig. 9.12 Effect of control edge to jet edge clearance on input impedance.

attributed to the pressure forces on the higher-pressure side, a more complete explanation includes the reduction of pressure in the opposite pocket and its dependence on jet position. Foss' results show that smaller values of Y (Fig. 9.14) increase the jet deflection. This effect is the primary mechanism involved in the patents of Pan [13] and Manion [14].

In Manion's device close walls are used. Attachment is, however, prevented by keeping the effective length of the wall, to which attachment might occur, short. This is done by moving the splitter in very close (three nozzle widths) and providing vents to the surroundings.

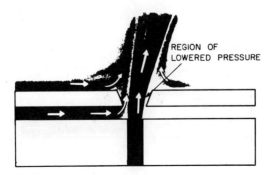

Fig. 9.13 Control edge with short wall.

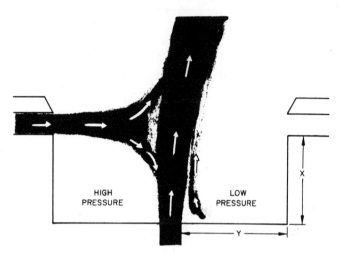

Fig. 9.14 Pockets added to increase gain.

A device with an interaction region shaped as in Fig. 9.15 has been designed by Pavlin [15]. The input impedance of this amplifier can be made very high at small control pressures since it depends on the downstream opening size with respect to the nozzle width. Although its pressure recovery and its pressure gain are only fair, the device has very good flow gain.

An amplifier that combines the attributes of Manion's high-pressure recovery amplifier and Pavlin's amplifier as reported by Howland [16] incorporates two sets of vents (Fig. 9.16). The primary purpose of the lower pair of vents is to prevent attachment to the closely placed walls as in Manion's device, but Howland indicates that this pair may also be used as a second set of controls. Using the lowest pair of openings (that is, the conventional controls), pressure

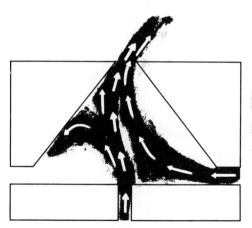

Fig. 9.15 High input impedance control region.

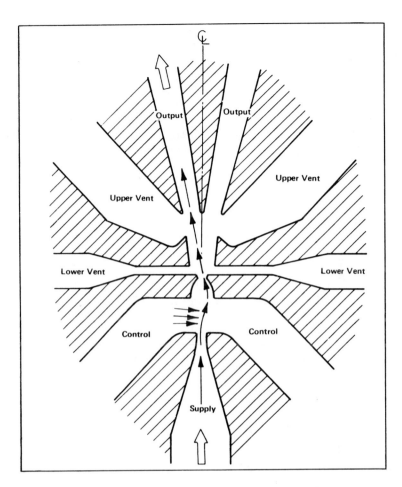

Fig. 9.16 High input impedance high gain amplifier.

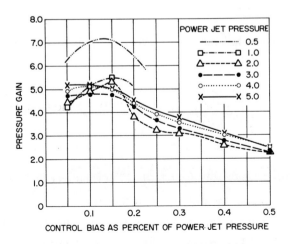

Fig. 9.17 Gain as a function of control bias.

gains between 16 and 20 have been obtained. Use of the lower vents as controls results in a gain of approximately 3.

The data show (Fig. 9.17) that the gain of jet deflection amplifiers is affected by the dc control bias (the average of the left and right control pressures), but to date no experiment has been made to determine the reason. The fact that the gain is dependent on the pressure in the interaction region is used as a method of varying the gain in a jet deflection amplifier patented by Evans [17]. The pressure in the interaction region of this device is varied by inserting flow, which is obtained from a separate source, through an additional inlet (see Fig. 9.18).

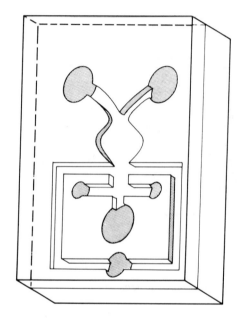

Fig. 9.18 Variable gain amplifier.

9.4.1 Analytical Formulation of the Interaction Region for Open and Closed Controls

As the jet proceeds downstream it entrains flow so that the volume flow increases. At $x = x_1$ (the control width) of Fig. 9.10, we find from Eqs. (9.1)–(9.3) that the volume flow $[h \int_{-\infty}^{+\infty} u \, dy]$ is

$$q(x_1) = hbu_o(1 + 3x_1/b\sigma_e)^{1/2} \quad \text{for turbulent flow} \quad (9.4a)$$

and

$$q(x_1) = hbu_o[1 + 36x_1/N_R b] \quad \text{for laminar flow} \quad (9.4b)$$

The flow entrained q_{ent} in the distance x_1 is therefore

$$q_{\text{ent}} = hbu_o[(1 + 3x_1/b\sigma_e)^{1/2} - 1] \quad \text{for turbulent flow} \quad (9.5a)$$

$$q_{\text{ent}} = hbu_o[(1 + 36x_1/N_R b)^{1/3} - 1] \quad \text{for laminar flow} \quad (9.5b)$$

9.4 Analysis of Input Region

When the jet is undeflected, half of this flow is entrained through each control. Let us calculate the lowering of pressure near the jet caused by this flow when the control nozzles are open to atmospheric pressure p_a and when the control nozzles are closed (our results are only approximations because in the vicinity of the nozzle the jet models we are using do not have the same entrainment characteristics as the actual jets).

To make this calculation we need to know the width of the jet at $x = x_1$. In order to relate the jet width to the nozzle width we define the jet width $b_1(x_1)$ as

$$b_1(x_1) \equiv [q(x_1)/q(0)]b \qquad (9.6)$$

From Eq. (9.4) and the definition given in Eq. (9.6),

$$b_1(x_1) = b(1 + 3x_1/b\sigma_e)^{1/2} \qquad \text{for turbulent flow} \qquad (9.7a)$$

$$b_1(x_1) = b[1 + 36x_1/N_R b]^{1/3} \qquad \text{for laminar flow} \qquad (9.7b)$$

We can also write the entrained flow for both cases as

$$q_{\text{ent}} = hu_o(b_1 - b) \qquad (9.8)$$

To get some idea of the rate of turbulent jet spreading consider that experiments have shown that, for the confined jet, $\sigma_e \cong 10$. Then, if x_1 is approximately equal to b,

$$b_1(x_1) = 1.14b$$

Thus the width changes about 14% in moving one nozzle width downstream. The laminar jet spreading depends on the Reynolds number. For example, when $N_R = 100$ and $x_1 = b$, $b_1(x) = 1.10b$. As N_R increases, the jet width decreases.

We are now in the position to calculate the pressure p_1 on the sides of the jet (Fig. 9.10). If we ignore energy losses, then the flow q_c into each control due to the difference between the ambient pressure p_a and the internal pressure p_1 is given by

$$q_c = x_1 h[2(p_a - p_1)/\rho]^{1/2} \qquad (9.9)$$

If the vent region is also at ambient pressure and the jet region near the control is below ambient because of entrainment, the edge flow q_e moves from the vent into the interaction region (Fig. 9.10). The edge flow may, therefore, be expressed as

$$q_e = [(D - b_1)/2]h[2(p_a - p_1)/\rho]^{1/2} \qquad (9.10)$$

where D must be greater than b_1. From Eqs. (9.9) and (9.10) we may observe that the clearance between the control edge and the jet, $(D - b_1)/2$, and the control width x_1 determine q_e and q_c at the same pressure difference. This situation is analogous to two resistances in parallel. Now

$$q_e + q_c = q_{\text{ent}}/2 \qquad (9.11)$$

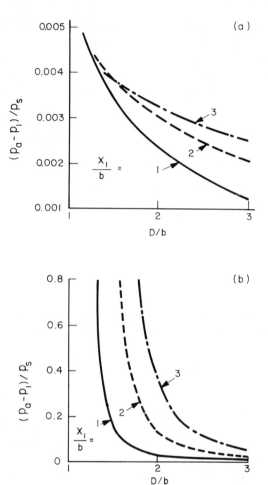

Fig. 9.19 Jet-side pressure as a function of offset: (a) open controls (turbulent jet), (b) closed controls (turbulent jet).

so that we may combine Eqs. (9.9)–(9.11) to obtain

$$p_a - p_1 = \rho q_{ent}^2/2h^2[D - b_1 + 2x_1]^2, \qquad D > b_1 \qquad (9.12)$$

The substitution of Eq. (9.8) into (9.12) leads to

$$\frac{p_a - p_1}{p_s} = \left(\frac{b_1 - b}{D - b_1 + 2x_1}\right)^2, \qquad D > b_1 \qquad (9.13)$$

where $p_s \cong \rho u_o^2/2$. When the controls are closed the entrained flow equals the edge flow, and Eq. (9.13) becomes

$$\frac{(p_a - p_1)}{p_s} = \left(\frac{b_1 - b}{D - b_1}\right)^2, \qquad D > b_1 \qquad (9.14)$$

9.4 Analysis of Input Region

For turbulent flow we may use Eq. (9.7a) in conjunction with Eqs. (9.13) and (9.14) so that

$$\frac{p_a - p_1}{p_s} = \left[\frac{(1 + 3x_1/\sigma_e b)^{1/2} - 1}{(D + 2x_1)/b - (1 + 3x_1/\sigma_e b)^{1/2}}\right]^2$$

$$D > (1 + 3x_1/\sigma_e b)^{1/2} \quad \text{(open controls)} \qquad (9.15a)$$

$$\frac{p_a - p_1}{p_s} = \left[\frac{(1 + 3x_1/\sigma_e b)^{1/2} - 1}{D/b - (1 + 3x_1/\sigma_e b)^{1/2}}\right]^2$$

$$D > (1 + 3x_1/\sigma_e b)^{1/2} \quad \text{(closed controls)} \qquad (9.15b)$$

Equations (9.15) are shown plotted in Fig. 9.19 for $\sigma_e = 10$. When the controls are open (Fig. 9.19a) the pressure along the jet edges is only slightly below atmospheric. The pressure approaches atmospheric as the setback distance increases. For a given setback the pressure p_1 decreases with increases in control width.

When the controls are closed (Fig. 9.19b) p_1 once again approaches atmospheric pressure for large setbacks. However, the pressure difference increases without limit for small setbacks. This, of course, does not happen. The anomalous behavior occurs because we have treated the jet as a solid barrier and assumed that entrainment is independent of the pressure in the jet vicinity. Thus Eq. (9.15) is reasonably accurate only when the setback ratio D/b is greater than 2.

9.4.2 Analytical Formulation of the Interaction Region for Pressurized Controls

The deflection of the jet in a beam deflection amplifier is accomplished in general by both pressure and momentum; however, when the edges of the controls are close to the jet (i.e., when the offset is small) pressure effects become predominant. Small clearances between the jet edges and the adjacent control edges result in high input resistance, which is usually desirable. If the clearances are made sufficiently small, however, and in particular for turbulent jets, the operation of the amplifier becomes extremely sensitive to small variations in these clearances so that it is difficult to obtain the same results from two amplifiers built according to the same pattern, and in general the jet will be biased off-center when the control pressure difference is zero.

Because laminar jets entrain less flow and are much less noisy, the clearances between the jet edges and the control edges can be made small with an appreciable reduction in the problems of operation that result from using small clearances with a turbulent jet.

The advantages of small clearances are not only an increased gain but also a flat saturation characteristic. The pressure difference Δp_r appearing across the entrance to the output of a beam deflection amplifier is related to the deflection y_o by a curve of the form of Fig. 9.20. In a conventional amplifier, the deflection is approximately proportional to the input pressure difference Δp_c, so that the

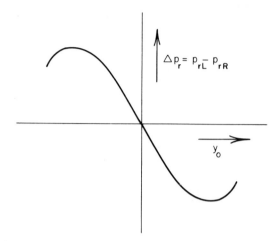

Fig. 9.20 Output pressure difference as a function of jet deflection.

output pressure difference as a function of input pressure difference also has the form of Fig. 9.20. This type of characteristic is not desirable. A flat saturation characteristic (one that maintains the maximum output difference for large values of input difference) is much preferable.

If the control edges are placed at the correct distance, part of the power jet will be deflected into the opposite control when the deflection corresponding to maximum output is reached. At that point, slight deflections of the jet into the right control by raising the pressure in the left control will cause the pressure in the right control also to rise. Hence, all further increases in pressure difference across the controls cause only slight changes in output pressure difference and yield an essentially flat saturation curve.

Let us consider the deflection of the jet caused by the application of control pressure. The jet deflection may be due to pressure forces, momentum forces, or a combination of both. To determine which forces predominate, note that for a control edge distance D the clearance between the control edge and the jet is $\frac{1}{2}(D - b_1)$.

Now if $D - b_1(x_1)$ is large compared to x_1, the pressures near the jet (p_1, p_2) will both be approximately equal to the vent pressures, and hence Δp will be small where $\Delta p \equiv p_2 - p_1$ (see Fig. 9.21). Thus, the deflection of the jet will be almost entirely due to momentum.

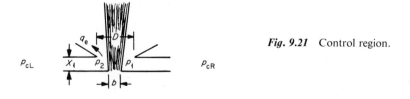

Fig. 9.21 Control region.

9.4 Analysis of Input Region

On the other hand, if $D - b_1$ is small compared to x_1, the momentum effects will be negligible and the deflection will be primarily due to Δp.

We will examine the latter case (the pressure-controlled amplifier) in some detail in the following development.

When the momentum is negligible the deflection y_{o1} in the vicinity of the control under the application of a steady pressure difference Δp is [from Eq. (5.20a)]

$$y_{o1} = x_1^2 \, \Delta p / 2 b \rho u_o^2 \qquad (9.16)$$

under the assumption that the jet width and velocity are approximately constant over the distance x_1.

If p_2 and p_1 are the pressures in the interaction region (Fig. 9.21) then we have

$$p_s - p_3 = \tfrac{1}{2} \rho u_o^2 \qquad (9.17a)$$

where p_s is the supply pressure and

$$p_3 \cong (p_1 + p_2)/2 \equiv \text{control bias pressure} \qquad (9.17b)$$

Since $p_3 \ll p_s$ we will assume here that

$$p_s = \tfrac{1}{2} \rho u_o^2 \qquad (9.18)$$

so that Eq. (9.16) can be rewritten as

$$y_{o1} = (x_1^2/4b) \, \Delta p/p_s \qquad (9.19)$$

which agrees with the formulation presented for the jet barrier in Eq. (2.80).

When the jet is deflected, the distance d_L between the left control edge and the jet is

$$d_L = \tfrac{1}{2}(D - b_1) + y_{o1} \qquad (9.20a)$$

whereas for the right control edge

$$d_R = \tfrac{1}{2}(D - b_1) - y_{o1} \qquad (9.20b)$$

Our objective now is to express the control pressure difference $(p_{cL} - p_{cR})$ in terms of the pressure difference across the jet $(p_2 - p_1)$ and the geometric parameters (x_1, D, b). As an interim step, however, we first try to relate p_{cL} with p_2 and p_{cR} with p_1. We continue then by summing the flows on each side of the jet in the interaction region. Thus

$$q_{cL} = q_{eL} + q_{ent}/2 \qquad (9.21a)$$

and

$$q_{cR} = q_{eR} + q_{ent}/2 \qquad (9.21b)$$

The edge flow and the control flow can be in either direction, but the entrained flow is always out of the interaction region. We must, therefore, distinguish three cases for each control.

Case 1 *Flows in from the Controls and out through the Edges.* The pressure conditions are

$$p_{cL} > p_2 > 0, \qquad p_{cR} > p_1 > 0$$

where the ambient pressure is taken as zero. In this case all the pressures are above ambient. The control pressures exceed the side pressures. As a result the control and edge flow are in the same direction. For this case we may express the left-side flows that are compatible with Eq. (9.21a) as

$$q_{cL} = (x_1 h)[2(p_{cL} - p_2)/\rho]^{1/2} \qquad (9.22\text{a})$$

$$q_{eL} = (d_L h)[2(p_2)/\rho]^{1/2} \qquad (9.22\text{b})$$

where the vent pressure is assumed to be at zero (ambient). The right-side flows may be represented similarly.

Case 2 *Flows in from the Controls and in from the Edges.* The pressure conditions are

$$p_2 < 0, \quad p_1 < 0, \qquad p_{cL} > p_2, \quad p_{cR} > p_1$$

Note that the control pressures may be below ambient as long as they are above the side pressures. In this case the control and edge flows are in opposite directions. Thus, to remain compatible with the directions selected in (9.21a) the left-side flows must be expressed as

$$q_{cL} = (x_1 h)[2(p_{cL} - p_2)/\rho]^{1/2} \qquad (9.23\text{a})$$

$$q_{eL} = (d_L h)j[2(p_2)/\rho]^{1/2} \qquad (9.23\text{b})$$

The control flow expression is the same as for Case 1 but the edge flow expression requires the factor $j = (-1)^{1/2}$. Once again we may formulate the right-side flows in a similar manner.

Case 3 *Flow out from the Controls and in from the Edges.* The pressure conditions are

$$p_{cL} \leq p_2 < 0, \qquad p_{cR} \leq p_1 < 0$$

The pressures are all below ambient. The control and edge flows are again in the same direction, but in this case the flow direction is opposite to that of Case 1. The left-side flows that we may use in conjunction with Eq. (9.21a) are

$$q_{cL} = (x_1 h)j[2(p_{cL} - p_2)/\rho]^{1/2} \qquad (9.24\text{a})$$

$$q_{eL} = (d_L h)j[2(p_2)/\rho]^{1/2} \qquad (9.24\text{b})$$

The edge flow is the same as in Case 2, but now the control flow also requires the factor j.

Let us now determine the relation between side pressure, control pressure, and the geometric parameters. We illustrate the procedure for the left-side variables only. In all cases similar relations hold also for the right-side variables. The

9.4 Analysis of Input Region

substitution of Eqs. (9.22)–(9.24) into (9.21) yields (after squaring) the following expressions for the three cases under consideration:

$$(1 + d_L^2/x_1^2)p_2 + (d_L/x_1)p_{ent}^{1/2}p_2^{1/2} + (p_{ent}/4 - p_{cL}) = 0 \quad \text{(Case 1)} \quad (9.25a)$$

$$(1 - d_L^2/x_1^2)p_2 + j(d_L/x_1)p_{ent}^{1/2}p_2^{1/2} + [(p_{ent}/4) - p_{cL}] = 0 \quad \text{(Case 2)} \quad (9.25b)$$

$$(1 + d_L^2/x_1^2)p_2 - j(d_L/x_1)p_{ent}^{1/2}p_2^{1/2} - [(p_{ent}/4) + p_{cL}] = 0 \quad \text{(Case 3)} \quad (9.25c)$$

where

$$p_{ent} = (\rho q_{ent}^2)/(2x_1^2 h^2). \quad (9.25d)$$

We may solve Eqs. (9.25) as a quadratic in $p_2^{1/2}$ and then square the result to obtain p_2. This procedure yields

$$p_2 = \left[p_{ent}D_M + 4p_{cL}D_p - 2\frac{d_L}{x_1}(4D_p p_{cL} p_{ent} - p_{ent}^2)^{1/2}\right]\bigg/4D_p^2 \quad \text{(Case 1)}$$

$$\text{for} \quad p_{cL} \geq \frac{(b_1 - b)^2 p_s}{4x_1^2 D_p} \quad (9.26a)$$

$$p_2 = -\left[p_{ent}D_p + 4p_{cL}D_M - 2\frac{d_L}{x_1}(4D_M p_{cL} p_{ent} + p_{ent}^2)^{1/2}\right]\bigg/4D_M^2 \quad \text{(Case 2)}$$

$$\text{for} \quad -\frac{(b_1 - b)^2}{4x_1^2 D_M} \leq \frac{p_{cL}}{p_s} \leq \frac{(b_1 - b)^2}{4x_1^2 D_p} \quad (9.26b)$$

$$p_2 = -\left[p_{ent}D_M - 4p_{cL}D_p + 2\frac{d_L}{x_1}(-4D_p p_{cL} p_{ent} - p_{ent}^2)^{1/2}\right]\bigg/4D_p^2 \quad \text{(Case 3)}$$

$$\text{for} \quad \frac{p_{cL}}{p_s} \leq -\frac{(b_1 - b)^2}{4x_1^2 D_M} \quad (9.26c)$$

where

$$D_p = d_L^2/x_1^2 + 1, \quad D_M = d_L^2/x_1^2 - 1 \quad (9.26d)$$

The above equations for the three cases all give p_2 in terms of p_{cL} and d_L, but d_L is itself dependent on p_{cL}, p_{cR}, p_2, and p_1 so that we must simultaneously consider both the left and right controls.

For the cases of most interest, a bias control pressure would be considered and the left and right pressures would be allowed to vary antisymmetrically, about this bias pressure. This necessitates solving a fifth-degree equation. For purposes of illustration, however, we will consider a much simpler case, namely one in which the right control is kept at ambient pressure and the left control pressure is allowed to vary.

We thus assume that $p_{cR} = 0$ (corresponding to Case 2 for the right control) and take into account the three cases for the left control.

For the right control, Case 2 for $p_{cR} = 0$ can easily be obtained from Eq. (9.26b):

$$p_1 = -\frac{p_{ent}}{4[(d_R/x_1) + 1]^2} \quad (9.27)$$

We now define Case 1,2 as corresponding to the left control at Case 1 and the right control at Case 2. Similarly in Case 2,2 and Case 3,2 the left digit refers to the case on the left side and the right digit refers to Case 2 ($p_{cR} = 0$).

Case 1,2 Now from Eqs. (9.26a) and (9.27) we may formulate the side pressure difference as

$$\Delta p \equiv p_2 - p_1 = \frac{[p_{\text{ent}} D_M + 4 p_{cL} D_p - 2(d_L/x_1)(4 D_p p_{cL} p_{\text{ent}} - p_{\text{ent}}^2)^{1/2}]}{4 D_p^2}$$

$$+ \frac{p_{\text{ent}}}{4[(d_R/x_1) + 1]^2} \qquad (9.28)$$

Equations (9.8) and (9.25d) give the relation between p_{ent} and the jet parameters as

$$p_{\text{ent}} = p_s \left(\frac{b_1 - b}{x_1}\right)^2 \qquad (9.29)$$

If we insert (9.29) into Eq. (9.28) the result is

$$4 D_p^2 \left(\frac{\Delta p}{p_s} - \frac{(b_1 - b)^2}{4 x_1^2 (d_R/x_1 + 1)^2}\right) = D_M \left(\frac{b_1 - b}{x_1}\right)^2 + 4 D_p \frac{p_{cL}}{p_s}$$

$$- 2 \frac{d_L}{x_1} \left(\frac{b_1 - b}{x_1}\right) \left[4 D_p \frac{p_{cL}}{p_s} - \left(\frac{b_1 - b}{x_1}\right)^2\right]^{1/2}$$

$$(9.30)$$

We wish to know how $\Delta p/p_s$ changes as p_{cL}/p_s changes. Since d_L and d_R are functions of $\Delta p/p_s$, the equation above is of higher degree in $\Delta p/p_s$ than in p_{cL}/p_s; hence, we solve for p_{cL}/p_s as a function of $\Delta p/p_s$.

We therefore write Eq. (9.30) in the form

$$A_{11} = A_{21} + A_{31}(p_{cL}/p_s) - A_{41}[A_{31}(p_{cL}/p_s) - A_{51}]^{1/2} \qquad (9.31a)$$

where

$$A_{11} = 4 D_p^2 \left[\frac{\Delta p}{p_s} - \frac{(b_1 - b)^2}{4 x_1^2 (d_R/x_1 + 1)^2}\right] \qquad (9.31b)$$

$$A_{21} = D_M \left(\frac{b_1 - b}{x_1}\right)^2 \qquad (9.31c)$$

$$A_{31} = 4 D_p \qquad (9.31d)$$

$$A_{41} = 2 \frac{d_L}{x_1} \left(\frac{b_1 - b}{x_1}\right) \qquad (9.31e)$$

$$A_{51} = \left(\frac{b_1 - b}{x_1}\right)^2 \qquad (9.31f)$$

9.4 Analysis of Input Region

The solution of quadratic equation (9.31a) is

$$\frac{p_{cL}}{p_s} = \frac{2(A_{11} - A_{21}) + A_{41}^2 \pm A_{41}[4(A_{11} - A_{21}) + A_{41}^2 - 4A_{51}]^{1/2}}{2A_{31}} \quad (9.32)$$

for

$$p_{cL}/p_s \geq (b_1 - b)^2/4x_1^2 D_p$$

where the positive root must be used.

Case 2,2 From Eqs. (9.26b) and (9.27) the side pressure difference Δp is

$$\Delta p = \frac{-[p_{ent} D_p + 4p_{cL} D_M - 2(d_L/x_1)(4D_M p_{cL} p_{ent} - p_{ent}^2)^{1/2}]}{4D_M^2} + \frac{p_{ent}}{4[(d_R/x_1) + 1]^2} \quad (9.33)$$

In a similar manner to Case 1,2 we find that in this case

$$\frac{p_{cL}}{p_s} = A_{12} + \left(\frac{b_1 - b}{2x_1}\right)^2 - \left(\frac{d_L}{x}\right)\left(\frac{b_1 - b}{x_1}\right)(A_{12}/D_M)^{1/2} \quad (9.34a)$$

where

$$A_{12} = D_M \left[\frac{(b_1 - b)^2}{4x_1^2 (d_R/x_1 + 1)^2} - \frac{\Delta p}{p_s}\right] \quad (9.34b)$$

Case 3,2 From Eqs. (9.26c) and (9.27) the side pressure difference Δp in this case is

$$\Delta p = \frac{-[p_{ent} D_M - 4p_{cL} D_p + 2(d_L/x_1)(-4D_p p_{cL} p_{ent} - p_{ent}^2)^{1/2}]}{4D_p^2} + \frac{p_{ent}}{4[(d_R/x_1) + 1]^2} \quad (9.35)$$

The solution of Eq. (9.35) is

$$\frac{p_{cL}}{p_s} = \frac{-(A_{43}^2 + 2A_{13} - 2A_{23}) \pm A_{43}(A_{43}^2 - 4(A_{23} - A_{13}) - 4A_{53})^{1/2}}{2A_{33}} \quad (9.36a)$$

where

$$A_{13} = -4D_p^2 \left[\frac{\Delta p}{p_s} - \frac{(b_1 - b)^2}{4x_1^2[(d_R/x_1) + 1]^2}\right] \quad (9.36b)$$

$$A_{23} = D_M \left(\frac{b_1 - b}{x_1}\right)^2 \quad (9.36c)$$

$$A_{33} = 4D_p \quad (9.36d)$$

$$A_{43} = 2\frac{d_L}{x_1}\left(\frac{b_1 - b}{x_1}\right) \quad (9.36e)$$

$$A_{53} = \left(\frac{b_1 - b}{x_1}\right)^2 \quad (9.36f)$$

To evaluate Cases 1,2, 2,2, and 3,2 note from Eqs. (9.19), (9.10), and (9.26b) that

$$D_p = \left(\frac{D-b_1}{2x_1} + \frac{x_1 \Delta p}{4b\, p_s}\right)^2 + 1 \tag{9.37a}$$

$$D_M = \left(\frac{D-b_1}{2x_1} + \frac{x_1 \Delta p}{4b\, p_s}\right)^2 - 1 \tag{9.37b}$$

$$\frac{d_R}{x_1} = \left(\frac{D-b_1}{2x_1} - \frac{x_1 \Delta p}{4b\, p_s}\right) \tag{9.37c}$$

Hence, for given values of Δp, p_{cL} can be evaluated.

9.4.3 The Input Characteristic

The input characteristic is the relation between the control pressure and the control flow. This is usually measured with the opposite control open. The condition of the opposite control is of little effect when the offset is relatively large (so that the controls are decoupled) but does have an effect for the close wall case.

Once again we must consider the three cases treated in Section 9.4.2, namely

(1) flow *in* to control, *out* of edge;
(2) flow *in* to control, *in* to edge;
(3) flow *out* of control, *in* to edge.

Case 1 From Eqs. (9.22a) and (9.22b) we find that

$$p_{cL} = (\rho q_{cL}^2/2x_1^2 h^2) + (\rho q_{eL}^2/2d_L^2 h^2) \tag{9.38a}$$

where as in Eq. (9.26a)

$$p_{cL} \geq (b_1 - b)^2 p_s / 4x_1^2 D_p \tag{9.38b}$$

If we substitute for the edge flow q_{eL} the expression given in Eq. (9.21a), Eq. (9.38a) becomes

$$\frac{\rho}{2h^2} q_{cL}^2 \left(\frac{1}{x_1^2} + \frac{1}{d_L^2}\right) - \frac{q_{cL}}{d_L^2} q_{ent} \frac{\rho^{1/2}}{2h^2} + \frac{\rho q_{ent}^2}{8h^2 d_L^2} - p_{cL} = 0 \tag{9.39}$$

We may eliminate q_{ent} from Eq. (9.39) by using Eq. (9.8). The result is

$$q'_{cL} \equiv \frac{\rho^{1/2} q_{cL}}{bh p_s^{1/2}} = \frac{\dfrac{\sqrt{2}\, b(b_1-b)}{2d_L^2} + \left[2b^2\left(\dfrac{1}{x_1^2} + \dfrac{1}{d_L^2}\right)\dfrac{p_{cL}}{p_s} - \dfrac{b^2(b_1-b)^2}{2d_L^2 x_1^2}\right]^{1/2}}{b^2[(1/x_1^2) + (1/d_L^2)]} \tag{9.40}$$

Now d_L is a function of Δp, and from Eq. (9.32) we may relate Δp and p_{cL}. Thus we can find d_L as a function of p_{cL}. For each value of p_{cL} we may then calculate q_{cL} from Eq. (9.40) and obtain the input characteristic.

9.4 Analysis of Input Region

Case 2 From (9.32a) and (9.32b) we obtain

$$p_{cL} = (\rho/2h^2)[(q_{cL}^2/x_1^2) - (q_{eL}^2/d_L^2)] \tag{9.41a}$$

The range of p_{cL} in this case is given following Eq. (9.26b):

$$-(b_1 - b)/4x_1^2 D_M \le p_{cL}/p_s \le (b_1 - b)/4x_1^2 D_p \tag{9.41b}$$

Now as in Case 1 we combine Eqs. (9.8), (9.21a), and (9.41a) to get

$$\frac{\rho^{1/2} q_{cL}}{bh p_s^{1/2}} = \frac{-\dfrac{\sqrt{2}\, b(b_1 - b)}{2 d_L^2} + \left[2b^2 \left(\dfrac{1}{x_1^2} - \dfrac{1}{d_L^2} \right) \dfrac{p_{cL}}{p_s} + \dfrac{b^2 (b_1 - b)^2}{2 d_L^2 x_1^2} \right]^{1/2}}{b^2[(1/x_1^2) - (1/d_L^2)]} \tag{9.42}$$

Equation (9.42) is the input characteristic for Case 2 and can be used in conjunction with Eq. (9.34a) to obtain a relation between q_{cL} and p_{cL}.

Case 3 If we square Eqs. (9.24a) and (9.24b) and add the results together,

$$p_{cL} = -(\rho/2h^2)[(q_{cL}^2/x_1^2) + (q_{eL}^2/d_L^2)] \tag{9.43a}$$

where from (9.26c)

$$p_{cL}/p_s \le -(b_1 - b)^2/4x_1^2 D_M \tag{9.43b}$$

and similarly as before, we obtain from (9.8), (9.21a), and (9.43a)

$$\frac{\rho^{1/2} q_{cL}}{bh p_s^{1/2}} = \frac{\dfrac{\sqrt{2}\, b(b_1 - b)}{2 d_L^2} - \left[-2b^2 \left(\dfrac{1}{x_1^2} + \dfrac{1}{d_L^2} \right) \dfrac{p_{cL}}{p_s} - \dfrac{b^2 (b_1 - b)^2}{2 d_L^2 x_1^2} \right]^{1/2}}{b^2[(1/x_1^2) + (1/d_L^2)]} \tag{9.44}$$

The combination of Eqs. (9.36a) and (9.44) provides a relation between q_{cL} and p_{cL} for this case.

The input characteristic with q_{cL} given in normalized form, i.e.,

$$q'_{cL} = \rho^{1/2} q_{cL}/hb p_s^{1/2}$$

is plotted in Figs. 9.22a–d for turbulent flow and for laminar flow at three Reynolds numbers.

Since the analysis does not consider the effects that occur when the edge of the jet touches the control edge, the calculation is cut off for d_L or $d_R < 0$. It is obvious, however, that at sufficiently negative control pressures the flow must become increasingly negative, and an estimated curve is continued (as a dashed line) for the negative portions for which the analysis no longer holds.

The major difference among the curves is between the turbulent and the laminar cases. This difference is primarily due to the greater flow in the turbulent case. The shapes of the curves are quite similar.

Note that for the turbulent case because of the strong entrainment there is positive flow, even when the control pressure is somewhat negative. As a result, making the flow zero (by blocking the left control, for example) will cause the jet to deflect completely to the left.

For the laminar case, two jet positions are possible for zero flow so that the jet

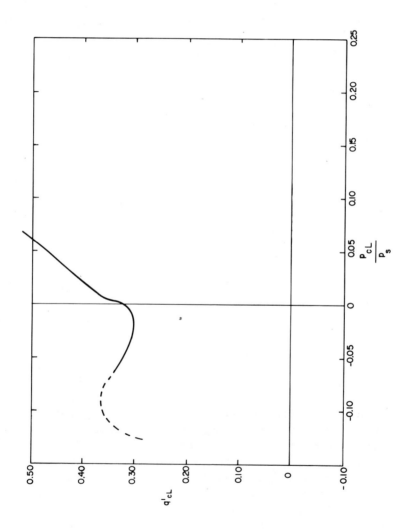

Fig. 9.22a Input flow as a function of left control pressure: (a) turbulent flow, $\sigma_e = 10$, $D = 2$, $x_1 = 4$; (b) laminar flow, $D = 2$, $x_1 = 4$, $N_R = 500$; (c) laminar flow, $D = 2$, $x_1 = 4$, $N_R = 1000$; (d) laminar flow, $D = 2$, $x_1 = 4$, $N_R = 1500$. Parts (b), (c), and (d) are on following pages.

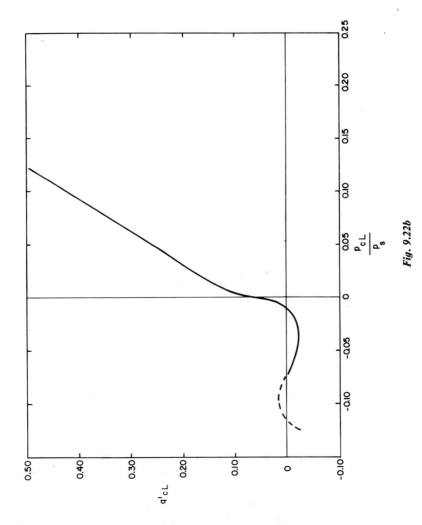

Fig. 9.22b

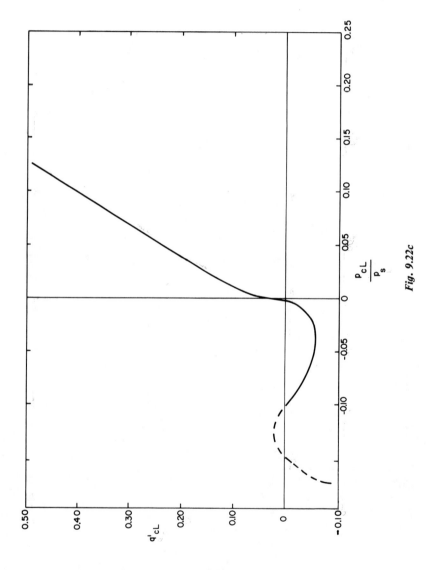

Fig. 9.22c

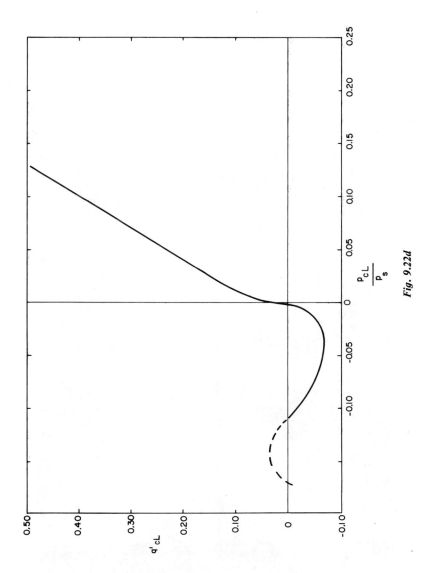

Fig. 9.22d

will stay in either of two positions with the left control blocked and the right control open.

These theoretical curves depart from experimental results for two reasons: First, even when the vents are open to ambient, the vent pressure near the jet is slightly less than ambient because of jet entrainment. As a result, the maximum flow at negative left control pressures is considerably reduced. Second, the analysis assumes that the jet is two-dimensional and that the clearance between the edge of the jet and the control edge becomes zero at a sharply defined boundary. If the deflection is increased past this point, a portion of the jet will be peeled off and tend to raise the pressure p_2 in the control region. In actual fact, however, the jet spreads more near the top and bottom plates than it does in the midplane. As a result, flow begins to peel from the portions of the jet near the top and bottom plates at a deflection for which edge flow can still pass in the mid-

Fig. 9.23a Flow pattern for fully open outlet.

9.4 Analysis of Input Region

Fig. 9.23b Flow pattern for blocked outlet.

plane because the jet width is not as great in the midplane. This tends to flatten out the curve in the negative pressure region; that is, instead of the well-defined maximum and minimum indicated in Figs. 9.22, the curvature is very slight (for slight pressurization of the vents) so that a region of negligible change of flow with pressure (very high input resistance) is obtained. Increased pressurization of the vents both increases the curvature and moves the entire curve to the right; consequently, by pressurizing the vents sufficiently on a laminar amplifier, the curve can be moved over sufficiently so that the device will become bistable.

The flow pattern in a laminar close-wall amplifier is shown in Figs. 9.23a and 9.23b corresponding to fully open and blocked outlets, respectively. The jet has a tendency to deflect to one side or the other unless control flow exists. Note that the control flow causes a slight pinching of the power stream as opposed to our analytical model which assumes an inflexible barrier.

We emphasize the important point that, for $\Delta p/p_s$ small, the two input controls are decoupled; that is, the flow in one control has almost no effect on the other. For larger values of $\Delta p/p_s$ or if the control edges are sufficiently close together, the deflection of the stream can appreciably affect the flow past the control edge. As a result of this modulation of the clearance between the edge of the control and the edge of the jet, it is possible to obtain a larger pressure difference across the jet than is applied across the controls. This is a positive feedback effect that can measurably increase the deflection of the jet and thereby the gain for a given applied pressure and, as our example shows, results in a negative input resistance.

A further point to note is that for turbulent two-dimensional flow the jet angle of spread is independent of the momentum, and the turbulent jet within the amplifier, although not really two-dimensional, also acts in a somewhat similar manner. If the supply pressure is lowered, however, to the point where the flow becomes laminar, the spread angle becomes a function of the Reynolds number. The clearance between the jet and the control edge will then depend on the supply pressure.

9.5 EFFECT OF THE VENTS

The vents affect the gain of the amplifier. We have already noted that if the vents of Fig. 9.2 are completely closed off or are small and the jet is then deflected (for example, toward the right), the right cavity will build up a pressure, and there will be return flow toward the power jet, which will decrease the jet deflection at low frequencies where the difference between the time of the return flow signal and the original control signal results in the return signal being sufficiently out of phase. At a high enough frequency, however, the return signal will be in phase with the control signal, resulting in increased deflection at the higher frequencies.

We have also noted that pressurizing the vents shifts the input pressure–flow characteristic to the right and tends to increase the negative resistance on small-offset devices.

Where the primary concern is with low frequencies, it may be desirable to convert the negative feedback in the vents to a positive feedback by Horton's crossover technique [1] already described. On the other hand, positive feedback may be obtained as shown by Dexter [18] without crossover by designing the vent so that the flow out of it tends to entrain additional flow so as to decrease the pressure on the side of the amplifier toward which the jet has been deflected, as shown in Fig. 9.24. This, in general, means that the vent opening must be large enough to allow the fluid from the jet to pass through without appreciable restriction but small enough to prevent fluid from the surroundings from leaking past the jet into the cavity, and that the receiver widths must be small so that the jet flow into the vent has appreciable momentum.

In spite of the increased gain possible at low frequencies, this effect to date has not been sufficiently well understood to make it useful.

9.5 Effect of the Vents

Fig. 9.24 Gain increase by vent and receiver design.

It is found in general that momentum effects of this kind (and others) give rise to circulations (vortices) within the vent which tend to produce low-frequency oscillations.

These momentum effects will, of course, exist even though the vent opening is large rather than partially closed off as in Fig. 9.24; however, partial closing of the vent increases the circulation.

Another closely related momentum effect that causes problems is the reflection of a jetlike stream from the receivers. When a receiver is blocked or partially blocked, the power jet does not simply stagnate against the receivers, but instead a portion of the jet is reversed and streams back into the vent region thus contributing to vortices and oscillations as discussed by Brown and Humphrey in Part 2 of Ref. [19]. A set of diagrams from their paper showing this oscillation is reproduced in Figs. 9.25a–c.

A vent design that reduces these circulations was conceived by Griffin [20] and was incorporated into the laminar amplifier designed by Manion and Mon [8] (Fig. 9.26). The vent consists of an upper and a lower portion. Any reflections from the receivers tend to strike the divider between the upper and lower vents so that they affect at most the upper portion of the power jet and cause negligible jet deflection. The lower vent allows flow from the atmosphere to supply the entrainment needs of the jet. Thus, in operation, flow will usually move inward along the lower vents and outward along the upper vents for turbulent amplifiers without the mixing and the circulations that occur when the divider is not present (Fig. 9.25). In the case of laminar amplifiers almost no motion occurs in the lower vent.

Let us now consider for analysis the more common configuration of Fig. 9.27.

When all the flow does not go into the receivers, that which does not constitutes the spilled or returned flow. A problem arises in determining how much flow spills into the right vent and how much into the left vent. The simplest assumption is that all flow in the right half of the jet that does not enter the

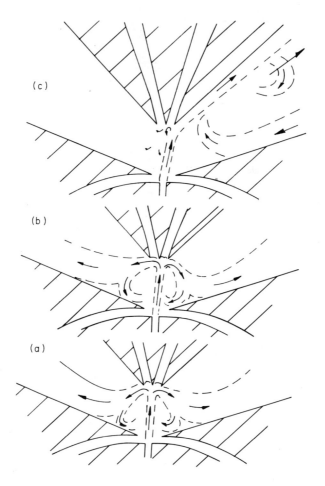

Fig. 9.25 Oscillation caused by vent vortices: (a) jet centered, (b) jet moving to right, and (c) jet at maximum amplitude of oscillation. (From Brown and Humphrey [19].)

right receiver is spilled into the right vent and similarly for the left half. For small deflections this should be approximately correct, particularly when the splitter between the receivers projects sufficiently far upstream as in Fig. 9.23.

If, for example, the jet is deflected to the right, then the flow in the right half of the amplifier in the vicinity of the splitter is

$$q_R = h \int_0^\infty u(l, y - y_o)\, dy = h \int_{-y_o}^\infty u(l, y)\, dy \qquad (9.45)$$

Of this flow some amount q_{oR} will move into the receiver and the rest will go into the vent. If the vent opening is sufficiently large the vent pressure will remain close to ambient so that the returned flow will have only a minor effect on the jet

9.5 Effect of the Vents

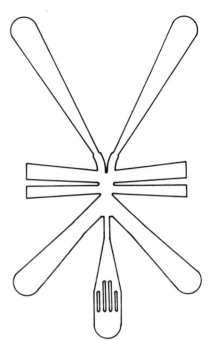

Fig. 9.26 Laminar beam deflection amplifier.

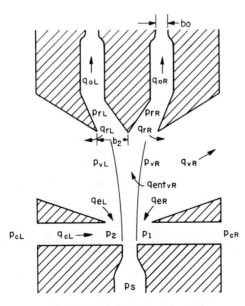

Fig. 9.27 Open vent configuration.

deflection. However, if there is appreciable resistance to vent flow, there can be either a higher or lower pressure than ambient in the vents.

The pressure can be less than ambient because of jet entrainment so that for a large enough receiver width a pressure below ambient can be obtained. For a receiver designed for maximum pressure gain, or if the receiver is partially blocked, there will be spilled flow that will increase the pressure in the vent to above ambient. The returned flow on the right q_{rR} will depend on the jet loading and on the jet deflection. We therefore write

$$q_{rR} = k_5 y_o(l) \tag{9.46}$$

where k_5 depends on the load but is also a weak function of the deflection. For small deflections k_5 should be approximately constant for a given load. From Fig. 9.27 we see that four flows must be considered in the right vent region, the flow q_{entvR} entrained in the vicinity of the vent (from x_1 to l), the spilled flow q_{rR}, the edge flow q_{eR}, and the sum of these flows that makes up the vent flow q_{vR}.

The right vent pressure is then given by

$$p_{vR} = K_v q_{vR}^2$$

where K_v is the vent resistance coefficient.

A pressure Δp_v will then in general exist across the jet in the vicinity of the vent. For large vents this pressure will usually be quite small except at the resonant frequencies.

Some amplifiers include a vent called a center dump between the two receivers. This vent prevents flow spilled from one receiver from entering the other. It also tends to prevent instability under blocked load conditions.

9.6 ANALYSIS OF THE OUTPUT REGION

As was initially pointed out in the analysis of Peperone et al. [5] previously mentioned, the width of the receivers affects the gain characteristics, the pressure recovery, and the energy recovery. It is obvious that the wider the receiver apertures the greater the energy recovered; however, wide apertures decrease the pressure (i.e., the energy per unit mass) recovered. It follows that the receiver widths and the loading of the receivers must be considered as important factors in designing a jet deflection amplifier.

One can take into account the effect of the receiver loading on the flow and hence on the gain by a method given by Manion [21], by Camarata [22], and by Rupert [23], based on the previous efforts of Olson and Camarata [24] and of Kallevig [25].

For a given velocity distribution, Olson and Camarata point out (Fig. 9.28) that the flow actually entering the duct may be either more or less than that portion of the profile intercepted by the receiver. They indicate, however, and

9.6 Analysis of the Output Region   353

data are given by Kallevig [25], that the pressure acting on the receiver is that due to the portion of the dynamic pressure intercepted by the receiver, even when the receiver is completely blocked.

On this basis, Manion [21] assumes that the actual flow into the receiver may be found by determining the load characteristic of the duct (receiver) and then assuming the source pressure to be that obtained by averaging over the dynamic pressure intercepted by the duct.

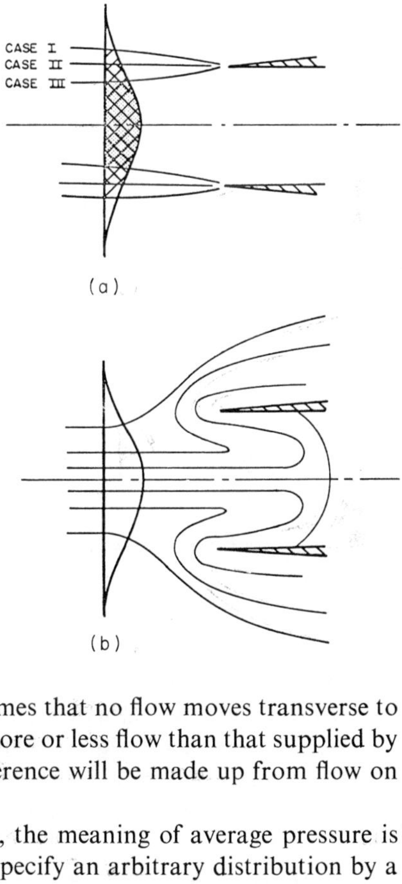

Fig. 9.28 Effect of load on flow accepted by duct: (a) finite load (shaded portion of velocity profile contains mass flow entering diffuser), (b) blocked (infinite) load.

Manion considers two receivers and assumes that no flow moves transverse to the splitter so that, if the receiver requires more or less flow than that supplied by interception of the velocity profile, the difference will be made up from flow on the same side of the splitter.

For any distribution other than uniform, the meaning of average pressure is ambiguous because one cannot in general specify an arbitrary distribution by a single number.

Most authors writing in the fluidic literature have obtained the dynamic pressure by averaging over the square of the velocity $\overline{u^2}$. Some have used the average velocity squared $\bar{u}^2$.

Since the pressure is actually the energy per unit mass, both of which (energy and mass) are conservative quantities, the most meaningful expression to use is the average of the cube of the velocity divided by the average velocity $\overline{u^3}/\bar{u}$; this expression has been used in the fluidics literature only rarely because a good theory is not yet available. Rupert [23] did consider the kinetic energy but was able to obtain a formulation which was valid only for conditions close to that of Case II (Fig. 9.28), i.e., when the flow entering the duct is approximately equal to that part of the profile intercepted by the duct so that loading effects are minimal.

The theoretical calculation of the output pressure is usually done by averaging $\rho u^2(l, y)/2$ over the receiver width where $u(l, y)$ is the velocity distribution in the vicinity of the receiver. For the blocked receiver, however, Reid [26] and others have shown that this method gives too low a value.

A general theoretical description that is adequate for all types of receiver loading has not yet been achieved, although Rupert [23] has obtained good agreement for flows near the design condition. As Fig. 9.28 illustrates, however, for the general case the flow entering the duct can be greater than or less than the design condition. In particular, if the duct is blocked no flow enters the duct; i.e., all the flow that enters is reversed and is spilled.

When flow is neither entrained nor spilled, the total pressure at the jet input is given by Rupert as the kinetic energy entering the duct divided by the mass flow into the duct. Since pressure is by definition the kinetic energy per unit mass, this procedure is more reasonable than the commonly used procedure of assuming that the pressure at the receiver is given by integrating the profile dynamic pressure at the receiver width. However, the latter procedure is simpler and gives results that are acceptable even for blocked loads. Because a better procedure accurate for all cases does not exist, we will use this method.

9.6.1 The Transfer Characteristic

The transfer characteristic gives the output pressure difference as a function of the pressure difference across the controls.

Assume that the undeflected jet velocity distribution at the splitter is $u(l, y)$. Let each receiver be of width b_2 (Fig. 9.27). The actual flow into the receiver is not necessarily determined by integrating the velocity profile over the receiver width since the receiver impedance must be taken into account. Thus the flow into the duct depends essentially on the pressure produced by the flow profile and on the receiver loading.

When the receivers are partially blocked, some of the jet flow will spill. If, for example, the jet is deflected toward the right receiver, then the pressure in the vicinity of the right receiver will be greater than that near the left receiver, and some of the flow spilled from the right receiver will be forced toward the left side of the amplifier. This effect tends to cause the output pressure to vary in a nonlinear manner with input pressure. This transverse motion can be minimized if the splitter is made long enough (Fig. 9.26). A center dump may also be used.

In the following analysis we assume that the splitter is long enough so that spilled flow does not move across it.

9.6 Analysis of the Output Region

The method usually used is to define an average pressure at each receiver as

$$p_{rR} = \frac{\rho}{2b_2} \int_0^{b_2} u^2(l, y) \, dy \tag{9.47}$$

where p_{rR} is the average pressure at the right outlet.
For the undeflected jet $p_{rL} = p_{rR}$ and $\Delta p_r = (p_{rL} - p_{rR}) = 0$,

$$\Delta p_r = 0 = \frac{\rho}{2b_2} \int_{-b_2}^{0} u^2(l, y) \, dy - \frac{\rho}{2b_2} \int_0^{b_2} u^2(l, y) \, dy \tag{9.48a}$$

If the jet is deflected by an amount y_o, then

$$\Delta p_r = \frac{\rho}{2b_2} \int_{-b_2}^{0} u^2(l, y - y_o) \, dy - \frac{\rho}{2b_2} \int_0^{b_2} u^2(l, y - y_o) \, dy \tag{948b}$$

If the distribution is of the hyperbolic secant form given in Eq. (9.1), i.e., if

$$u(l, y) = k_3(l) \, \text{sech}^2 \, k_4(l) y \tag{9.49}$$

then Eq. (9.48b) becomes

$$\Delta p_r = \frac{\rho k_3^2}{2b_2} \int_{-(y_o + b_2)}^{-y_o} \text{sech}^4 \, k_4 y \, dy - \frac{\rho k_3^2}{2b_2} \int_{-y_o}^{(b_2 - y_o)} \text{sech}^4 \, k_4 y \, dy \tag{9.50}$$

where k_3 and k_4 are defined for turbulent jets and laminar jets in Eqs. (9.2) and (9.3) with $x = l$. After integration, Eq. (9.50) becomes

$$\Delta p_r = \frac{-\rho k_3^2}{2b_2 k_4} \{ 2 \tanh k_4 y_o - \tfrac{2}{3} \tanh^3 k_4 y_o + \tfrac{1}{3} \tanh^3 k_4(y_o + b_2)$$
$$+ \tanh k_4(b_2 - y_o) - \tfrac{1}{3} \tanh^3 k_4(b_2 - y_o) - \tanh k_4(y_o + b_2) \} \tag{9.51}$$

This gives us the pressure difference at the receiver as a function of the deflection y_o and the receiver width b_2. Figure 9.20 shows that Δp_r initially increases as y_o increases but reaches a maximum for some value of y_o after which a further increase in y_o causes Δp_r to decrease. This is in general an undesirable property in an amplifier. For this reason various methods have been used to arrange the amplifier in such a way that the deflection cannot become greater than that value of y_o corresponding to the maximum. For example, Griffin [20] built a momentum-deflected amplifier that has the controls at an acute angle to the amplifier so that the right control stream forms a jet aimed at the left receiver if no power jet is present (and the left control is similarly pointed at the right receiver). As a result the power stream cannot be deflected past the receivers regardless of the control pressure.

Although Manion and Mon have built a laminar pressure controlled amplifier somewhat similar in appearance to Griffin's amplifier, the flat saturation characteristic is obtained not by momentum but by bringing the control edges in close to the power jet so that, if the left control pressure, for example, is large, the right edge of the jet strikes the right control edge, peeling off some of the power jet and causing the pressure to rise in the right control region thus tending to prevent further deflection.

Equation (9.51) relates the pressure difference across the receivers to the jet deflection. To determine the transfer characteristic we must use this relation in conjunction with those developed for the deflection dependence on control pressure difference [Eqs. (9.32), (9.34a), and (9.36a)].

9.6.2 Optimization of Gain

We will discuss the optimization of the pressure difference across the receiver in some detail; however, it should be noted that the results will be different if one optimizes with respect to power or to flow.

We consider only optimization for small deflections and for steady state conditions.

If output pressure difference is to be maximized with respect to control pressure difference and with the output receiver width b_2 as the parameter to be optimized, then we require that

$$\frac{\partial}{\partial b_2}\left(\frac{\partial \Delta p_r}{\partial \Delta p}\right) = 0 \qquad (9.52)$$

where $\partial \Delta p_r/\partial \Delta p = G_p$ is the pressure gain but

$$\frac{\partial \Delta p_r}{\partial \Delta p} = \frac{\partial \Delta p_r}{\partial y_o}\frac{\partial y_o}{\partial \Delta p} \qquad (9.53)$$

Since $\partial y_o/\partial \Delta p$ does not depend on b_2, we consider the deflection sensitivity G_d where

$$G_d \equiv \partial \Delta p_r/\partial y_o \qquad (9.54)$$

To find an expression for the deflection sensitivity G_d we return to Eq. (9.48), although the results can also of course be obtained from Eq. (9.51).

From Eq. (9.48) we obtain

$$\frac{\partial \Delta p_r}{\partial y_o} = \frac{\rho}{2b_2}\int_{-b_2}^{0}\frac{\partial u^2}{\partial y_o}(l, y - y_o)\,dy - \frac{\rho}{2b_2}\int_{0}^{b_2}\frac{\partial u^2}{\partial y_o}(l, y - y_o)\,dy \qquad (9.55)$$

but, for any function $f(y - y_o)$,

$$-\partial f/\partial y_o = \partial f/\partial y$$

hence, Eq. (9.55) reduces to

$$G_d \equiv \partial \Delta p_r/\partial y_o = (-\rho/2b_2)[u^2(l, y_o) - u^2(l, -b_2 - y_o) - u^2(l, b_2 - y_o) + u^2(l, -y_o)] \qquad (9.56)$$

For $y_o = 0$

$$G_{do} \equiv \frac{\partial \Delta p_r}{\partial y_o}\bigg|_{y_o=0} = \frac{-\rho}{2b_2}[2u^2(l, 0) - u^2(l, -b_2) - u^2(l, b_2)] \qquad (9.57)$$

9.6 Analysis of the Output Region

since $u(l, y)$ is an even function

$$G_{do}(b_2) = (-\rho/b_2)[u^2(l, 0) - u^2(l, b_2)] \tag{9.58}$$

We now find the value of b_2 that maximizes the deflection sensitivity G_{do} by taking the derivative of Eq. (9.58) and setting $\partial G_{do}/\partial b_2 = 0$. From Eq. (9.49) we find that the maximizing relation is

$$(4b_2 k_4 \tanh k_4 b_2 + 1) \operatorname{sech}^4 k_4 b_2 = 1 \tag{9.59}$$

The first root of this yields

$$b_2 = 0.78/k_4(l) \tag{9.60}$$

This width, which holds for both turbulent and laminar two-dimensional jets, results in the maximum output pressure difference change with respect to deflection for small deflections.

Let us now consider the optimization of the nozzle-to-splitter distance l. We write G_p in terms of the variable involved, where again we concern ourselves only with small deflections about the splitter,

$$G_p = \frac{\partial \Delta p_r}{\partial \Delta p} = \frac{\partial \Delta p_r}{\partial y_o} \frac{\partial y_o}{\partial \Delta p} = G_{do} \frac{\partial y_o}{\partial \Delta p} \tag{9.61}$$

If we assume the vent pressure difference is zero, then from the result obtained in Problem 5.1

$$\partial y_o/\partial \Delta p = (1/2b\rho)(2\tau_1 \tau_l - \tau_1^2) \tag{9.62a}$$

where

$$\tau_1 = \int_0^{x_1} \frac{dx}{u_c(x)} = \int_0^{x_1} \frac{dx}{k_3(x)} \tag{9.62b}$$

$$\tau_l = \int_0^l \frac{dx}{u_c(l)} = \int_0^l \frac{dx}{k_3(x)} \tag{9.62c}$$

and u_c is the centerline velocity and is equal to $k_3(x)$. Hence, using the value of G_{do} from Eq. (9.58), the velocity distribution of Eq. (9.49), and the optimum value of b_2 from Eq. (9.60), we may express the pressure gain as

$$G_p = -0.273 k_3^2(l) k_4(l) (2\tau_1 \tau_l - \tau_1^2)/b \tag{9.63}$$

Let us first consider turbulent flow, in which case we take k_3 and k_4 from Eq. (9.2) with $x = l$. If we then evaluate τ_1 and τ_l through the use of Eq. (9.49), Eq. (9.63) becomes

$$G_p = -\frac{0.121 \sigma_e [2(l + x_o)^{3/2} - x_o^{3/2} - (x_1 + x_o)^{3/2}][(x_1 + x_o)^{3/2} - x_o^{3/2}]}{b(l + x_o)^2} \tag{9.64}$$

Taking the derivative of G_p with respect to l and setting it equal to zero, we find that the value of l that maximizes G_p for small deflections is

$$l = [2x_o^{3/2} + 2(x_1 + x_o)^{3/2}]^{2/3} - x_o \tag{9.65}$$

where
$$x_o = b\sigma_e/3$$

In momentum-type amplifiers, $x_1 \sim b$; hence, if we assume $\sigma_e = 10$, we find

$$l = 6.37b \tag{9.66}$$

for which, from Eqs. (9.60) and (9.2),

$$b_2 = 0.733b \tag{9.67}$$

If we go through the same process for laminar flow we find that no maximum exists and G_p asymptotically approaches its maximum as l approaches infinity.

This result points up an important point. Although the gain increases for very small deflections, as the distance l increases, the pressure recovered at the receivers decreases. Thus, to obtain output signals of usable magnitude we must restrict l to values that will permit a reasonable recovery pressure.

It should also be kept in mind that these results can at most be considered qualitative aids since three-dimensional effects due to the top and bottom plates have been neglected.

The calculation so far has allowed us to optimize the receiver width and the nozzle-to-splitter distance. We have also obtained the pressure difference across the power jet. In order to obtain the transfer characteristic, all that is necessary is to use the previously obtained equations that relate the pressure across the jet to the pressure across the controls.

9.7 THE OUTPUT CHARACTERISTICS

The pressure and flow one can obtain at the output of a proportional amplifier depend on the jet deflection, so that every possible jet deflection gives another member of a family of output flow versus output pressure characteristics. The measurements are made for a fixed pressure difference across the controls by changing valve openings on both output loads by the same amount. In practice the procedure is to change the valve settings with zero pressure difference across the control so that the pressure and flows are the same in both output lines. The control pressure difference is then changed by an amount Δp_c and the pressures and flows read in both outputs. The control pressure difference is then changed to $2\Delta p_c$ and more readings are taken, and so on. The output valve settings are then changed with zero pressure difference across the control and the process is repeated. A member of the family of output curves consists of the output pressure versus output flow curve for a given pressure difference across the control.

Thus, for some given pressure difference across the controls, a pressure p_{rR} will appear at the right receiver. The relation between pressure and flow at the right outlet is given by

$$p_{rR} - p_{\text{load}} = K_o q_{oR}^2 \tag{9.68a}$$

9.7 The Output Characteristics

where

$$K_o = \rho/2b_o^2 h^2 c_{dc}^2 \tag{9.68b}$$

where $b_o h$ is the outlet area (Fig. 9.27) and c_{dc} is the outlet discharge coefficient.

The output characteristic is obtained by allowing p_{load} to vary and finding q_{oR} as a function of p_{load}.

9.7.1 The Pressure Recovery

The pressure recovery characteristic of a fluidic element is determined by measuring (as a function of supply pressure) the pressure recovered at one receiver when the jet is centered in that receiver; hence,

$$p_r = \frac{\rho}{2b_2} \int_0^{b_2} u^2(l, y - b/2)\, dy = \frac{\rho}{2b_2} \int_{-b_2/2}^{b_2/2} u^2(l, y)\, dy$$

whence from (9.49)

$$p_r = \frac{\rho k_3^2}{2b_2 k_4} \left\{ \tanh \frac{k_4 b_2}{2} - \frac{1}{3} \tanh^3 \frac{k_4 b_2}{2} \right\}$$

If the pressure gain has been optimized, then, from (9.60), $k_4 b_2 = 0.78$ and

$$p_r = 0.227 \rho k_3^2$$

Hence for turbulent flow [from Eq. (9.2)]

$$\frac{p_r}{p_s} = \frac{0.340 b\sigma_e}{1 + b\sigma_e/3}$$

where it is assumed that

$$p_s \cong \rho u_o^2/2$$

and for laminar flow

$$\frac{p_r}{p_s} = 0.047 \left(\frac{b N_R}{1 + b_R N/36} \right)^{2/3}$$

where

$$N_R = b u_o/\nu$$

For $l = 6.4b$, we find for the turbulent case with $\sigma_e = 10$

$$p_r/p_s = 0.35$$

For $l = 6.4b$, $N_R = 1000$, we find for the laminar case

$$p_r/p_s = 0.45$$

Of course, if p_r is found as a function of p_s, then for small values of p_s the jet will be laminar and have a small Reynolds number. As p_s is increased, the Reynolds number will increase and the pressure recovery will increase. Eventually, however, the jet will become turbulent and the recovery pressure will become only weakly dependent on the Reynolds number. (If the jet were actually two-dimensional, it would be completely independent of Reynolds number when it is turbulent.)

9.8 ASPECT RATIO

In some of the very early work at Harry Diamond Laboratories aspect ratios of 10, 20, and 30 were tried on the same amplifier, and it was found that the lower the aspect ratio the higher the gain. Later Van Tilburg et al. [9] examined the effect of aspect ratio on gain and found that, for the units tested, the gain was greatest at an aspect ratio of 2 (Fig. 9.29). These results contradict one's

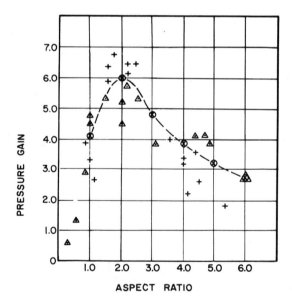

Fig. 9.29 Effect of aspect ratio on pressure gain: +, Test 1; △, Test 2.

intuitive feeling that, because of friction caused by the top and bottom plates, the gain should increase with height monotonically, eventually reaching some asymptotic value. The reason that this did not happen in the units tested is not known, but one would expect that the eventual decrease in gain as the plate distance h is increased is due to swirls that can occur more easily when h is large and are damped out at smaller values of h.

9.9 INTRODUCTION TO DYNAMIC ANALYSIS

Although at a number of laboratories some attempts were being made experimentally to determine transfer functions for proportional devices, nothing of any note was published until 1964, when Brown [27] considered a transfer matrix representation of the amplifier and Belsterling and Tsui [28] developed an equivalent circuit representation of the amplifier by use of transport delays plus lumped impedance concepts. A similar type of lumped parameter plus delay analysis including some experimental confirmation was published almost simultaneously by Boothe [29].

Brown has subsequently considerably amplified his original work and experimentally determined the various matrix elements [30]; Belsterling, too, has revised and improved his original concepts [31–33].

Frequency response characteristics of the proportional amplifier were published in 1965 by Roffman and Katz [34]. In spite of considerable scatter in the data, the presence of resonant peaks at high frequencies due to reflection within the amplifier chamber and in the output lines was apparent.

The jet deflection amplifier was simulated on the analog computer and discussed by Manion in a paper [21] which takes into account both the receivers and the vents in terms of their effective impedances. This simulation included the nonlinear (orifice-type) resistances in the controls, the receiver, and the venting passageways.

Analytic approximations to the jet deflection amplifier internal dynamics have been made using flow control volumes. Healey [35] reported an amplifier control volume analysis for blocked output channels, and presented experimental frequency data for an amplifier operating with a liquid. Agreement between analysis and experiment (shown in Fig. 9.30) indicates the merit of the control volume model. Manion [36] used a similar analysis.

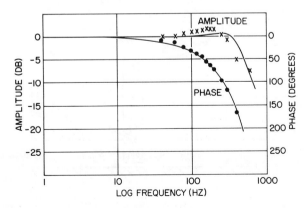

Fig. 9.30 Agreement of theory with experimental data: ——, theoretical; ×, experimental amplitude; ●, experimental phase.

In both papers the results indicate that without receiver channel influence the interaction chamber has a resonance due to coupling of jet capacitance and the vent impedance. This is further complicated by feedback effects from the outlets.

Manion [37] also obtained theoretical matrix coefficients based on the modeling given in his previous paper [36], but including the output channel admittances.

Brown and Humphrey have presented [19] in a two-part paper the most comprehensive amplifier dynamic test results yet published.

A dynamic analysis of the proportional amplifier is presented in Sections 9.10–9.12, which follow. In Sections 9.10 and 9.11 we formulate the input impedance and transfer function of the amplifier in terms of circuit components (impedances). Then Section 9.12 shows how these impedances are related to the geometric parameters of the amplifier.

9.10 THE INPUT IMPEDANCE

For the dynamic analysis of the amplifier we will assume that the changes of pressure are very small so that a linear analysis holds. All of the variables are then given in terms of deviation from the steady state value. Momentum is assumed negligible.

Consider Fig. 9.31 in which the flows and pressures shown are functions of time and the transformed variables P and Q are functions of s. (The entrained flows are assumed independent of time and have been omitted.)

The equivalent dynamic circuit is given in Fig. 9.32 for the interaction region where the returned flow Q_r is shown as originating from sources whose strengths are proportional to the deflection. The control volumes of Fig. 9.31 become the nodes of Fig. 9.32.

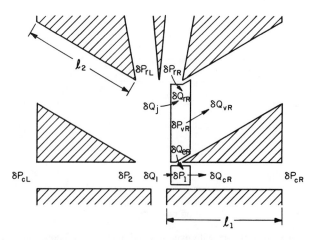

Fig. 9.31 Dynamic analysis configuration.

9.10 The Input Impedance

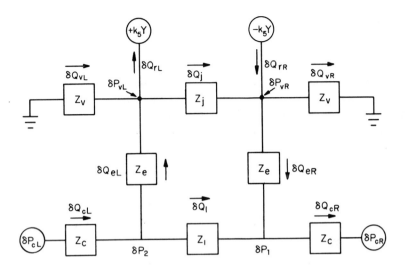

Fig. 9.32 Equivalent dynamic circuit.

We use upper case P's and Q's to designate the Laplace transforms of the respective p's and q's.

The symbol δ is used to indicate small changes.

It will be assumed that the control pressures are applied antisymmetrically, i.e.,

$$\delta P_{cL} = -\delta P_{cR} \tag{9.69a}$$

As a result it follows that

$$\delta P_2 = -\delta P_1 \tag{9.69b}$$

$$\delta Q_{cL} = \delta Q_{cR} = \delta Q_c \tag{9.69c}$$

Our objective is to determine the input impedance Z_{in}, which is defined as

$$Z_{in} \equiv (\delta P_{cL} - \delta P_{cR})/\delta Q_c \tag{9.70}$$

Since $Z_c = (\delta P_{cL} - \delta P_2)/\delta Q_{cL} = (\delta P_1 - \delta P_{cR})/\delta Q_{cR}$ and $\delta Q_{cL} = \delta Q_{cR} = \delta Q_c$ the control impedance Z_c (see Fig. 9.32) may be expressed as

$$2Z_c = [(\delta P_{cL} - \delta P_{cR}) - (\delta P_2 - \delta P_1)]/\delta Q_c \tag{9.71}$$

We may eliminate the control pressure by combining Eqs. (9.70) and (9.71). The result is

$$Z_{in} = 2Z_c + (\delta P_2 - \delta P_1)/Q_c \tag{9.72}$$

Thus the input impedance is the control line impedance in series with another impedance. We may determine this other impedance by noting that

$$\delta Q_c = \delta Q_e + \delta Q_1 \tag{9.73}$$

The flow across the control edges is

$$\delta Q_e = [(\delta P_2 - \delta P_1) - (\delta P_{vL} - \delta P_{vR})]/2Z_e \qquad (9.74\text{a})$$

where we have assumed that $\delta Q_{eL} = \delta Q_{eR} = \delta Q_e$. This assumption is, however, correct only if the jet deflection in the vicinity of the edge is small compared to the clearance between the control edge and the jet edge. If the deflection is not small, then $Z_{eL} \neq Z_{eR}$ and Eq. (9.74a) is not valid.

Except for resonant frequencies of the vents,

$$\delta P_{vR} - \delta P_{vL} \ll \delta P_2 - \delta P_1$$

hence, for nonresonant frequencies,

$$\delta Q_e = (\delta P_2 - \delta P_1)/2Z_e \qquad (9.74\text{b})$$

The flow across the jet depends on the jet impedance Z_1 and is

$$\delta Q_1 = (\delta P_2 - \delta P_1)/Z_1 \qquad (9.75)$$

If Eqs. (9.74b) and (9.75) are substituted into Eq. (9.73) we obtain

$$(\delta P_2 - \delta P_1)/\delta Q_c = 2Z_1 Z_e/(Z_1 + 2Z_e) \qquad (9.76)$$

Through the application of Eq. (9.76), the input impedance given in Eq. (9.72) becomes

$$Z_{in} = 2Z_c + 2Z_1 Z_e/(Z_1 + 2Z_e) \qquad (9.77)$$

As we might have expected, the input impedance is the sum of the control impedances in series with the parallel combination of $2Z_e$ and Z_1.

It is worth pointing out that, for large Z_e (small edge clearances), Eq. (9.77) becomes

$$Z_{in1} = 2Z_c + Z_1 \qquad (9.78\text{a})$$

whereas for small Z_e (large clearances)

$$Z_{in2} = 2Z_c \qquad (9.78\text{b})$$

Thus the impedance of the jet in the interaction region has negligible effect when the clearance is large and an appreciable effect when the clearance is small. We must emphasize that the above results hold for small deflections and, where the conditions $\delta Q_{cL} = \delta Q_{cR}$, $\delta P_{cL} = -\delta P_{cR}$ are fulfilled.

9.11 THE TRANSFER FUNCTION

We now find how the change in pressure difference at the receivers, $\delta P_{rL} - \delta P_{rR}$ ($=\Delta P_r$), is related to change in pressure difference across the controls $\delta P_{cL} - \delta P_{cR}$ ($=\Delta P_c$). Once again we make use of the jet deflection to provide the

9.11 The Transfer Function

connection between output and input variables. Thus, a limited transfer function may be separated into the product of two differentials and has the form

$$d(\Delta P_r)/d(\Delta P_c) = [d(\Delta P_r)/d(Y)][d(Y)/d(\Delta P_c)] \tag{9.79a}$$

To obtain the complete transfer function, we must relate the pressure difference across the receivers ΔP_r to the pressure difference across the output apertures ΔP_o $(= \delta P_{oL} - \delta P_{oR})$. This relationship depends on the output impedance of the amplifier Z_o and the load impedance provided by the component connected to the amplifier Z_L. Since

$$\Delta P_o = [Z_L/(Z_o + Z_L)] \Delta P_r \tag{9.79b}$$

the complete transfer function for the amplifier is

$$\frac{d(\Delta P_o)}{d(\Delta P_c)} = \left[\frac{d(\Delta P_r)}{d(Y)}\right]\left[\frac{d(Y)}{d(\Delta P_c)}\right]\left(\frac{Z_L}{Z_o + Z_L}\right) \tag{9.79c}$$

To determine the transfer function of Eq. (9.79c) we will first find the relation between Y and ΔP $[=(P_2 - P_1)]$. Then, since from Eqs. (9.70) and (9.72) we know that

$$\Delta P/\Delta P_c = 1 - 2Z_c/Z_{in} \tag{9.80}$$

we will be able to relate Y and ΔP_c. For the case of a uniform pressure gradient $\Delta P/b$ in the control region (0 to x_1) and another uniform pressure gradient $\Delta P_v/b$ in the vent region (x_1 to l) we may adapt Eq. (5.11) so that

$$Y(\tau, s) = \frac{1}{\rho b}\int_0^{\tau_1}(\tau - \tau')\exp[-s(\tau - \tau')]\,\Delta P\,d\tau'$$
$$+ \frac{1}{\rho b}\int_{\tau_1}^{\tau}(\tau - \tau')\exp[-s(\tau - \tau')]\,\Delta P_v\,d\tau' \tag{9.81}$$

where

$$\Delta P = \delta P_2 - \delta P_1, \qquad \Delta P_v = \delta P_{vL} - \delta P_{vR}$$

$$\tau_1 = \int_0^{x_1}\frac{dx}{u_c(x)}, \qquad u_c(x) = k_3(x)$$

$$\tau_1 \le \tau \le \tau_l, \qquad \tau_l = \int_0^l\frac{dx}{u_c(x)}$$

l is the distance from the power jet nozzle to the splitter.

Upon integration, Eq. (9.81) becomes

$$Y(\tau, s) = (\Delta P/\rho b s^2)\{\exp[-s(\tau - \tau_1)][1 + s(\tau - \tau_1)] - e^{-s\tau}(1 + s\tau)\}$$
$$+ (\Delta P_v/bs)\{1 - \exp[-s(\tau - \tau_1)][1 + s(\tau - \tau_1)]\} \tag{9.82}$$

If we assume that the velocity $u_c(x)$ is constant and is the nozzle velocity u_o, we obtain

$$\tau = x/u_o, \qquad \tau_1 = x_1/u_o, \qquad \tau_l = l/u_o$$

and we may rewrite Eq. (9.82) for the deflection at the receiver as

$$Y(l, s) = (1/\rho b s^2)\{B_1 \Delta P_v + B_2 \Delta P\} \tag{9.83a}$$

where

$$B_1 = 1 - (1 + \alpha_2 - \alpha_1) \exp[-(\alpha_2 - \alpha_1)] \tag{9.83b}$$

$$B_2 = (1 + \alpha_2 - \alpha_1) \exp[-(\alpha_2 - \alpha_1)] - (1 + \alpha_2) \exp(-\alpha_2) \tag{9.83c}$$

and

$$\alpha_1 = x_1 s/u_o, \qquad \alpha_2 = ls/u_o$$

In Eq. (9.83a) we have a relation between the deflection Y and the pressure differences ΔP and ΔP_v. Since we desire only the dependency of Y on ΔP we must now find an expression that links ΔP_v with ΔP.

For the volume involving ΔP_{vR} of Fig. 9.31 (or the junction point on the equivalent circuit diagram of Fig. 9.32) we obtain

$$\delta Q_v + \delta Q_e = \delta Q_r + \delta Q_j \tag{9.84}$$

We continue by writing each of the flows in Eq. (9.84) in terms of known quantities. The edge flow δQ_e is already known from Eq. (9.74a). The vent flow δQ_v is simply

$$\delta Q_v = (\delta P_{vR} - \delta P_{vL})/2Z_v = \Delta P_v/2Z_v \tag{9.85}$$

where Z_v is the impedance of the vent.

Now the transverse jet flow δQ_j corresponding to the jet deflection is given by

$$\delta Q_j = h \int_{x_1}^{l} s Y(\tau, s) \, dx \tag{9.86}$$

From Eqs. (9.86) and the deflection given in Eq. (9.82) we find that

$$\delta Q_j = (h u_o/\rho b s^2)\{A_1 \Delta P_v + A_2 \Delta P\} \tag{9.87}$$

where

$$A_1 = \alpha_2 - \alpha_1 + 2[\exp[-(\alpha_2 - \alpha_1)] - 1] + \alpha_2 \exp[-(\alpha_2 - \alpha_1)]$$
$$- \alpha_1 \exp[-(\alpha_2 - \alpha_1)]$$

$$A_2 = -2[\exp[-(\alpha_2 - \alpha_1)] - 1] - (\alpha_2 - \alpha_1) \exp[-(\alpha_2 - \alpha_1)]$$
$$+ 2[\exp(-\alpha_2) - \exp(-\alpha_1)] + \alpha_2 \exp(-\alpha_2) - \alpha_1 \exp(-\alpha_1)$$

We assume for the present that for small deflections the returned flow δQ_r is proportional to the deflection:

$$\delta Q_r = k_5 Y(l, s) \tag{9.88}$$

9.11 The Transfer Function

where k_5 depends on the output impedance and therefore on the frequency. We will show in Section 9.12 that Eq. (9.88) is a valid representation of the return flow. At that time we will also evaluate k_5.

Now the substitution of the flows given in Eqs. (9.74a), (9.85), (9.87), and (9.88) into Eq. (9.84) yields in conjunction with Eq. (9.83a)

$$\Delta P_v = \frac{(1/2Z_e) - (hu_o A_2/\rho b s^2) - (k_5 B_2/\rho b s^2)}{(1/2Z_v) + (1/2Z_e) + (hu_o A_1/\rho b s^2) + (k_5 B_1/\rho b s^2)} \Delta P \qquad (9.89)$$

Equation (9.89) provides us with a relation between ΔP_v and ΔP. This enables us to utilize the expression for edge flow given in Eq. (9.74a) rather than the reduced form of Eq. (9.74b). We may also use Eq. (9.89) to eliminate ΔP_v from Eq. (9.83a) so that

$$Y(l, s) = \frac{\Delta P}{\rho b s^2} \left\{ B_2 + \frac{B_1[(1/2Z_e) - (hu_o A_2/\rho b s^2) - (k_5 B_2/\rho b s^2)]}{(1/2Z_v) + (1/2Z_e) + (hu_o A_1/\rho b s^2) + (k_5 B_1/\rho b s^2)} \right\} \qquad (9.90)$$

We insert the relation between ΔP_c and ΔP [Eq. (9.80)] and obtain the relationship between the input pressure difference ΔP_c and the jet deflection $Y(l, s)$ in the vicinity of the splitter:

$$\frac{d(Y(l, s))}{d(\Delta P_c)} = \frac{1 - 2Z_c/Z_{in}}{\rho b s^2} \left\{ B_2 + \frac{B_1[(1/2Z_e) - (hu_o A_2/\rho b s^2) - (k_5 B_2/\rho b s^2)]}{(1/2Z_v) + (hu_o A_1/\rho b s^2) + (k_5 B_1/\rho b s^2) + (1/2Z_e)} \right\} \qquad (9.91)$$

The differential relation between the pressure difference at the entrance to the receivers and the jet deflection has already been presented for the static case in Eq. (9.58). If we assume that the relation also holds dynamically, Eq. (9.58) becomes

$$\frac{d(\Delta P_r)}{d(Y(l, s))} = \frac{-\rho}{b_2} [u^2(l, 0) - u^2(l, b_2)] \qquad (9.92)$$

From Eqs. (9.79a), (9.91), and (9.92) the limited transfer function for the amplifier is

$$\frac{d(\Delta P_r)}{d(\Delta P_c)} = \frac{(1 - 2Z_c/Z_{in})[u^2(l, 0) - u^2(l, b_2)]}{b b_2 s^2}$$

$$\times \left\{ B_2 + \frac{B_1[(1/2Z_e) - (hu_o A_2/\rho b s^2) - (k_5 B_2/\rho b s^2)]}{(1/2Z_e) + (1/2Z_v) + (hu_o A_1/\rho b s^2) + (k_5 B_1/\rho b s^2)} \right\} \qquad (9.93a)$$

and the complete transfer function from (9.79c) is

$$\frac{d(\Delta P_o)}{d(\Delta P_c)} = \frac{d(\Delta P_r)}{d(\Delta P_c)} \left(\frac{Z_L}{Z_o + Z_L} \right) \qquad (9.93b)$$

9.12 EVALUATION OF THE IMPEDANCES AND OF k_s

We will now indicate briefly the evaluation of the impedances that appear in the transfer function [Eq. (9.93)].

In calculating the resistance portion of each impedance, the approach to be used for the proper small pressure fluctuation approximations depends on the geometry associated with the particular impedances. For shapes that are essentially parallel wall ducts, we can use the Hagen–Poiseuille resistance (as modified for rectangular ducts).

We linearize around the nonlinear resistance associated with an orifice. The Hagen–Poiseuille resistance associated with a short duct terminated with an orifice can usually be neglected in comparison with the orifice resistance, but if the duct is sufficiently long, both resistances should be considered.

In the following we assume that only the control edge resistance and the output resistance are of the orifice type, but in practice the particular configuration must be taken into account.

(a) *The Control Duct Impedance*

Since the control line capacitance is small, the impedance Z_c is given by

$$Z_c = R_c + sL_c \tag{9.94a}$$

where

$$R_c \cong 12\mu l_1(x_1^2 + h^2)/x_1^3 h^3 \tag{9.94b}$$

$$L_c = \rho l_1/x_1 h \tag{9.94c}$$

where l_1 is the length of the control

(b) *The Vent Impedance*

The vent impedance for a wide open vent as in Fig. 9.31 is also an inertance in series with a resistance and is evaluated similarly:

$$Z_v = R_v + sL_v \tag{9.95a}$$

$$R_v = 12\mu l_{2e}(b_v^2 + h^2)/(b_v h)^3 \tag{9.95b}$$

$$L_v = \rho l_{2e}/b_v h \tag{9.95c}$$

where l_{2e} is the effective length of the vent and b_v is its effective width.

For vents at right angles to the amplifier axis, the effective length l_{2e} is then equal to the vent wall length l_2, but obviously when the angle is less than a right angle $l_{2e} < l_2$. When the vents are partially blocked (Fig. 9.24) we must also consider the capacitance of the vent.

(c) *The Control Edge Impedance*

The control edge impedance is essentially a resistance and can be evaluated in terms of the clearance between the jet and the control edge:

$$Z_e = R_e = \frac{q_{eo}}{[(D - b_1)/2]^2 h^2} \tag{9.96}$$

9.12 Evaluation of the Impedances and of k_5

where q_{eo} is the flow past the control edge when the jet is undeflected. Because Z_e depends on Q_{eo}, it is a function of the bias pressure $(p_{cL} + p_{cR})/2$.

(d) *The Jet Impedance*

The impedance of the jet in the vicinity of the controls has already been evaluated in Eq. (5.38) as

$$Z_1 = \frac{\rho b s}{h[x_1 + (2u_o/s) \exp(-sx_1/u_o) - (2u_o/s) + x_1 \exp(-sx_1/u_o)]} \quad (9.97)$$

and by Taylor Series approximation for the exponential terms we obtain for low frequencies

$$Z_1 = R_1 + 1/sC_1 \quad (9.98a)$$

where

$$R_1 = 3\rho b u_o / h x_1^2 \quad (9.98b)$$

$$C_1 = h x_1^3 / 6 \rho b u_o^2 \quad (9.98c)$$

(e) *The Output Impedance*

The output impedance Z_o depends on the resistance R_o, capacitance C_o, and inertance L_o of the output receiver. In general the output impedance has the form

$$Z_o = R_o + sL_o + 1/sC_o \quad (9.99)$$

Appropriate values of R_o, L_o, and C_o may be estimated by applying the concepts for passive components presented in Chapter 2 to the geometry of the output receiver.

Belsterling and Tsui [28] have shown that the output impedance exerts a large influence on the transfer function of the amplifier. The reason for this is shown in the evaluation of k_5.

(f) *The Evaluation of the Coefficient k_5*

In Eq. (9.88) we assumed that $\delta Q_r = k_5 Y$.

In order to find k_5 we must recognize that the dynamic change in spilled (or returned) flow δq_{rR}, for example, is equal and opposite to the change in output flow δq_{oR}; that is,

$$\delta q_r = -\delta q_o \quad (9.100a)$$

or

$$\delta Q_r = -\delta Q_o \quad (9.100b)$$

Further, in keeping with our previous definition

$$\Delta P_r \equiv \delta P_{rL} - \delta P_{rR} = 2\delta P_r \quad (9.101)$$

Now the relation between the pressure δP_r and flow δQ_r depends on the driving point impedance Z_d of the receiver and load combination. Thus, we may write

$$Z_d \equiv \delta P_r / \delta Q_r = Z_o + Z_L \quad (9.102)$$

It follows then from Eqs. (9.100), (9.101), and (9.102) that

$$Z_d = \Delta P_r / 2\delta Q_r \quad (9.103)$$

The relation between ΔP_r and Y is given by Eq. (9.92). Inserting it into Eq. (9.103),

$$\delta Q_r = \frac{\rho[u^2(l, 0) - u^2(l, b_2)] Y}{2b_2 Z_d} \tag{9.104}$$

and hence, from Eqs. (9.88) and (9.104),

$$k_5 = \frac{\rho[u^2(l, 0) - u^2(l, b_2)]}{2b_2 Z_d} \tag{9.105}$$

We find that the constant k_5 is therefore inversely proportional to the driving point impedance.

Let us determine the driving point impedance for a simple case. Suppose that the output receivers are considered as lossless lines, each terminated by an orifice. Then the driving point impedance Z_d is given in the frequency domain by Eqs. (3.80) and (3.82). In the Laplace domain we may obtain Z_d from Eq. (3.78) as

$$Z_d \equiv \frac{\delta P_{rL}}{\delta Q_{oL}} = \frac{R_{or} \cosh s (L_o C_o)^{1/2} + (L_o/C_o)^{1/2} \sinh s (L_o C_o)^{1/2}}{R_{or}(C_o/L_o)^{1/2} \sinh s (L_o C_o)^{1/2} + \cosh s (L_o C_o)^{1/2}} \tag{9.106}$$

where L_o is the inertance of the outlet, C_o is the capacitance of the outlet, and R_{or} is the variational resistance ($= 2k_{or} q_o$) of the orifice.

For low frequencies, Eq. (9.106) may be approximated by

$$Z_d = \frac{R_{or} + sL_o + (s^2 R_{or} L_o C_o/2) + (s^3 L_o^2 C_o/6)}{1 + sR_{or} C_o + s^2 C_o L_o/2} \tag{9.107}$$

If R_{or} is small, that is, if the orifice opening is not much smaller than the tube, then at sufficiently low frequencies Eq. (9.107) can be simplified to

$$Z_d = R_{or} + sL_o \tag{9.108}$$

Thus for this case we obtain from Eqs. (9.105) and (9.108) that

$$k_5 = \frac{\rho[u^2(l, 0) - u^2(l, b_2)]}{2b_2(R_{or} + sL_o)} \tag{9.109}$$

9.13 STAGING OF AMPLIFIERS *

9.13.1 Desired Characteristics

The important characteristics of linear amplifiers include:

(a) A flat frequency response over the bandwidth of interest.

(b) A phase shift proportional to frequency over the bandwidth of interest. (It can be shown for minimum phase networks that a flat frequency response over a bandwidth of $3f_c$ essentially guarantees a phase shift proportional to frequency over the bandwidth f_c.)

---

* Reference [8].

9.13 Staging of Amplifiers

(c) A relatively small phase shift over the entire bandwidth of interest (if the amplifier is part of a feedback loop).

(d) A large dynamic range where the dynamic range is defined as the smallest input signal resulting in an output above noise divided into the largest input signal that does not saturate the amplifier.

(e) A high input impedance.

(f) A low output impedance.

(g) High gain.

(h) A reasonably broad bandwidth for which a, b, c, and d hold.

(i) A flat saturation characteristic; that is, the output should be proportional to the input over some range of amplitudes (the dynamic range) and should give approximately a constant output for input signals larger than those within the dynamic range.

9.13.2 Options in Staging

When amplifiers are staged, the above characteristics should be kept in mind. In general one finds that a change in geometry that improves one characteristic may cause another to become worse. Similarly the staging of amplifiers changes certain overall characteristics (with respect to the characteristic of a single stage) in a different manner than it changes others. To some extent the designer has control over the way the characteristics are affected.

9.13.3 Dynamic Range

As an example of the various options, consider the dynamic range. The dynamic range of staged amplifiers is at most equal to the dynamic range of the amplifier in the cascade having the least dynamic range. All amplifiers in the cascade should therefore deflect through the same angle and reach saturation together since, if one saturates before the others, the full dynamic range of the others is not being used.

Now the input to the first stage is, of course, smaller than the input to the second stage (and so on) since the second stage input includes the effect of the gain of the first stage.

If the power jets are to deflect through the same angle for all stages, then the power jet velocities must also increase in succeeding stages.

The output pressure of the ith stage is the input pressure to the $i + 1$st stage. Considered as the source of the $i + 1$st stage, the ith stage has a source resistance R_{oi}, where R_{oi} is the output resistance of the ith stage. The output is applied to the input of the $i + 1$st stage that has an input resistance $R_{c(i+1)}$.

The pressure appearing at the $i + 1$st input is therefore

$$p_{c(i+1)} = \frac{R_{c(i+1)} p_{ri}}{R_{oi} + R_{c(i+1)}} \tag{9.110}$$

where p_{ri} is the receiver pressure of the ith stage.

Now the receiver pressure of the $i + 1$st stage will be

$$p_{r(i+1)} = G_{pb} p_{c(i+1)} = \frac{G_{pb} R_{c(i+1)} p_{ri}}{R_{oi} + R_{c(i+1)}} \quad (9.111)$$

where G_{pb} is the pressure gain when the output is blocked.

It follows that if the pressure gains of the ith and $i + 1$st stages are equal

$$\frac{p_{r(i+1)}}{p_{ri}} = \frac{G_{pb} R_{c(i+1)}}{R_{oi} + R_{c(i+1)}} = \frac{p_{c(i+1)}}{p_{ci}} \quad (9.112)$$

If the deflections of all stages are to be the same, the power supply pressures must be proportional to the control pressures; consequently,

$$\frac{p_{c(i+1)}}{p_c} = \frac{p_{s(i+1)}}{p_{si}} = \frac{G_{pb} R_{c(i+1)}}{R_{oi} + R_{c(i+1)}} \quad (9.113)$$

9.13.4 Bandwidth

The bandwidth of a beam deflection amplifier depends on the transport time of the power jet particles so that, for a given design and size, the less the power jet velocity the smaller the bandwidth.

Now the bandwidth of staged amplifiers is given to a good approximation by

$$1/f_c = (1/f_{c1}^2 + 1/f_{c2}^2 + \cdots + 1/f_{cn}^2)^{1/2} \quad (9.114)$$

where f_{ci} is the bandwidth of the ith stage and f_c is the overall bandwidth.

It is seen that the overall bandwidth is less than that of the individual stage having the smallest bandwidth. If the bandwidths are all equal, the overall bandwidth decreases as the square root of the number of stages; however, if one stage has a bandwidth appreciably less than the others, the rest have only a small effect. It follows that, if all stages are of the same length, the overall bandwidth will be determined by the bandwidth of the first stage since the power jet velocity is least for that stage. On the other hand, the designer does have the option of successively increasing the amplifier length without a drastic loss in bandwidth.

9.13.5 Overall Phase Shift

The maximum phase shift for any stage depends on the transport time for the particles. The overall phase shift is the sum of the phase shifts throughout the stages. The overall phase shift is important as far as stability is concerned if the staged amplifiers are to be used with feedback. This consideration suggests that

9.13 Staging of Amplifiers

it is more advantageous to increase the overall gain by increasing the gain per stage rather than increasing the number of stages (providing that the per stage gain can be increased without decreasing the bandwidth).

9.13.6 Reynolds Number Consideration

Because amplifiers using laminar flow are much less noisy than those using turbulent flow, appreciably larger dynamic ranges may be obtained with laminar amplifiers for small signals than with turbulent amplifiers. Manion and Mon [8] have, therefore, investigated laminar amplifiers.

The spreading of a two-dimensional laminar jet over a distance x_1 is given in Eq. (9.7b) as

$$b_1(x_1)/b = (1 + 36x_1/bN_R)^{1/3}$$

where the Reynolds number is based on nozzle width.

Manion and Mon [8], however, have found that, for aspect ratios of the order of unity or less, the spread Reynolds number is dependent on h, the distance between the top and bottom plates, and that the spread rate remains approximately constant if the product $u_o h$ is kept constant.

These results on the jet spread rate explain those obtained by Kelley and Boothe [38], who found (as shown in Fig. 9.33) that for aspect ratios less than or equal to two, a single gain curve is obtained if the Reynolds number (which they designate as a stretched Reynolds number) is defined as

$$N_{Rh} \equiv (\rho u d_e/\mu)h/b$$

where d_e is the hydraulic diameter, that is four times the nozzle area divided by its circumference.

Manion and Mon found that the flow remained laminar up to Reynolds numbers $(\rho u h/\mu)$ of about 1600. Furthermore, they found for their design that the spreading became excessive for Reynolds numbers (based on h) less than 700. Since the clearance they used between the control edge and the jet was very small, excessive spreading caused the power jet to deflect into the controls, even at fairly small input signals, thereby cutting down the dynamic range. Since the power jet velocity in successive stages must increase and since the first stage has a minimum Reynolds number of 700, the Reynolds number in succeeding stages must increase if the stages are identical, so that the maximum Reynolds number (1600) for laminar flow would be exceeded in the third stage if the pressure gain per stage is about 10 (a velocity increase per stage of approximately 3). Manion and Mon, therefore, decrease the height in succeeding stages to keep the Reynolds number within bounds.

Identical plan forms are used for all stages, only the aspect ratio being changed. This simplifies the fabrication problem, since the same outline may be used for all stages.

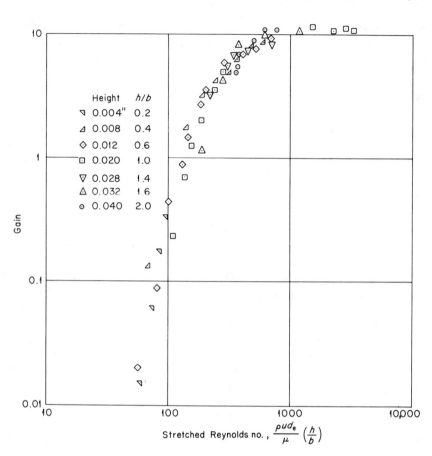

Fig. 9.33 Gain versus stretched Reynolds number. Nozzle width is 0.020 in. (From Kelley and Boothe [38].)

Manion and Mon arbitrarily selected the maximum (first stage) and minimum (last stage) heights as 2 and 0.1 mm, respectively. The ratio of maximum supply pressure to minimum supply pressure possible for a Reynolds number range from 700 to 1600 is then about 2500. The limits on aspect ratio thus limit the maximum pressure gain for a cascade to 2500 times the gain of the last stage, or about 25,000 (assuming that the last stage has a gain of 10).

9.13.7 Input and Output Impedances

Since the previous arguments indicate that it is advantageous in general to preserve the same plan form while changing the aspect ratio in succeeding stages, the input and output impedances are more or less determined for all stages once the plan form pattern has been decided. Some flexibility is possible

9.13 Staging of Amplifiers

however because any particular stage could consist of a number of units in parallel. The parallel use of units reduces the effective output impedance at the expense of also reducing the effective input impedance for the parallel set.

9.13.8 Pressure Gain

The actual pressure gain G_{pi} from the ith stage depends on the output resistance of the stage and the input resistance of the succeeding stage $(i+1)$. Thus,

$$G_{pi} = \frac{G_{pb} R_{c(i+1)}}{R_{oi} + R_{c(i+1)}} \quad (9.115)$$

The maximum gain for each stage thus occurs when $R_{oi} \ll R_{c(i+1)}$. As a result for maximizing pressure gain it is not desirable to parallel units in successive stages.

The overall pressure gain G_p of a stage amplifier is

$$G_p = \prod_{i=1}^{n} G_{pi} \quad (9.116)$$

For example, a three-stage cascade would have a pressure gain $G_p = G_{p1} G_{p2} G_{p3}$.

9.13.9 Power Gain

In general, if the resistances are of the form $R_c = A_c q^n$, where A_c is a constant and q is the volume flow, then it can be shown that maximum power transfer occurs when

$$\frac{R_{c(i+1)}}{R_{oi} + R_{c(i+1)}} \equiv f_R = \frac{n+1}{n+2} \quad (9.117)$$

If the resistances are independent of flow, then $n = 0$ and $f_R = 0.5$. For orifice resistors, $n = 1$ and $f_R = 0.66$.

In general, n will lie between zero and unity so that f_R will be between 0.5 and 0.66.

9.13.10 Flow Gain

As the number of elements in parallel fed by a single element is increased, the flow gain increases but the effective pressure gain decreases and the power gain decreases. In general the number of parallel units fed by a single unit should be limited so as to keep the effective pressure gain greater than or equal to unity; that is, from Eq. (9.115) with $G_p = 1$,

$$\frac{R_{c(i+1)}}{R_{oi} + R_{c(i+1)}} \geq \frac{1}{G_{pb}} \quad (9.118)$$

9.13.11 Ratio of Output-to-Input Resistance

The ratio of output-to-input resistance of a single element of the small edge-clearance type is usually about unity. If f_R as defined in Eq. (9.117) is 0.5 for all stages, then the output-to-input resistance ratio for all the stages will also be about unity. For $f_R > 0.5$ for each stage, the output-to-input resistance ratio for many stages will be increased over the single-stage ratio and, for $f_R < 0.5$, the resistance ratio will be decreased. Thus maximizing pressure gain decreases this ratio, whereas maximizing flow gain increases it.

PROBLEMS

9.1 The bias level is given by $(p_{cL} + p_{cR})/2$. Discuss the effect of raising the bias level on the input impedance of close-wall devices.

9.2
(a) Find the pressure gain for small deflections for a laminar amplifier. Plot G_p as a function of l/b for $N_R = 1000$, $x_1 = 4b$, assuming that the receiver width b_2 is varied as l is varied so as to optimize the pressure gain for small deflections.

(b) Plot the pressure difference Δp_r as a function of y_o for a laminar jet (with b_2 optimized) for $l = 8b$ and for $l = 16b$ with $N_R = 1000$.

9.3 Equation (9.25a) actually yields two roots for $p_2^{1/2}$. Show that the correct root is the negative one given in Eq. (9.26a).

9.4 Show that, if the receiver width is sufficiently great compared to the jet width, the receivers may draw more flow than the receiver edges subtend of the flow profile. Explain the result of this effect on the vent pressures and therefore on the performance of the device.

9.5 Assume that, for steady (dc) control pressure difference applied to the jet, the deflection at the distance x_1 where x_1 is the control width is given by

$$y_{o1} = (x_1/2b\rho u_o^2)(x_1 \Delta p + \Delta M)$$

where ΔM is the difference in left and right control momenta applied to the jet.

Discuss the effect of changing the edge clearance on the relative values of $x_1 \Delta p$ and ΔM.

9.6 (or term paper) Discuss a method for determining the deflection of the jet at the receivers if the edge clearance is large.

9.7 Find the pressure recovery characteristic (pressure recovery versus supply pressure) for a range of supply pressures resulting in Reynolds numbers from 0 to 2000, where $N_R = bu_o/v$, $p_s = \frac{1}{2}\rho u_o^2$, $l = 7b$, $b_2 = 1.5b$, $b = 1$ mm, $h = 0.5b$, and the medium is air.

Assume that the jet is laminar for $N_R < 1600$ and turbulent for $N_R > 1600$.

9.8 Discuss the effect of a change in temperature on the gain of a small clearance laminar amplifier.

9.9 Discuss the power optimization of an amplifier configuration.

9.10 Assume that the amplifier has a center dump of width b, i.e., equal to the power jet width. Optimize the receiver widths for maximum pressure gain.

9.11 Calculate the edge flow at zero deflection q_{eo} as a function of control pressure bias level for a Reynolds number of 1000, $D = 2b$, $x_1 = 4b$.

9.12 Prove Eq. (9.117).

SUGGESTED TERM PAPERS

9.1
(a) Determine Δp as a function of Δp_c for a close-wall laminar amplifier if $D = 2$, $x_1 = 4$, $N_R = 1000$, and the left and right control pressures are changed antisymmetrically about control bias level pressures of 0, $0.05 p_s$, $0.10 p_s$ and $0.15 p_s$; i.e., let $(p_{cL} + p_{cR})/2$ remain constant at $0.05 p_s$, for example, as p_{cL} and p_{cR} are changed.
(b) Find the input characteristic for the above case.
(c) Find the input characteristic if the control bias level is zero and the vents are pressurized to $0.01 p_s$, $0.02 p_s$, and $0.05 p_s$.

9.2 Analyze in detail the steady state characteristics of an amplifier for which the clearances are large.

9.3 Assume that the downstream edge of the controls includes a short portion parallel to the jet axis instead of ending in a point. Discuss the effect on the input characteristic analytically.

SUGGESTED RESEARCH PROJECTS

9.1 Attempt to find a formulation for recovery pressure that agrees with Reid's [26] results for blocked loads. The research should concern itself with rectangular receivers and should include the effects of receiver width and receiver loading on the pressure recovered.

9.2 Find a three-dimensional representation of the jet spreading characteristics between plates.

9.3 Examine the effect of the friction due to the top and bottom plates on the frequency response of an amplifier.

NOMENCLATURE

b	Power jet nozzle width
b_v	Vent width
b_1	Power jet width at distance x_1
b_2	Output receiver width
c_{dc}	Control nozzle discharge coefficient
c_{dv}	Vent discharge coefficient
d_e	Equivalent (hydraulic) diameter
d_L	Clearance between left control edge and power jet
d_R	Clearance between right control edge and power jet
D	Distance between left and right control edges
D_M, D_p	Eq. (9.26d)
G_d	Deflection sensitivity
G_p	Pressure gain
G_{pb}	Pressure gain at blocked load
h	Slit height (distance between top and bottom plates)
J	Power jet momentum flux per unit depth
$k_3(x), k_4(x)$	Coefficients in jet velocity profile
k_5	Eq. (9.46) and Eq. (9.109)
K_c	Resistance coefficient for control nozzle
K_o	Resistance coefficient for output orifice
K_v	Resistance coefficient for vent
l	Distance between power jet nozzle and splitter
L_c	Control inductance
L_o	Output inductance
L_v	Vent inductance
N_R	Reynolds number
N_{Rh}	Stretched Reynolds number
p_a	Ambient pressure
p_b	Bubble pressure
p_c	Control pressure
p_{ent}	Equivalent entrainment pressure [Eq. (9.25d)]
p_o	Output pressure
p_{load}	Load pressure
p_{rR}	Pressure at the right receiver
Δp_r	Pressure difference across receivers $= p_{rL} - p_{rR}$
p_s	Supply pressure
p_{vR}	Right vent pressure adjacent to power jet
Δp_v	Pressure difference across vents $= p_{vL} - p_{vR}$
p_1, p_2	Pressures in interaction region
Δp	Pressure difference in interaction region $= p_2 - p_1$
p_3	$= (p_1 + p_2)/2$, Average pressure within interaction region (control bias pressure)
P	$= \mathscr{L}\{p\}$, Laplace transform of pressure
q	Power jet volume flow
q_c	Control flow
q_e	Flow between the control edge and jet edge
q_{ent}	Entrained flow
q_j	Transverse flow due to jet deflection
q_r	Return flow
q_v	Vent flow
Q	$= \mathscr{L}\{q\}$, Laplace transform of flow

R_c	Control resistance
R_e	Edge clearance resistance
R_o	Output resistance
R_{or}	Orifice resistance
R_v	Vent resistance
s	Laplace transform variable
t	Time
u_c	Centerline velocity
u_o	Power jet nozzle exit velocity
$u(x,y)$	Velocity profile
x	Axial distance
x_o	Virtual origin of power jet
x_1	Control width acting on power jet
y	Transverse distance
y_o	$= y_o(l)$, Jet deflection at $x = L$
y_{o1}	$= y_o(x_1)$, Jet deflection at $x = x_1$
Y	$= \mathscr{L}\{y_o(l)\}$, Laplace transform of jet deflection at L
Z_c	Control nozzle impedance
Z_d	Driving point impedance [Eq. (9.102)]
Z_e	Impedance of clearance between control edge and jet
Z_{in}	Input impedance
Z_L	Load impedance
Z_v	Vent impedance
Z_1	Jet impedance in the interaction region
α_1	$x_1 s / u_o$
α_2	ls/u_o
ν	Kinematic viscosity
ρ	Density
ϕ	Jet spread factor
σ_e	Jet spread factor
τ_1	Time for jet centerline particle to move the distance x
τ_l	Time for jet centerline particle to move the distance l
τ	Time for jet centerline particle to move the distance x

REFERENCES

1a. J. M. Kirshner and F. M. Manion, The jet-deflection proportional amplifier. *Fluidics Quart.* **2**, No. 4 (1970).

1b. B. M. Horton, Pressure Equalized Amplifier, U.S. Patent No. 3,282,280, 1 November 1966.

2. R. R. Palmisano, Fluid amplifier demonstration vehicle. *Proc. Fluid Amplification Symp.* I (October 1962).

3. G. L. Roffman, Staging of closed proportional amplifiers. *Proc. Fluid Amplification Symp.* III (May 1964).

4. F. T. Brown, Pneumatic Pulse Transmission with Bistable-Jet-Relay Reception and Amplification. Sc.D. Thesis, MIT, Dept. of Mech. Eng., May 1962.

5. S. J. Peperone, S. Katz, and J. M. Goto, Fluerics 4. Gain Analysis of the Proportional Amplifier, AD-296513, HDL, 30 October 1962.

6. J. F. Douglas and R. S. Neve, Investigation into the behavior of a jet interaction proportional amplifier. *Proc. Cranfield Fluidics Conf.*, 2nd (January 1967).

7. A. Powell, Characteristics and control of free laminar jets. *Proc. Fluid Amplification Symp.* HDL (October 1962).

8. F. M. Manion and G. Mon, Fluerics 33. Design and Staging of Laminar Proportional Amplifiers, HDL-TR-1608, June 1972.
9. R. W. Van Tilburg, R. H. Bellman, and W. L. Cochran, Fluerics 21. Optical Fabrication of Fluid Amplifiers and Circuits. AD-636842, HDL, May 1966.
10. C. Bourque and B. G. Newman, Reattachment of a two dimensional incompressible jet to an adjacent flat plate. *Aeronaut. Quart.* **XI**, 201 (1960).
11. F. M. Manion, "Jet Interaction in a Defined Region." M.M.E. Thesis, Catholic Univ., Washington, D.C., March 1962.
12. J. F. Foss, Flow characteristics of the defined region geometry for high-gain proportional amplifiers. *Advan. Fluidics* (May 1967), ASME.
13. C. H. T. Pan, "Fluid Amplifiers," U.S. Patent No. 3,240,221, 15 Mar. 66.
14. F. Manion, "Vented Pure Fluid Analog Amplifier," U.S. Patent No. 3,283,768, 8 Nov. 66.
15. C. Pavlin, Experimental study of a proportional fluid amplifier. *Proc. Fluid Amplification Symp.* **III** (October 1965), HDL. Also C. F. Pavlin—Fluid Amplifiers, U.S. Patent No. 3,456,665, 22 July 1969.
16. G. R. Howland, A high-gain proportional amplifier. *Bendix Tech. J.* **1**, No. 4 (1969). Also J. M. Hyer and J. M. Eastman—"Deflector Fluidic Amplifier," U.S. Patent No. 3,486,520, 30 Dec. 69.
17. R. A. Evans, "Control Apparatus," U.S. Patent No. 3,451,408, 24 June 1969.
18. E. M. Dexter, An Analog Pure Fluid Amplifier. *Symp. Fluid Jet Contr. Devices* ASME (28 November 1962).
19. F. T. Brown and R. A. Humphrey, Dynamics of a proportional amplifier (Part I), ASME Paper 69-WA/Flcs-2, and (Part 2) 69-WA/Flcs-3, 1969. *J. Basic Eng. Ser. D* **92**, (1970).
20. W. S. Griffin, A fluid-jet amplifier with flat saturation characteristics. *J. Basic Eng. ASME* 734 (1969).
21. F. Manion, Proportional amplifier simulation. *Advan. Fluidics, ASME* (May 1967).
22. F. J. Camarata, Analytical procedure for predicting performance of single-stage momentum exchange proportional amplifiers. *Advan. Fluidics, ASME* (May 1967).
23. J. G. Rupert, Analysis of the pressure flow characteristics of a submerged jet-receiver-diffuser-load system. *Nat. Conf. Fluid Power, 23rd Annu. Meeting, 19–20 October 1967*.
24. R. E. Olson and F. J. Camarata, Pressure recovery characteristics of compressible two-dimensional free jet flows. *Proc. Fluid Amplification Symp.* **I** (October 1965), HDL.
25. J. A. Kallevig, Effect of receiver design on amplifier performance and jet profile of a proportional fluid amplifier. *Proc. Fluid Amplification Symp.* **II** (October 1965), HDL. Also J. M. Kirshner, Fluerics 35. Jet Dynamics and Its Application to the Beam Deflection Amplifier, HDL-TR-1630, March 1973.
26. K. N. Reid, An experimental study of the static interaction of an axisymmetrical fluid jet and a single receiver-diffuser. *Fluid Amplification Symp. Proc.* **IV** (October 1965), HDL.
27. F. T. Brown, On the stability of fluid systems. *Proc. Fluid Amplification Symp.* **I** (May 1964), HDL.
28. C. A. Belsterling and K. C. Tsui, Application techniques for proportional pure fluid amplifiers. *Proc. Fluid Amplification Symp.* **II** (May 1964), HDL.
29. W. A. Boothe, A Lumped Parameter Technique for Predicting Analog Fluid Amplifier Dynamics. J. A. C. C., June 1964.
30. F. T. Brown, "Stability and Response of Fluid Amplifiers and Fluidic Systems," NASA CR 72192, DSR 5169-3, 1 October 1967. Prepared under Contract NASA 3-5203.
31. C. A. Belsterling, "Development of Techniques for the Static and Dynamic Analysis of Fluid State Components and Systems," U.S. AVLABS, Tech. Rep. 66-16, February 1966.
32. C. A. Belsterling, "Designing Fluidic Systems," Control Engineering, April 1966.
33. C. A. Belsterling, "Fluidic Systems Design Manual," U.S. Army AVLABS, Rep. 67-32, July 1967.
34. G. L. Roffman and S. Katz, Predicting closed loop stability of fluid amplifiers from frequency tests. *Fluid Amplification Symp. Proc.* **I**, 297–313 (1965).

References

35. A. J. Healey, Vent effects on the response of a proportional fluid amplifier, ASME Paper No. 67-WA/FE-12, 1967. *J. Basic Eng. Ser. D* **90**, 90 (1968).
36. F. M. Manion, "Dynamic Analysis of Flueric Proportional Amplifiers," ASME Paper No. 68-FE-49, 1968.
37. F. M. Manion, "Development of a Flueric Amplifier Transfer Matrix," 1968, Wescon Tech. Papers, Session 15, 20–23 August 1968.
38. L. R. Kelley and W. A. Boothe, "Hydraulic Fluids," ASME Paper 68-WA/FE-26, 1968.

Chapter 10

THE BISTABLE SWITCH

10.1 EARLY HISTORY

Early in 1959 after Horton suggested that amplification might be achieved by deflecting one fluid jet with another, a number of experiments with round jets indicated that the gain was rather small. In an effort to increase the gain, Raymond Warren and Romald Bowles decided to use slitlike nozzles and to surround the jets with top and bottom plates and walls. In so doing they rediscovered the wall attachment effect. After the initial dismay at being unable to obtain proportional deflection, they realized that they had found something that could be more important in some ways than a proportional device and the investigations of the bistable device began.

Subsequently it was learned that the deflection of a stream toward an adjacent wall had first been noted as early as 1800 by Thomas Young, and had been named the Coanda Effect in 1948 by Professor A. Metral, and had even been used for switching a jet stream by auxiliary jets within the housing of an aircraft engine by Kadosch et al. [1].

A survey of the early history of this effect has been given by Kadosch [2, 3] as follows:

In 1800, Thomas Young reported on some experiments he had carried out [4]:

> The lateral pressure which urges the flame of a candle towards the stream of air from a blow pipe is probably exactly similar to that pressure which causes the inflexion of a current of air near an obstacle. Mark the dimple which a slender stream of air makes on the surface of water. Bring a convex body into contact with the side of the stream and the place of the dimple will immediately show the current is deflected toward the body; and if the body be at liberty to move in every direction it will be urged toward the current.

The phenomenon was rediscovered by Chilowski in connection with experiments on projectiles, and was applied by Lefay [5] (who named it the Chilowski effect) and by Ravelli [6] in 1929.

Henri Bouasse and Abbott Carriere repeated Young's experiments and Bouasse [7] described them in 1931.

In the keynote address delivered at the Third HDL Fluid Amplification Symposium in October 1965, Dr. Henri Coanda related how he became interested in aviation in 1906 and built a jet engine aircraft that he showed at the Paris Air Show in 1910. The flames from the jet attached themselves to mica plates (that had been installed to keep the flames away from the pilot and fuselage) and the plywood plane was burned.

Other things held his attention for many years, but in 1930 Coanda noticed the way droplets of water from a faucet attached to and moved down his arm and was reminded of the way that the jet had attached to the mica plates. It was then that he began experimenting with jets [8] on what later was named by Metral and Zerner [9] the Coanda Effect.

10.2 PRINCIPLES OF OPERATION

10.2.1 Attachment

To see how entrainment can be used to produce a switching device, consider a two-dimensional turbulent jet leaving a slitlike nozzle, as in Fig. 10.1, shortly after it has been turned on. The jet entrains fluid on both sides. Because of the presence of the walls, some of the molecules are evacuated between the jet and each wall, causing a low-pressure region into which there is a counterflow down the wall. As a result of any perturbation, such as the turbulent fluctuations of the jet, the jet may momentarily bend toward one wall, thus tending to cut down the counterflow, and thereby lowering the pressure between the jet and that wall more than that along the other wall. This mechanism therefore quickly causes the jet to attach to the wall, as shown in Fig. 10.2. The recirculating region between the attached jet and the wall is called the low-pressure bubble.

Attachment occurs only if the side walls are sufficiently long. Short walls, however, may cause the jet to bend without attachment. The necessary minimum length of wall for attachment when a jet is turned on initially has not received a great deal of consideration. The problem is complicated by the fact that an existing jet may be moved into attachment with a wall and remain there, although the wall cannot cause attachment when the jet issues initially into the stagnant region between the walls. In all the cases we consider, the walls are sufficiently long for spontaneous attachment.

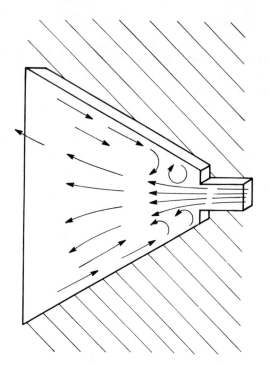

Fig. 10.1 Initial flow.

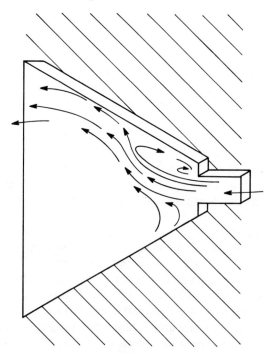

Fig. 10.2 Attachment.

10.2 Principles of Operation

10.2.2 Effect of Controls

The wall attachment effect enables us to make a bistable device (Fig. 10.3). The jet will, at random, attach to either wall. In either case we can shift it to the opposite wall by use of the controls.

If the jet is attached to the right wall, as in Fig. 10.3, and both controls are open, the pressure in the bubble (on the attached side) is lower than that on the opposite side of the jet. Fluid is being entrained through both controls.

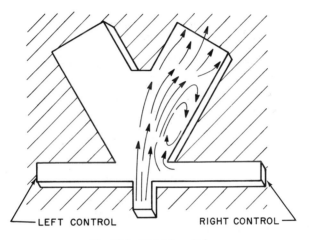

Fig. 10.3 Bistable switch.

If sufficient flow is inserted momentarily through the right control, the jet will flip to the left wall. The narrowest portion of the device just downstream of the controls is called the secondary throat. The operation of the device is critically affected by the ratio of the width of the secondary throat to that of the power jet nozzle.

If this ratio is not too large, temporarily closing off the flow through the left control (or applying a negative pressure of sufficient magnitude in any other manner) will lower the pressure on that side to below that in the bubble and will cause the jet to switch from the right wall to the left.

10.2.3 Effect of Blocking an Outlet

Figure 10.4 shows the effect of complete blockage on a bistable switch having high stability. Stability in this case implies that the distance from the nozzle to the splitter is long enough so that the forces holding the jet to the wall are sufficient to keep it attached to the blocked side, even though the fluid itself is leaving from the opposite receiver. An appreciable static pressure will exist in the blocked receiver and the jet will again flow out of that side once the blockage is removed.

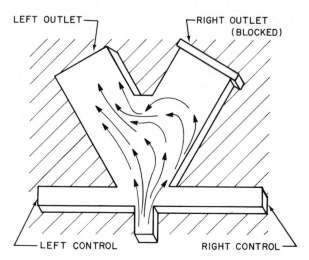

Fig. 10.4 Effect of blocking the outlet.

If the splitter distance (from the nozzle) is small, blocking the outlet will cause the jet to detach and the static pressure in the blocked receiver will be low.

If the splitter distance is made short enough the fluid will tend to flow into both outlets.

10.2.4 The Use of Vents or Bleeds

Although loading increases the pressure at the outlets in the case of complete or partial blockage, it may also decrease the pressure when the load is another bistable switch with a large entrainment. Figure 10.5 shows two bistable switches connected together in tandem with all the flow from unit 1 entering the controls of unit 2. If the units are the same size and are operating at the same supply pressure, the flow from unit 1 may be sufficient to remove the bistability from

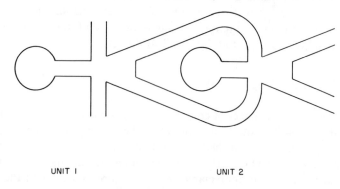

Fig. 10.5 Bistable switch cascade.

10.2 Principles of Operation

unit 2. This defect can be corrected by using a larger supply pressure for unit 2. However, if the supply pressure of unit 2 becomes too large, the state of unit 1 can be changed through increased entrainment. In some cases an oscillating system may result for specific values of supply pressure in each unit.

If we wish to achieve a power device, unit 2 may be made larger than unit 1. Here again there are restrictions to the supply pressure in each unit. If, however, we are concerned with computation devices, we would like to use units all approximately the same size. To accomplish this as well as to minimize the effects of loading in general, bleeds that vent the excess flow (or supply additional flow if needed) are used, as shown in Fig. 10.6.

In a device with bleeds, the pressure on the side opposite the bubble tends to be approximately ambient regardless of the load on the outlets. If the units are operated with air as the medium in an air environment and if there is no reason to seal the units off from the environment, the bleeds are left open to atmospheric pressure. In other cases (such as in a contaminated atmosphere), the bleeds are connected to a filtered reservoir.

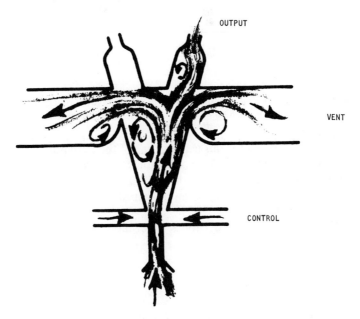

Fig. 10.6 Vented bistable switch.

10.2.5 Splitter Vent Effects

The first bistable devices used pointed wedges as splitters (Fig. 10.7), but it was subsequently found that blunt splitters and cusp-shaped splitters (Fig. 10.7) increase the stability of wall attachment and that a faster rise time is obtained at

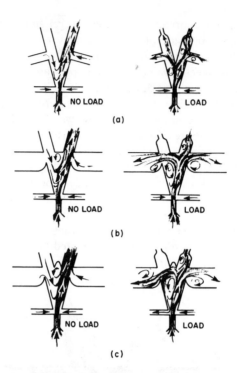

Fig. 10.7 Effects of splitters and bleeds: (a) Knife edge or rounded splitter upstream of bleed, (b) blunt splitter downstream of bleed, and (c) cusp splitter downstream of bleed.

the output. The reason for this is that some of the output flow is peeled off by a cusped or blunt splitter and caused to flow in a vortex (Fig. 10.7) so as to increase the forces acting to hold the jet to the wall.

The above advantages are, however, gained at the expense of an increase in total switching time.

Present theories for the dynamics of bistable switches consider only sharp splitters.

10.2.6 Curved Attachment Walls

Advantages and disadvantages have also been demonstrated for curved walls as opposed to straight walls.

In a curved wall device the attachment bubble may be extremely small or even nonexistent. A wall curvature can be found that maximizes the pressure difference holding the jet to the wall so that maximum stability is obtained. This optimum curvature (one giving the lowest pressure) is a convex logarithmic spiral [10]. The jet entrainment is, however, appreciably increased over that near a straight wall, so that the maximum velocity and hence the jet energy decays faster [11]. The greater transverse pressure difference across the jet in the case of a curved wall means that the nozzle-to-splitter distance can be reduced as compared to that necessary for the straight wall case so that the greater decay rate of the curved jet does not necessarily result in a lower pressure recovery and possibly can increase the pressure recovery if properly designed.

Although a number of papers discussing the use of curved walls have appeared in the literature, [12–16], the results have been somewhat contradictory. Two effects have made it difficult to compare the results obtained by various authors. The low pressure near the attachment wall results in an increased flow from a given nozzle for the same supply pressure. This increased flow is greatest for the curved wall designed to give the lowest wall pressure. This effect also distorts the profile out of the nozzle.

The second effect is that of the receivers and vent design. The width of the receivers, their position relative to the centerline (as affected by the splitter, particularly if the splitter is not sharp), their distance from the nozzle, and the position of the vents all affect the recovery pressure and make comparisons based purely on whether or not the wall is curved almost worthless. The straight-walled bistable switch has received a great deal of attention, so that the effects of various parameters are now reasonably well understood. More experimental and theoretical work is necessary before the effects of curved walls can be integrated into the theory.

10.2.7 Figures of Merit and Dimensional Parameters

One figure of merit for a bistable switch is the pressure recovery, which is defined as the total pressure captured at an outlet divided by the power jet pressure. Another figure of merit is the fan-out ratio, which is defined as the number of similar units that can be switched by the output of a single unit.

The dimensions are usually given in terms of the power jet nozzle width, to which all lengths are usually normalized. An aspect ratio (ratio of channel height to nozzle width) of 2 is common in molded devices, but the aspect ratio is often about unity in etched devices.

10.3 WALL ATTACHMENT THEORIES

The presence of a wall adjacent to a two-dimensional jet leaving a nozzle results in the jet bending toward the wall and flowing along it for some distance. This phenomenon is known as wall attachment or the Coanda effect. A number of simplified models have attempted to provide a theory to relate the geometrical and fluid parameters to the attachment point and the bubble pressure. Although the process is too complicated to allow a simple theory to give as good results as one might like, relatively good agreement has been obtained for certain cases.

10.3.1 Parallel Wall with Arbitrary Offset

We first consider the case of a parallel wall with an offset (Fig. 10.8). Most authors use modifications of the assumptions that Bourque and Newman [17] and Sawyer [18] ascribe to Dodds [19]. We shall discuss this method as published by Bourque and Newman.

The jet is identified in the model by three streamlines, as shown in Fig. 10.8. The streamline that proceeds from the edge of the nozzle a_1 to the point of attachment a_5 is called the attachment streamline. In the steady state this line

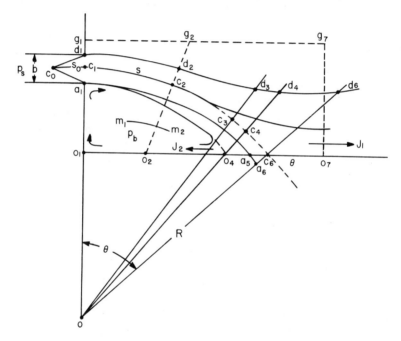

Fig. 10.8 Attachment to an offset parallel wall.

divides the jet flow from that circulating in the bubble. The bubble is the name given to the region $\overline{a_1 a_5 o_1 a_1}$ between the walls and the streamline, or alternatively to the region $\overline{c_1 c_6 o_1 c_1}$. (Note: Curves, whether open or closed, are designated by overlining the points through which the curve passes. Closed curves are indicated by the first and last point being identical.)

10.3.1.1 Jet Profile

The analysis assumes that the jet profile is the same as that of a free jet, and Goertler's profile is used to represent it, with the origin of the flow being placed a distance s_0 upstream of the nozzle with s_0 chosen such that the volume flow q_s out

10.3 Wall Attachment Theories

of the nozzle, as given by Goertler's equation at $s = 0$, is the same as that of a jet with a uniform profile:

$$q_s = h(Jb/\rho)^{1/2} = hb[2(p_s - p_a)/\rho]^{1/2} \tag{10.1}$$

where h is the distance between top and bottom plates, b is the nozzle width, J is the momentum flux out of the nozzle per unit depth $[=2(p_s - p_a)b]$, ρ is the density, p_s is the supply pressure, and p_a is the ambient pressure. Now Goertler's equation [Eq. (4.144)] may be written as

$$u = [3J\sigma_e/4\rho(s + s_o)]^{1/2} \operatorname{sech}^2[\sigma_e y/(s + s_o)] \tag{10.2}$$

where s is the distance along the jet centerline, y is the distance perpendicular to the jet centerline, and σ_e is the spread parameter. We may now determine the equation of a streamline in the same way as we did in Chapter 4 [Eq. (4.88)].

Since volume flow cannot cross a streamline, for incompressible flow, streamlines are also lines of constant volume flow. The equation of a streamline is therefore as given in Eq. (4.35):

$$q_1 = \text{const} = h \int_0^{y_1} u \, dy \tag{10.3}$$

where q_1 is the volume flow contained between $y = 0$ and $y = y_1$. The substitution of Eq. (10.2) into (10.3) yields

$$q_1 = h[3J(s + s_o)/4\rho\sigma_e]^{1/2} \tanh[\sigma_e y_1/(s + s_o)] \tag{10.4}$$

For the particular case of the attachment streamline ($y_1 = y_a$ and $q_1 = q_s/2$), Eq. (10.4) becomes

$$t_a^2 = b\sigma_e/3(s + s_o) \tag{10.5}$$

where

$$t_a = \tanh[\sigma_e y_a/(s + s_o)]$$

From the free jet assumption we may find s_o by noting that at $s = 0$ the streamline distance y_a approaches infinity and so t_a approaches unity. Therefore, from Eq. (10.5) and as already indicated in Chapter 4,

$$s_o = \sigma_e b/3 \tag{10.6}$$

The equation of the attachment streamline [Eq. (10.5)] then becomes

$$(3s/b\sigma_e) + 1 = 1/t_a^2 \tag{10.7}$$

Equation (10.7) is shown plotted in Fig. 10.9 for several values of the jet spread parameter σ_e. The same result is shown in normalized form in Fig. 4.8 at $\dot{m}/\dot{m}_f = 1.00$. The jet spreads more as σ_e decreases. Note also that as downstream distance increases the attachment streamline position is more sensitive to changes in the spread parameter.

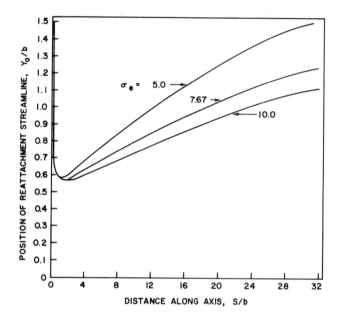

Fig. 10.9 The effect of spread factor on attachment streamline.

10.3.1.2 Attachment Point Theory

Bourque and Newman [17] use two different assumptions in considering momentum flux. The first is called the attachment point theory. It assumes that the momentum flux is conserved in a volume surrounding the attachment point a_5 (Fig. 10.8), i.e., that the effects of variations in pressure are negligible, and that the average angle of the input momentum flux J is essentially the same as the angle made by the centerline extended (θ). Under these assumptions

$$J \cos \theta = J_1 - J_2 \tag{10.8}$$

It is further assumed that J_1 and J_2 are given, respectively, by

$$J_1 = \int_{-\infty}^{y_a} \rho u^2(s_6, y) \, dy \tag{10.9a}$$

$$J_2 = \int_{y_a}^{\infty} \rho u^2(s_6, y) \, dy \tag{10.9b}$$

where, from Fig. 10.8, $s_6 = \overline{c_1 c_6}$ at which point $y_a = \overline{c_6 a_6}$. Thus J_1 is considered as being equal to the momentum flux that would lie between y_a (the attachment streamline) and infinity if the jet were a free jet, and J_2 is taken as the rest of the momentum in the jet.

The assumption that J_1 and J_2 may be calculated [Eq. (10.9)] as if the jet were a free jet is difficult to justify for the following reasons:

10.3 Wall Attachment Theories

In the steady state the mass flow within the bubble of the attached jet is a constant so that the flow circulates without the mass flow increasing or decreasing. It is apparent that at least some momentum must also circulate. If the angular momentum associated with the circulation of flow within the bubble is to remain constant at any given point (as it must in the steady state), the momentum entering must be equal to the frictional losses. In particular, if the process were completely lossless, no momentum would enter the bubble. Intuitively one would suppose that, if the wall geometry allows the bubble to be approximately circular, the losses will be relatively small because the fluid would rotate roughly as a rigid body, whereas in a long narrow bubble the velocity gradients (across the line $\overline{m_1 m_2}$ in Fig. 10.8) are very large and will result in relatively large losses. The jet velocity profile must therefore be different for different conditions of momentum loss so that in all cases the momentum entering the bubble is equal to that lost.

In considering the circulating momentum it is apparent that the momentum flux is greatest in the vicinity of the point of attachment a_5 (Fig. 10.8) but decreases for several reasons as it goes around the loop. First, there are losses to the wall; second, there are losses due to the interaction of the circulating flow on one side of the dividing arc $\overline{m_1 m_2}$ with the circulating flow on the opposite side of $\overline{m_1 m_2}$; and finally, some of the momentum is converted into pressure both in the vicinity of $\overline{m_1 m_2}$ and at the base wall. It is intuitively clear that the shorter $\overline{m_1 m_2}$ is the more the recirculation takes on the property of rigid-body rotation and thus the momentum flux loss should be less. In general, $\overline{m_1 m_2}$ will become shorter as the offset distance $B\,(=\overline{o_1 a_1})$ increases.

The recirculating momentum is minimum in the vicinity of a_1 (Fig. 10.8) and increases as it moves back toward the attachment point along the upper side of the bubble, since as it moves along that region it takes up momentum from the jet until it has achieved its maximum value when it reaches the vicinity of the attachment point.

The arguments above indicate that only some fraction of the momentum calculated on the attachment point model should actually reenter the bubble; yet, at least for those cases where the wall is parallel to the nozzle, the calculation shows reasonably good agreement with experimental results. On the basis of the above arguments this could be true only if an appreciable part of this momentum is lost in the recirculation process or if an error due to one of the other assumptions tends to compensate.

The second reason that the assumption that the profile is approximately that of a free jet is difficult to justify is because it is known that the entrainment of a jet curving over a convex surface is greater than that of a free jet, whereas the entrainment of a jet curving within a concave surface is lower than that of a free jet (Fig. 10.10).

Sawyer [20] has pointed this out, adding that if we combine these two effects we get essentially the flow pattern of an attached jet (Fig. 10.8) where the half of the jet on the bubble side corresponds to Fig. 10.10b and the opposite half to Fig. 10.10a. Sawyer's calculation shows that thickness of each half of the jet varies by only about $1\frac{1}{2}\%$ from that of a free jet; however, this small difference

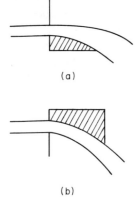

Fig. 10.10 The effect of jet curvature on entrainment: (a) flow over a convex surface—entrainment increased, (b) flow deflected by a concave surface—entrainment decreased.

may make a relatively large difference in determining the point of attachment since essentially all of the momentum associated with this $1\frac{1}{2}\%$ is subtracted from the reverse momentum J_2 and is added to the forward momentum J_1. Since J_1 is itself usually less than 25% of the total momentum, subtracting from it even this small percentage of the total momentum makes an appreciable change.

It is important to note that the point at which the momentum is considered to divide is also chosen by assumption. Other points can be used (and have been used by other authors) with at least equal justification.

Our objective now is to use the attachment point theory to determine the attachment distance x_R ($\overline{o_1 c_6}$ on Fig. 10.8). There are five geometric variables: θ, y_a, x_R, s_6 ($=\overline{c_1 c_6}$), and the radius of curvature R. We need five equations relating these variables. One equation is the equation of the attachment streamline [Eq. (10.7)] with s replaced by s_6. Another equation can be obtained from the attachment point model [Eqs. (10.8) and (10.9)] by using Goertler's jet profile [Eq. (10.2)]. The result is

$$\cos\theta = \frac{3}{2}\left\{\tanh\frac{\sigma_e y_a}{s_6 + s_o} - \frac{1}{3}\tanh^3\frac{\sigma_e y_a}{s_6 + s_o}\right\} \qquad (10.10)$$

The remaining three equations come from geometric relations. It is assumed that the curvature of the jet is essentially constant (i.e., that the jet moves in the arc of a circle) and that the radius of curvature R is related to the bubble pressure p_b and the ambient pressure p_a by

$$p_a - p_b = J/R \qquad (10.11)$$

where $R = \overline{oc_6} = \overline{oc_1}$.

10.3 Wall Attachment Theories

The bubble pressure is not actually uniform, but if p_b is defined as the mean bubble pressure, then experiments by Wada and Shimizu [21] indicate that this assumption is approximately correct.

From Fig. 10.8 it is seen that

$$R(1 - \cos \theta) = B + b/2 \qquad (10.12)$$

where $B = \overline{o_1 a_1}$ and is called the offset. Equation (10.12) is the first of three geometric relations. The second relation is

$$s_6 = R\theta \qquad (10.13)$$

and finally the third geometric relation is

$$x_R + y_a/\sin \theta = R \sin \theta \qquad (10.14)$$

We may find the variables θ, y_a, x_R, s_6, and R by solving Eqs. (10.7), (10.10) and (10.12)–(10.14) simultaneously. Figure 10.11 shows the attachment distance

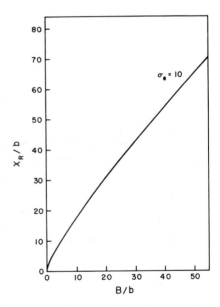

Fig. 10.11 Attachment distance for parallel offset.

as a function of the offset. Bourque and Newman's data and theory are in good agreement [22] for $\sigma_e = 10$.

Sometimes it is of interest to calculate the bubble pressure p_b. This is now easily done since the solution of the five simultaneous equations also provide the radius of curvature R. Thus, Eqs. (10.1) and (10.11) can be combined to give

$$(p_b - p_a)/(p_s - p_a) = -2b/R$$

and p_b may now be calculated.

10.3.1.3 Control Volume Assumption

The preceding calculations were based on the attachment point assumption [Eq. (10.8)]. Bourque and Newman [17] also carried out a similar calculation using a different control volume. In their second assumption, the control volume is taken as $\overline{o_1 g_1 g_7 o_7 o_1}$ of Fig. 10.8, and it is assumed that the pressure p_b exists everywhere in the region bounded by $\overline{o_1 c_1 c_6 o_1}$. Then from momentum considerations

$$J - J_1 = (p_a - p_b)(B + b/2) \tag{10.15}$$

Using Eqs. (10.11) and (10.12) this becomes

$$J \cos \theta = J_1 \tag{10.16}$$

This assumption thus results in ignoring the effect of J_2 since the other assumption resulted in

$$J \cos \theta = J_1 - J_2 \tag{10.8}$$

The primary reason for this discrepancy lies in the assumption that the bubble pressure is uniform throughout the bubble region. If momentum stagnates against the wall or is caused to turn by the wall $\overline{o_1 a_1}$ (Fig. 10.8), the pressure along that wall must obviously be greater than at other points within the bubble. Measurements within the bubble actually show the distribution to vary appreciably and to be greatest (closest to ambient) near the wall. One can thus remove the discrepancy between the results by accepting the wall pressure as an unknown. Then if the attachment point method is assumed correct, one can use the results of that calculation together with the control volume method to find the pressure along the base wall if one assumes that all of the momentum J_2 stagnates along the wall [23].

Since in truth not all of J_2 will reach the base wall and since at least part of it is turned by the wall instead of stagnating, the actual pressure at the wall conceivably could be either more than or less than that obtained if one assumes that J_2 stagnates against the wall. Perry [24] suggests that the ratio $(p_a - p_b)/(p_a - p_{b_1})$ be treated as an experimental constant where p_{b_1} is the base pressure (the pressure along the wall $\overline{o_1 a_1}$). In particular if $(p_a - p_b)/(p_a - p_{b_1})$ is given the value of 2, then instead of Eq. (10.16) one obtains Eq. (10.8) so that the attachment point method and the control volume method yield the same result. This, of course, in effect assumes that J_2 stagnates against the wall.

If momentum strikes the wall and stagnates it will not raise the pressure as much as if it strikes the wall and bounces off it to recirculate. On the other hand, as previously noted the momentum that circulates tends to cut down the amount of momentum entering the bubble from the jet thus providing a compensating error in the calculations.

10.3 Wall Attachment Theories

Kimura and Fukuzawa [25] discuss the attachment versus control model volume in some detail, pointing out that, for the general case (when the wall is inclined at an angle α), the control volume model may be made identical with the attachment point model by choosing the ratio

$$\frac{p_a - p_b}{p_a - p_{b_1}} \quad \text{as} \quad \frac{J\cos\alpha - J_1 + J_2}{J\cos\alpha - J_1} = \frac{2(\cos\alpha - \cos\theta)}{2\cos\alpha - 1 - \cos\theta} \quad (10.17)$$

where we have used the fact that

$$J = J_1 + J_2$$

This ratio is equal to 2 when $\alpha = 0$.

Bourque and Newman's experimental results show good agreement with their theory [22] [using Eq. (10.8) rather than Eq. (10.16)] over a large range of offset distances B if the spread factor σ_e is treated as an empirical constant and chosen as 10.

10.3.2 Inclined Wall

Bourque and Newman [17] also discuss the attachment to an inclined wall without an offset, again using both what they call the attachment point method and the control volume method. Neither method gives good agreement with experiment regardless of the value of σ_e chosen.

Levin and Manion [26] showed how the two geometries of Bourque and Newman can be combined so that the geometry of an offset wall at an angle can be treated. Their method, however, includes Bourque and Newman's assumptions for the inclined wall, and (except for small wall angles) the agreement with experiment is also poor regardless of the value chosen for σ_e.

One of the assumptions that is used in the control volume method by Bourque and Newman for the inclined wall and not for the parallel wall (and that was used by Levin and Manion as well as others) is that the momentum in the reversed flow at any arbitrary distance s from the origin may be calculated in the same manner as J_2; that is, Eq. (10.9b) can be used with any value of s substituted for $\overline{s_6}$; for example, if $\overline{o_2 g_2 g_7 o_7 o_2}$ of Fig. 10.8 is chosen as the control volume, then this assumption says that the momentum flux in both the forward and reversed flow can be calculated from Eq. (10.9b) with s_2 $(=\overline{c_1 c_2})$ substituted for $\overline{s_6}$. Now, whereas one might expect the momentum flux to change by only a small amount in turning in the vicinity of a_5, there is really no good reason to suppose that the momentum flux in the reversed flow at an arbitrary point s_2 will be the same as that in the forward flow within the bubble at s_2. In other words, this assumption implies that the momentum flux is dissipated in the reverse flow in the exact same way as the momentum flux grows in that portion of the jet entering the bubble. On the other hand, this assumption is not really radically different from the assumptions already used in the attachment point method

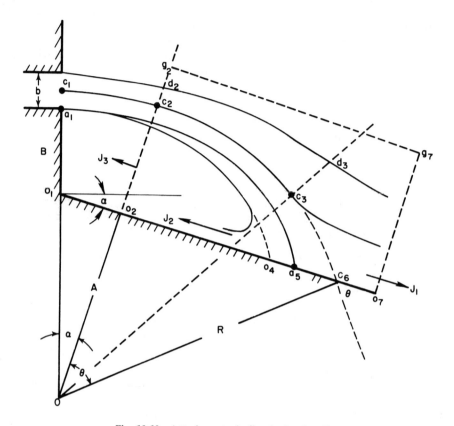

Fig. 10.12 Attachment of offset inclined wall.

since, as we have seen, the assumption of a free jet profile plus the attachment point method implies that the reverse momentum flux is dissipated at a rate such that it goes to zero in the vicinity of the nozzle.

The attachment point method used for the parallel wall of Fig. 10.8 can easily be generalized to the inclined wall of Fig. 10.12 since the basic changes occur in the geometric relations (10.12)–(10.14).

From Fig. 10.12 we see that instead of Eq. (10.12) we have

$$R\left(1 - \frac{\cos \theta}{\cos \alpha}\right) = B + b/2 \tag{10.18}$$

where α is the angle that the inclined wall makes with the nozzle axis. At $\alpha = 0$, Eqs. (10.12) and (10.18) are identical. Equations (10.13) and (10.14) for the inclined wall are

$$R(\alpha + \theta) = s_6 \tag{10.19}$$

and

$$x_R + y_a/\sin \theta = R(\sin \theta + \cos \theta \sin \alpha) \tag{10.20}$$

10.3 Wall Attachment Theories

Attachment point equation (10.8) becomes

$$J\cos(\theta + \alpha) = J_1 - J_2 \tag{10.21}$$

and for the control volume method (for volume $\overline{g_2 g_7 o_7 o_2 g_2}$) the momentum relation is

$$J\cos\alpha - J_3 = J_1 \tag{10.22}$$

where J_3 is calculated in the same way as J_2 is in Eq. (10.9b) but with s_2 substituted for s_6. This also means that s_2 must be used to locate the attachment streamline in the lower limit of the momentum integral.

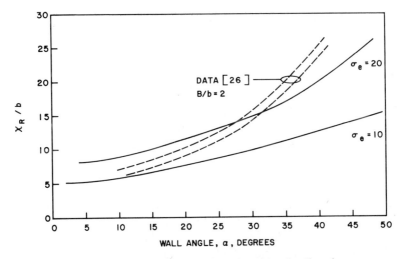

Fig. 10.13 Attachment as a function of wall angle.

Figure 10.13 shows the attachment distance versus wall angle at an offset of two nozzle widths. The attachment point method is used and σ_e is chosen as 10 and 20. Some data from Levin and Manion [26] is superimposed as a band on Fig. 10.13. The results show that for larger offsets or inclinations the value of σ_e must be greater than 20 before the theory approaches the experimental data.

10.3.3 Sawyer's Method

Sawyer [20] also considered separately the parallel offset wall and the inclined wall with no offset. His method is appreciably more complicated than that of Bourque and Newman but yields good agreement with experiment in both cases without the necessity of adjusting the value of σ_e. He obtains agreement using the value of σ_e measured for a free jet (about 7.7).

The first significant difference in his method is that he takes into account the difference in entrainment on the two sides of the attached jet. In addition he modifies Goertler's profile by representing the profile in the vicinity of the nozzle by a uniform velocity core bounded by mixing layer velocity distributions. Farther downstream he matches this distribution to a Goertler profile as the jet becomes fully developed. A third difference in Sawyer's assumptions from those of Bourque and Newman is in his choice of the point at which the momentum divides. Bourque and Newman use the point c_6 (Fig. 10.8) at which the jet centerline extended strikes the side wall. Sawyer selects a distance s_3 ($=\overline{c_1 c_3}$) along the jet such that the inner edge of the jet (which he defines as the distance from the jet axis for which $u/u_c = 0.1$, where u_c is the jet centerline velocity) meets the reversed flow (Fig. 10.8). The reversed flow profile is assumed to be identical to the forward flow profile in that part of the jet between the attaching streamline and the line for $u/u_c = 0.1$. (In Fig. 10.8, the $u/u_c = 0.1$ line—extrapolated to the wall—is indicated by $\overline{a_1 o_4}$. The distance s_3 is determined by the intersection of the line $\overline{od_3}$ with the centerline.)

It should be noted that all three of these assumptions are such as to decrease the momentum flux J_2 into the bubble, and that the last assumption (the choice of s_3) is the most arbitrary. Indeed, in Sawyer's earlier paper [18], the point at which the momentum divides was determined by the intersection of $\overline{od_4}$ with the centerline. One can therefore presume that this third assumption was probably made after initial calculations indicated that the results obtained by use of only the first two assumptions decreased the momentum flux into the bubble by an insufficient amount to give good agreement with experiment. Furthermore, the arbitrariness associated with selecting a line on the basis of setting u/u_c equal to some fraction is quite large. By choosing u/u_c sufficiently small, angle $a_1 o d_3$ and consequently the momentum flux J_2 can be made arbitrarily small.

The final major difference is that Sawyer takes into consideration the pressure variation in the vicinity of attachment by assuming that the pressure distribution is like that in a forced vortex.

Sawyer's results for both the offset parallel wall and the inclined wall agree reasonably well with experiment without the necessity of assuming σ_e to be an empirical constant. Sawyer, however, did not consider the case of a wall that was both offset and inclined.

10.3.4 Boucher's Modification

Boucher [27] has modified Sawyer's analysis and applied it to the case of a wall that is both offset and inclined. His assumptions are similar to Sawyer's [20] except that some simplification is obtained by using Simson's jet profile instead of a modified Goertler profile. This results in the profile in the vicinity of the nozzle being taken into account as it is in Sawyer's method, but the different rates of entrainment on both sides of the jet are not considered.

The distance at which the momentum flux divides is chosen in the same way as by Sawyer, namely in terms of the point at which the inner edge of the forward-moving jet meets the reversed flow. Although Simson's profile inherently provides a well-defined edge at which the velocity is zero, Boucher allows some

10.3 Wall Attachment Theories

flexibility by defining the inner edge of the jet as the line along which u/u_c is equal to some empirically determined fraction. The arbitrariness in this assumption in the case of Simson's profile is not nearly as significant as for Goertler's profile because the $u/u_c = 0.0$ line is only a relatively small distance from the $u/u_c = 0.1$ line in the case of Simson's profile (but is an infinite distance away in Goertler's profile). Boucher's results fit the experiment best if the inner edge is chosen such that $u/u_c = 0.05$. His theory is in best agreement with data for the parallel wall with offset and for the offset wall inclined at angles less than about 30 degrees. For the case of an inclined wall with zero offset, Boucher's results are in poor agreement with the data, although Sawyer's theory shows good agreement with the data, yet the methods of the two differ significantly only in that Boucher assumes a symmetric jet having a Simson profile whereas Sawyer uses a Goertler profile modified to take into account the uniform initial portion of the jet and the difference in entrainment on the two sides of the jet. Since the difference between a symmetric Goertler profile and the Simson profile is quite small, this seems to indicate that taking the entrainment difference into account is a significant factor; however, Boucher states that the agreement between data and theory for the offset parallel wall proves that the entrainment difference is not significant, and that the internal velocity is higher in the case of the smaller bubble associated with the inclined wall with no offset than for the other cases. This greater velocity according to Boucher results in larger pressure gradients within the bubble and greater deviation from a circular arc.

10.3.5 The Sinusoidal Assumption

Boucher's comments seem to be verified by Bourque's [28] work, which uses essentially the same assumptions as Bourque and Newman except that the assumption that the jet is deflected in a circular arc is replaced by the assumption that the attachment streamline is a sinusoidal curve. Another difference is that Bourque chooses the distance from the origin to the point of attachment as the point at which the momentum flux is considered to divide.

Bourque points out that the pressure is actually not constant within the bubble and that therefore one should not expect that the jet radius of curvature should be constant. Furthermore, he notes that for a plate without offset the data indicates (Fig. 10.14) that the reattachment distance becomes infinite as the plate angle of inclination approaches 67 degrees. (Figure 10.14 shows data of Bourque and Newman [17] for the case of an inclined wall of various lengths with no offset.)

The concept of an infinite attachment distance allows us to show more clearly that the jet curvature cannot be constant. For example, as s approaches infinity, Eq. (10.5) indicates that the hyperbolic tangent function approaches zero. Then, as a consequence of Eq. (10.10), $\cos \theta = 0$ and therefore $\theta = 90$ degrees. This shows, according to Bourque [28], that as the angle of inclination approaches 67 degrees the intersection of the jet with the plate must approach a right angle. However, it is obvious that this cannot happen if the jet curvature is constant.

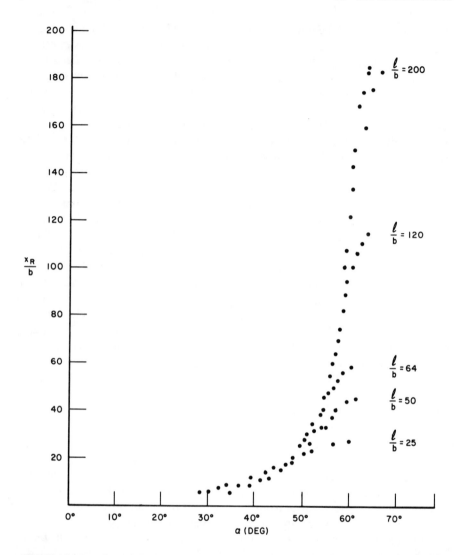

Fig. 10.14 Attachment distance versus wall angle for various length walls: l, wall length; b, nozzle width; x_R, attachment distance.

One can resolve this dilemma either by discarding Eq. (10.10) (and at least one of the assumptions that give rise to it), or by discarding the assumption of constant curvature [one can also weasel around the dilemma by assuming that Eq. (10.10) and constant curvature are simultaneously valid only for small angles of inclination]. Bourque therefore uses an assumption (which he attributes to Rodrigue [29]), namely

$$\frac{r}{b} = \frac{a}{b} \sin \frac{\pi}{2} \frac{\phi_1}{\phi_m} \qquad (10.23)$$

10.3 Wall Attachment Theories

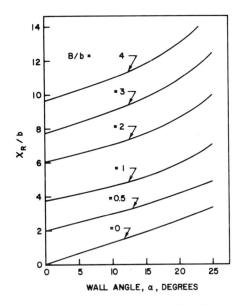

Fig. 10.15 Attachment distance versus wall angle for various offsets.

where r is the radius vector (the straight-line distance from point a_1 on Figs. 10.8 and 10.12 to the attachment streamline), ϕ_1 is the angle that the radius vector makes with the nozzle axis, and ϕ_m ($=67$ degrees) is the maximum possible value for ϕ_1. The constant a is evaluated by consideration of the geometry.

Bourque applies these assumptions to the inclined wall with an offset and gets reasonably good agreement over a large range of offsets and angles using a value of $\sigma_e = 10.5$. The method requires the solution of eight simultaneous equations. Two of these, the equation of the attachment streamline [Eq. (10.7)] and the attachment point momentum equation [Eq. (10.10)] are the same as the circular streamline case. The additional equations come from Eq. (10.23) and five geometrical relations that are different from the geometric relations previously used. The results are shown in Fig. 10.15. The attachment distance is plotted against wall angle for several values of offset that are commonly found in wall attachment devices. This theory (which agrees well with data) gives us a firm idea of the magnitude of the attachment distance.

10.3.6 Experimental Examination of the Attached Jet Assumptions

10.3.6.1 The Circular Arc

It has been pointed out that, whereas most analyses have assumed that the jet centerline moves in a circular arc, Bourque [28] has obtained good agreement with the experimental data for attachment distance by assuming a sinusoidal curve rather than the arc of a circle for the jet path.

The measurements of Wada and Shimizu [21] using various offsets but a single angle of inclination of 15 degrees show that the centerline curvature of the jet follows a circular arc quite closely except in the vicinity of the nozzle and in the vicinity of the attachment point. In the vicinity of the nozzle the jet centerline proceeds along a straight line for a distance whose length depends upon the offset. In the vicinity of the attachment point, the centerline bends away from, rather than towards, the attachment wall.

Their measurements of the relation between the jet radius of curvature and the mean bubble pressure verify the accuracy of the assumption relating the jet curvature to the difference of the mean pressure across it.

10.3.6.2 Jet Asymmetry and Spreading

Measurements of the velocity profile are also presented by Wada and Shimizu [21] that show that even near the point of attachment each half of the jet (both on the attached and the unattached side) has almost the same half-profile shape as a free jet. The spread factor is, however, different [30] for the two sides. Another figure [21] shows the asymmetry of the profiles near the point of attachment and the text states "the profile is extremely asymmetric near the attachment point because of the suppression of entrainment on the bubble side."

In connection with the discussion of his paper on time-dependent jet attachment, Sawyer [31] states that a direct measurement on the velocity profiles of attached jets has shown them to be symmetrical but with a transverse velocity at the velocity maximum, indicating enhanced entrainment on the outside and reduced entrainment along the inner edge. The average entrainment was slightly reduced so that σ_e was approximately equal to 10.

As discussed in Chapter 4, Trapani [32] and Raju and Kar [33] measured the spreading characteristics of two-dimensional jets bounded by top and bottom plates and found that, although the velocity profile was similar to a Goertler profile, the decay characteristics were different and other anomalies appeared. In particular, the bounded jet spreads less than a free jet. One can ascribe this decrease in spread to the restriction of lateral entrainment flow to the jet. The restriction is due to the friction of the air with the top and bottom plates. This process also tends to lower the pressure in the vicinity of the jet. Increasing the area of the top and bottom plates was shown by Trapani to change the spread factor from 10 to 10.8; i.e., the increased resistance caused the jet to spread even less than before.

Increased spreading in the attached jet is shown by the measurements of Wada and Shimizu [30], who also find that the spread rate is greatest near the nozzle ($\sigma_e \sim 6$) and then asymptotically approaches a constant value ($\sigma_e \sim 10$) farther downstream (for an aspect ratio of unity).

10.3.6.3 Choice of Velocity Profile and Discharge Coefficient

Another significant measurement discussed in the paper of Wada and Shimizu [21] shows that the attachment point depends on the boundary layer thickness when the jet leaves the nozzle and therefore depends on the nozzle length and

10.3 Wall Attachment Theories

configuration; that is, the nozzle discharge coefficient should be taken into account.

Bourque [34] has found the same effect and has shown that the actual attachment distance $[x_R/b]_a$ is

$$[x_R/b]_a = [x_R/b] Y_f \tag{10.24}$$

where

$$Y_f = \frac{1 - 2\delta_1/b}{1 - 2\delta_2/b}$$

and δ_1 is the displacement thickness, δ_2 the momentum thickness.

10.3.6.4 Two-Dimensionality

Most of the papers in the literature consider the jet to be two-dimensional; however, since fluid amplifiers incorporate top and bottom plates, an error can be introduced by this assumption. McRee and Moses [35] have examined the effect of aspect ratio on the attachment point for an offset parallel wall. Their data indicate that the aspect ratio begins to become important when it is less than about 9. For example, at a Reynolds number of 8000 an aspect ratio of 5 results in an increase of about 3% in attachment length from that of an infinite aspect ratio. An aspect ratio of 2 results in an increase of about 10% and an aspect ratio of 1 in about a 20% increase. These data are for offset ratios B/b all greater than 2. The difference in attachment length increases as the Reynolds number is decreased, and they find that at the lower Reynolds numbers an aspect ratio of 1 results in a 100% increase in attachment length.

McRee and Moses also point out that the attachment length depends on the distance from the top and bottom plates, so that the attachment distance is less for parts of the jet near the top and bottom plates than it is for portions of the jet near the midplane. They develop an equation to take into account this effect and obtain fairly good agreement with experiment.

McRee and Edwards [36] have extended the above analysis to inclined walls as well as carrying through experiments for a number of offsets and angles of inclination. Fair agreement between analysis and data was obtained.

10.3.7 Jet Attachment with Control Flow

Let us now consider the effect of control flow on the attachment distance. Figure 10.16a shows some typical mean flow streamlines for an attached jet with control flow q_c. The entrainment streamline (I) is displaced from the wall by the control flow. Thus the flow between the entrainment streamline (I) and the attachment streamline (II) is q_c. However, the control flow is most conveniently expressed as the difference between entrained flow q_{ent} and returned flow q_r. That is,

$$q_c = q_{ent} - q_r \tag{10.25}$$

The attached-side velocity profile shown in Fig. 10.16b will make this clear. The flow included between $y = y_E$ and $y = \infty$ is the flow entrained by the jet. The flow from $y = y_A$ to $y = \infty$ is the flow returned to the low-pressure bubble. The control flow is included between $y = y_E$ and $y = y_A$. Now we

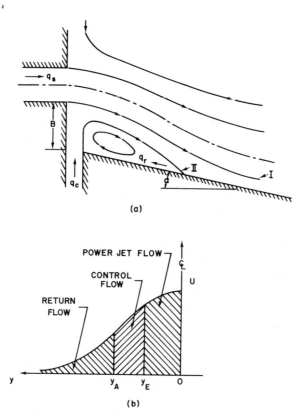

Fig. 10.16 Attachment with control flow: (a) mean flow streamlines with control flow, (b) attached-side velocity profile.

follow the procedure of Brown [37] to relate q_c and y_A. In this method q_{ent} and q_r are first determined. Then Eq. (10.25) is applied to find q_c. The entrained flow is the difference between the flow at some distance downstream and the flow which leaves the nozzle. The entrained flow on the attached side may therefore be expressed as

$$q_{ent} = h \int_0^\infty u\, dy \bigg|_{s=s_6} - q_s/2 \tag{10.26}$$

where h is the depth of the nozzle.

10.3 Wall Attachment Theories

If we use the velocity profile given in Eq. (10.2) and perform the integration, Eq. (10.26) becomes

$$\frac{q_{ent}}{q_s} = \frac{1}{2}\left[\left(\frac{s_6 + \sigma_e b/3}{b\sigma_e/3}\right)^{1/2} - 1\right] \quad (10.27)$$

The return flow (Fig. 10.16b) is

$$q_r = h \int_{y_A}^{\infty} u\, dy \quad (10.28)$$

and with the velocity profile of Eq. (10.2) this reduces to

$$\frac{q_r}{q_s} = \frac{1}{2}\left[\frac{s_6 + \sigma_e b/3}{b\sigma_e/3}\right]^{1/2}\left[1 - \tanh\frac{\sigma_e y_A}{s_6 + \sigma_e b/3}\right] \quad (10.29)$$

The substitution of Eqs. (10.27) and (10.29) into Eq. (10.25) yields

$$\frac{q_c}{q_s} = \frac{1}{2}\left[\left(\frac{s_6 + \sigma_e/3}{b\sigma_e/3}\right)^{1/2}\tanh\frac{\sigma_e y_A}{s_6 + b\sigma_e/3} - 1\right] \quad (10.30)$$

Equation (10.30) represents the equation of the attachment streamline with control flow. When there is no control flow, Eq. (10.30) reduces to Eq. (10.7) for the attachment streamline.

Now if we want to calculate the attachment distance x_R with control flow, we must solve Eqs. (10.18)–(10.21) and (10.30) simultaneously for the unknowns θ, y_A, x_R, s_6, and R. In this case we specify the offset B, the wall angle α, the spread factor σ_e, and the control flow ratio q_c/q_s. Based on these equations, Fig. 10.17

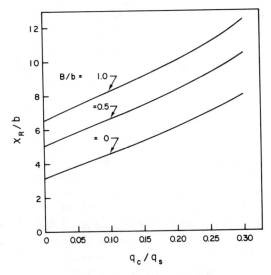

Fig. 10.17 The effect of control flow on attachment: $\alpha = 12$ degrees, $\sigma_e = 10$.

shows the effect of control flow on the attachment distance for $\alpha = 12$ degrees, $\sigma_e = 10$, and various values of offset. As might be expected the attachment distance increases with an increase in control flow.

10.4 THEORY OF THE BISTABLE SWITCH

10.4.1 Introduction

The effect of control flow on the attachment point was first analysed by Brown [37]. Sher [38] extended the analysis to include the effect of momentum in causing an initial deflection of the jet. On the basis of flow visualization studies by Keto [39] and those of his own, Warren [40] pointed out that there were three types of switching, which he named (a) terminated wall or bleed-type switching, (b) contacting-both-walls switching, and (c) splitter switching. Wilson [41] noted that since Bourque and Newman's [17] paper indicates a maximum attachment angle of 67 degrees, instability will occur and the jet will switch if the angle is made greater than 67 degrees, thus indicating a fourth type of switching.

The reason that 67 degrees is the upper stable limit for the attached jet in the case of an inclined wall with no offset is that, for angles greater than 67 degrees, the returned flow is greater than the entrained flow. Although the attachment angle at which instability occurs is probably different for the geometries of bistable switches than it is for the inclined wall, this effect (returned flow exceeding entrainment) may be a fourth contributor to the switching process.

If the offset is fairly large and the splitter farther downstream than the vents or bleeds, terminated wall switching usually occurs. In terminated wall switching, the gradual addition of control flow enlarges the separation bubble until the attachment point reaches the end of the wall, i.e., reaches the vent, at which time fluid enters from the vent into the bubble and the process becomes independent of the control flow.

If the offset is small, a slight addition of flow into the separation bubble will cause the opposite side of the jet to touch the opposite wall to which it will attach. The jet is then attached to both walls. The jet moves from one side to the other as the attachment bubble on the original side expands and that on the opposite wall contracts. The fact that the jet is attached to both walls does not mean that the process has become independent of control flow. If the control flow is removed at a sufficiently early stage, the jet will return to the original wall.

The switching sequence is illustrated [42] in Figs. 10.18a–e. Initially (Fig. 10.18a) the jet is attached to the left wall. Because of the very small offset of the particular device shown, there is no (or almost no) recirculation bubble on the wall. There is, however, a small vortex in the left control and a large vortex on the off side. The excess jet flow is being spilled out of the right vent.

10.4 Theory of the Bistable Switch

In Fig. 10.18b, control flow has been initiated. A recirculating bubble now exists along the left wall. The position of the off-side vortex is essentially unchanged.

In Fig. 10.18c the bubble has enlarged and moved downstream. The off-side vortex has moved to the vicinity of the splitter, a vortex has been initiated in the right control, and flow is spilling from both vents.

In Fig. 10.18d, the jet is attached to the right wall, although still leaving from the left outlet. The off-side (left) bubble has expanded still further, and the former right vortex is beginning to move out of the left outlet.

In Fig. 10.18e, the former right vortex has been swept downstream in the left outlet and the jet is completely on the right.

In splitter switching (Fig. 10.7) the vents are farther downstream than the splitter. When the growth of the separation bubble causes the jet to contact the splitter, part of the jet is peeled off by the splitter and the bubble opens up. If control flow ceases at this time, flow from upstream will enter the space between the wall and the jet (formerly the separation bubble), and the jet will continue to switch.

In the most general case the opposite wall, the vents, and the splitter are all involved to some extent in the switching process.

In general the switching time of a bistable device is a function of the amplitude of the control signal. If the amplitude is too small, the device will not switch. As the amplitude increases the switching time decreases. There is a minimum switching time which is approached asymptotically as the amplitude increases.

10.4.2 Steady State Modeling of Bistable Switch

In Chapter 6 we discussed the use of experimental static characteristic curves. These characteristic curves describe the operation of the bistable switch under a wide variety of possible conditions. In addition they provide us with information on the fan-out and pressure recovery referred to in Section 10.2.7. At this point then we will first present a brief review of the characteristic curves. Following this we indicate an approximate method of determining the curves prior to fabrication.

Figure 10.19 shows the basic input and output characteristics of a bistable switch (see Figs. 6.16 and 6.18). The input characteristics (for nonvented controls) begin at the point where the control is blocked ($q_c = 0$, $p_c = -p_g$). As flow is added to the control, the pressure–flow relation is virtually a straight line [43, 44]. The reason for this is that the internal pressure within the switch is increasing. Then at the switching point (q_{cs}, p_{cs}) there is a sudden decrease in control flow and the remainder of the input characteristic is displaced downward. The most important portion of the input characteristic, however, is the region up to the switching point. In the following section we will provide information of a

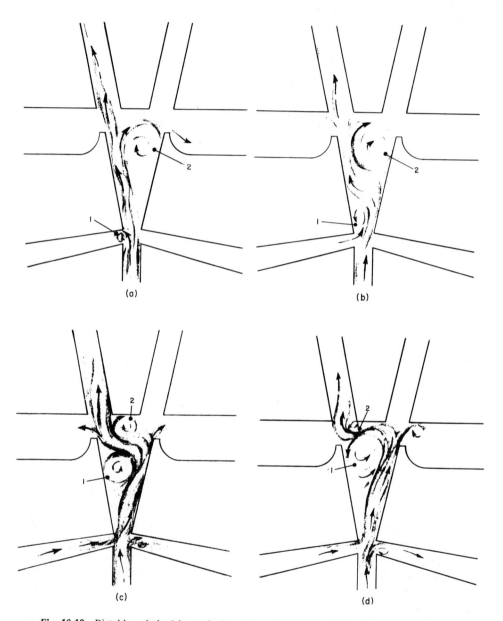

Fig. 10.18 Bistable switch: (a) attached position, (b) switching initiated—two-wall contact, (c) bubble vortex moves downstream, (d) bubble vortex reaches splitter, (e) bubble opens to vent (facing page).

10.4 Theory of the Bistable Switch

design nature so that the input characteristic can be approximated for a wide variety of geometric configurations.

The output characteristic from the active side (Fig. 10.19) is a typical source curve. A significant point on this curve is the point of maximum pressure recovery. For vented devices the maximum recovery occurs at zero output flow ($q_o = 0$, $p_o = p_{ob}$). As we draw flow from the bistable switch the output pressure decreases. The relation between output pressure and flow depends on the output impedance of the switch. Now since the fan-out ratio is the number n_F of similar units that can be switched by the output of a single unit, we may write

$$n_F = q_{os}/q_{cs} \tag{10.31}$$

where q_{os} is the output flow at the switching pressure p_{os}. Note that, when a bistable switch drives a number of identical switches, p_o must equal p_{cs} at the switching point. In Section 10.4.2.2 we describe a procedure that can be used to approximate the output characteristic for various geometric parameters.

Figure 10.20 shows the basic configuration of the bistable switch that we consider in the following sections. The control jet width is always equal to the power jet width b and the aspect ratio is 2. The designer may then select the offset B/b (from 0 to 1), the distance to the splitter l_s/b (from 8 to 15), and the wall angle α (from 10 to 30 degrees). The attached wall is rounded with radius b, where it joins the control port. The choice of these parameters then determines the attached wall length x_v that is suggested by Drzewiecki [43] as $4x_R$ (x_R may

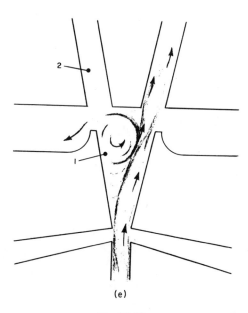

Fig. 10.18e

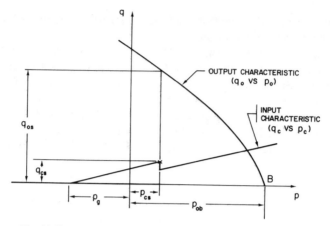

Fig. 10.19 Input and output characteristics of bistable switch.

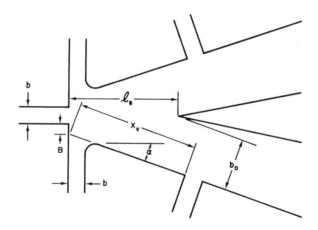

Fig. 10.20 Geometric configuration of bistable switch.

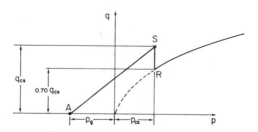

Fig. 10.21 Construction of input characteristic.

10.4 Theory of the Bistable Switch

be found from Fig. 10.15). The rationale for this suggestion is that the attachment length with controls open to atmosphere is about twice the attachment length for blocked controls or $2x_R$. Then to assure element stability the attached wall should be twice as long as open control port attachment or $4x_R$.

10.4.2.1 Approximating the Input Characteristic

Suppose that we have in mind a bistable switch of a particular geometry (b, B, α, l_s, x_v) and the known supply conditions (p_s, q_s). We want to determine the approximate input characteristic before we fabricate the unit. Figure 10.21 shows the information required and the method that has been worked out jointly by Drzewiecki [43] and the authors to construct the input characteristic.

The characteristic begins at the blocked control condition $(q_c = 0, p_c = p_g)$, point A on Fig. 10.21. It is therefore necessary to find the value of p_g that applies for our particular geometry. In Fig. 10.22 we show the pressure ratio p_g/p_s

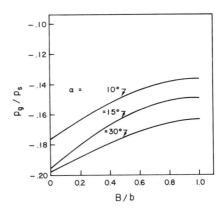

Fig. 10.22 Blocked bubble pressure versus offset.

plotted against offset for various wall angles. The information on this figure was obtained from the elaborate theoretical switch model of Goto and Drzewiecki [44] and agrees with experimental data within about 10%.

To determine the coordinates of switching point S on Fig. 10.21 we make use of extensive experimental data from Lush [45, 46], Moses and McRee [47], and Kimura [48]. These data show that when the attachment wall length x_v/b is 10 or larger, the flow to switch q_{cs} is independent of wall length. Figure 10.23 shows q_{cs} plotted against B/b for a wall angle of 15 degrees. At 12 degrees, Moses and McRee [47] find that the flow to switch is about 10% less than that at 15 degrees. The switching flow data may be closely approximated by the empirical relation

$$q_{cs}/q_s = 0.30 - 0.16[B/b - 1]^2 \tag{10.32}$$

From the data in Fig. 10.23 or Eq. (10.32), we know only the ordinate of switching point S in Fig. 10.21. However, Drzewiecki [43] has noted that the flow usually decreases by about 30% at the switching point. This means that

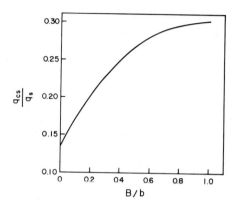

Fig. 10.23 Control flow to switch versus offset.

$0.70q_{cs}$ is the ordinate of point R (Fig. 10.21) on the unattached side characteristic. If we assume that the input characteristic on the unattached side is an orifice, then

$$q_c = c_{dc} A_c (2p_c/\rho)^{1/2} \tag{10.33}$$

where c_{dc} is the discharge coefficient and A_c the area of the control port. Now it follows from Eq. (10.33) and the condition $q_c = 0.70q_{cs}$ that

$$p_{cs} = 0.245 \rho q_{cs}^2 / c_{dc}^2 A_c^2 \tag{10.34}$$

Therefore, since we already know q_{cs}, we may determine p_{cs} from Eq. (10.34), or graphically if the control orifice characteristic is available. The point S is now located and we may draw a straight line from A to S (Fig. 10.21) to represent the attached side characteristic. After switching, the input characteristic follows the unattached side orifice characteristic. We demonstrate the procedure for a typical example.

Example 1 A bistable switch has an aspect ratio of 2 and control and power supply ports 1.0 mm wide. The offset B is 0.5 mm, the wall angle $\alpha = 15$ degrees, the wall length $x_v = 14b$, and the control nozzle discharge coefficient c_{dc} is 0.8. Determine the input characteristic when the supply pressure is 10 kN/m².

(a) Find the supply flow ($\rho = 1.2$ kg/m³)

$$q_s = A_s (2P_s/\rho)^{1/2} = 2 \times 10^{-2}[(2)(10)(10)^7/(1.2)]^{1/2}$$
$$q_s = 258 \quad \text{cc/sec}$$

(b) Find p_g. From Fig. 10.22 at $B/b = 0.5$ and $\alpha = 15$ degrees,

$$p_g/p_s = -0.162$$
$$p_g = (-0.162)(10) = -1.62 \quad \text{kN/m}^2$$

Thus the coordinates of point A (Fig. 10.21) are $q = 0$, $p = -1.62$ kN/m².

10. Theory of the Bistable Switch

(c) Find q_{cs}. From Fig. 10.23 or Eq. (10.32) at $B/b = 0.5$,

$$q_{cs}/q_s \cong 0.260$$
$$q_{cs} = (0.260)(258) = 67.1 \quad \text{cc/sec}$$

(d) Find p_{cs}. From Eq. (10.34),

$$p_{cs} = \frac{0.245 \rho q_{cs}^2}{c_{dc}^2 A_c^2} = \frac{(0.245)(1.2)(67.1)^2}{(0.64)(2)(10)^3} \frac{\text{kN}}{\text{m}^2}$$
$$= 0.52 \quad \text{kN/m}^2$$

The coordinates of point S (Fig. 10.21) are $q_{cs} = 67.1$ cc/sec, $p_{cs} = 0.52$ kN/m², and the coordinates of point R are $q = 0.70 q_{cs} = 47$ cc/sec, $p = 0.52$ kN/m².

10.4.2.2 Approximating the Output Characteristic

We approximate the output characteristic from the blocked output pressure recovery p_{ob} and the output impedance of the bistable device. Figure 10.24 shows the recovery predicted by the theory of Goto and Drzewiecki [44] for l_s/b between 8 and 16. Here the parameter $(p_{ob} l_s)/(p_s b)$ is plotted against offset for some typical

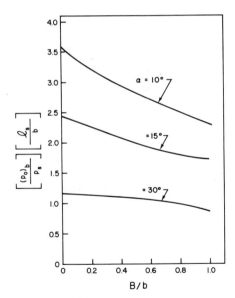

Fig. 10.24 Blocked output parameter versus offset for various wall angles.

wall angles. The recovery decreases with increasing offset, wall angle, and splitter distance. For example, when $l_s/b = 10$, $B/b = 0.5$ and $\alpha = 15$ degrees, p_{ob}/p_s from Fig. 10.24 is 0.20. If any of the geometric parameters l_s, B, or α are increased, p_{ob} decreases. However, the splitter distance is the most sensitive parameter.

Now Fig. 10.24 allows us only to locate point B (Fig. 10.19) on the output characteristic. It is a significant point though because we may now find the remainder of the characteristic to a good approximation by assuming a square law impedance for the output. That is,

$$p_o = p_{ob} - (\rho/2)(q_o/A_o)^2 \qquad (10.35)$$

where $A_o = b_o h$ and b_o may be obtained as a function of the other geometric parameters as (see Fig. 10.20):

$$b_o = (l_s^2 + [B + b/2]^2)^{1/2} \sin\left(\alpha + \tan^{-1}\frac{B + b/2}{l_s}\right) \qquad (10.36)$$

We may now return to the configuration of Example 1 to demonstrate the determination of the output characteristic.

Example 1 (*continued*) If the splitter is 12 nozzle widths downstream, find the output characteristic and the fan-out capabilities of the switch.

(e) Find p_{ob}. From Fig. 10.24 the recovery parameter $(p_{ob} l_s)/(p_s b)$ is 2.0. Thus,

$$p_{ob} = \frac{(2.0)}{(12)}(10) = 1.67 \quad \text{kN/m}^2$$

(f) Find b_o. From Eq. (10.36),

$$b_o = b[(12)^2 + (1)^2]^{1/2} \sin[15° + \tan^{-1}(1/12)] = 4.07 \quad \text{mm}$$

(g) Find the output characteristic. From Eq. (10.35),

$$p_o = p_{ob} - \frac{\rho}{2}\frac{q_o^2}{A_o^2} = 1.67 - \frac{1.2(10)^{-3}q_o^2}{2(4.07)^2(2)^2}$$

$$p_o = 1.67 - 0.0091(10)^{-3}q_o^2 \quad \text{kN/m}^2$$

where q_o is in cubic centimeters per second.

(h) Find the fan-out n_F. From Eq. (10.31),

$$n_F = q_{os}/q_{cs}$$

q_{os} is the flow when $p_o = p_{cs} = 0.52$ kN/m². Therefore

$$q_{os} = \left[\frac{1.67 - 0.52}{(0.0091)(10)^{-3}}\right]^{1/2} = 355 \quad \text{cc/sec}$$

and as a result

$$n_F = 355/67.1 = 5.29$$

and the fan-out is 5.

10.4 Theory of the Bistable Switch

10.4.3 Dynamic Modeling of the Bistable Switch

10.4.3.1 Initial (Lush) Model

The first noteworthy attempt to analyze the switching process in a realistic model was that of Lush [45, 46], who analytically considered terminated-wall switching.

Using a Goertler profile but with different entrainment rates in each half of the jet, Lush set up a quasi-steady model for the growth of the bubble volume and broke the switching process up into a stable and an unstable phase. In the stable phase (Fig. 10.25) an increase of control flow causes the bubble to grow to a size

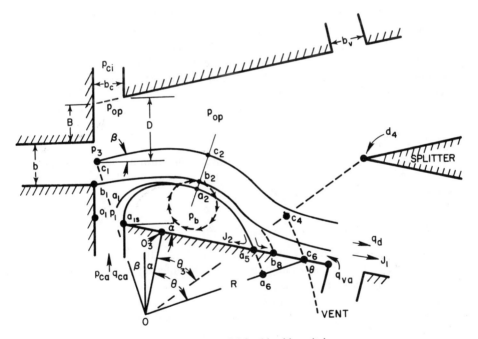

Fig. 10.25 Model for bistable switch.

dependent on the control flow, but if the control flow is kept constant at some value the bubble size will remain constant.

If the control flow is kept constant during the stable phase, the attachment streamline originates at the control corner a_{1s} of Fig. 10.25. However, at such times as the control flow is increasing during the stable phase, some of the control flow enters the bubble so that the attachment streamline passes through the point a_1 into the control nozzle. The process occurs as follows:

When the control flow is constant, none of it penetrates into the region bounded by the attachment streamline $\overline{a_{1s}a_2a_5}$. An increase in control flow results in flow into the bubble and an increase in the distance from the jet origin c_1 to the point of intersection of jet and wall c_6 and therefore in an increased entrainment. If the control flow is no longer increased, then in the stable phase the conditions again become steady and all of the control flow at its new higher level is carried downstream by the jet with its increased level of entrainment.

A point is reached in the bubble growth for which it becomes unstable, that is, its growth is no longer entirely due to control flow since the bubble is open to flow from the vent or from upstream.

For certain geometries an unstable phase arises because the flow carried downstream by entrainment is less than the control flow even when the control flow is constant, or conditions may arise (the attachment angle becomes large enough) that the returned flow is greater than the entrained flow thus causing the bubble to grow even in the absence of control flow. Ogzu and Stenning [49] report that for close-wall amplifiers of sufficiently small offsets, after the jet contacts the opposite wall, the angle of attachment on the original wall increases and instability occurs before the attachment streamline reaches the vent.

A pulse applied at the controls will be unable to cause switching if it is not large enough or of sufficient duration to cause the unstable phase to be initiated.

The bubble defined by Lush is the region bounded by the dividing streamline and the wall ($\overline{b_1 b_8 o_1 b_1}$ in Fig. 10.25).

Although Lush's approach was an important contribution, the switching times he obtained were a factor of about two too low compared with his experimental results.

10.4.3.2 *Epstein's Model*

Epstein [50] considered the combined jet formed from the power jet and control jet and chose as his bubble volume the region bounded by the walls and the jet centerline ($c_1 c_6 o_1 c_1$ of Fig. 10.25).

He considered the switching to take place in three phases instead of Lush's two phases. In the first phase the jet curvature changes due to the control flow but the point of attachment does not move. In the second phase the point of attachment moves to the end of the wall (to the vent) and the jet becomes unstable. His phase three is essentially the same as Lush's phase two. Some minor differences are Epstein's use of the sinusoidally shaped curvature originated by Bourque and his assumption of equal entrainment on both sides of the jet as opposed to the constant curvature and different entrainment on the two sides of the jet assumed by Lush. Like Lush, Epstein considered theoretically only the terminated wall type of switching. He obtained agreement with Lush's experimental results by adjusting his solution to fit the experimental minimum switch pressure and then solving for σ_e. The value required for σ_e was 31, a value that is inconsistent with experimental data.

10.4 Theory of the Bistable Switch

10.4.3.3 Opposite Wall Switching (Ogzu and Stenning)

A theoretical attack on the case of opposite wall switching was first done by Ogzu and Stenning [49] using a Simson jet profile. (However, they state that they also tried a Goertler profile and obtained similar results.) They use the bubble volume contained between the attachment streamline and the wall, as does Lush in determining the bubble growth. The quasi-steady assumption for the bubble growth was found to give only fair agreement, hence an additional term was added to take into account the effects of acceleration. Two methods of approximating this term are given, each of which yields a different result. These methods consider the acceleration of the attachment point rather than the more meaningful transverse acceleration of the jet. The use of the term obtained in one of these approximations results in a better agreement with the experimental data than its omission by about 20%.

Although most of the necessary equations for the analysis are derived from the assumptions, an empirical equation is required to relate the position of the separation point on the opposite wall to the attachment point on the initial wall.

Some important experimental results obtained by Ogzu and Stenning include:

1. The switching time and the control pulse duration required to cause switching were greater for the case of curved convex side-walls than for straight walls (except for high control flows when the switching times were about equal); however, the control flow required to cause switching was less for the curved wall amplifier.

2. The switching time was larger for an aspect ratio of unity than for an aspect ratio of 3 for straight wall amplifiers.

10.4.3.4 The Method of Goto and Drzewiecki

The analytical technique given by Goto and Drzewiecki [44] for bistable amplifier switching takes into account the effect of vents, splitter, and opposite wall, considers both laminar and turbulent jets, and includes the effect of input and output impedances. Results obtained agree well with experiment without the use of any empirical data and with σ_e chosen as the value 10, which is the value most experimenters give to it for the confined jet. The method, however, is very lengthy and requires the solution of over 60 simultaneous equations.

The authors do, however, provide a computer program so that various designs can be considered with relative ease.

The theoretical results agree well with published data for aspect ratios of 1 and offsets of 0.14 and 0.5 and with the experimental data of the authors who used aspect ratios of 2 and 3 and offset ratios (B/b) of 0.5 and 0.9.

Drzewiecki [51] has improved the formulation of the model by breaking the single opposite wall pressure region into two: one region upstream of the point of closest approach between the centerline and the wall and the other region downstream of that point. This allows a low-pressure region to occur upstream of the point of closest approach as the jet bubble grows and makes the model applicable to switching by a negative pressure pulse on the off side.

10.4.3.5 *The Quasi-Steady Assumption*

The theories of the bistable amplifier that we have discussed are reasonably satisfactory for engineering design purposes and for insertion in a circuit simulation program, in spite of the fact that certain of the assumptions are not quite justifiable. The agreement with data may not hold for designs that differ appreciably from those tested, however, so that caution should be used.

Probably the weakest assumption is that of quasi-steady flow.

In order for this assumption to be accurate, the jet profile should retain its shape under transverse motion and the forces due to acceleration must be small.

Studies of the jet in a high-frequency sinusoidal pressure field by Shields and Karamcheti [52] show that the profile distorts and that the spreading characteristics change under oscillatory motion. It is likely (but has not been shown) that the jet also distorts during a rapid switch.

Furthermore, if the quasi-steady assumption is to be valid, the time it takes a particle to move from the nozzle to the splitter should be short compared to the switching time, or more or less equivalently the pressure difference across the jet (internally) must be small compared to the power jet pressure.

In actual fact the particle transport time from nozzle to splitter *is* less than the switching time but not *much* less. The pressure difference across a jet in a close-wall amplifier ($B/b < 0.1$) may be as much as 40% or even more of the power jet pressure, although for offsets as large as 0.5 the pressure ratio may be less than 0.2.

If the addition of corrective terms to compensate for forces of acceleration improves the results, then these corrective terms are justifiable in a utilitarian sense purely on that basis rather than on the basis of better agreement of the assumption with fact. Although we know that the acceleration term is important only when the quasi-steady assumption is incorrect, we cannot abandon the quasi-steady assumption without getting involved in partial differential equations (i.e., the one-dimensional model would have to be discarded).

Thus we should look upon the current bistable switch theories as a tool that will enable us to predict to a reasonable approximation the characteristics (both static and dynamic) of a bistable device on the basis of its geometry and that gives us an approximate but by no means exact picture of the processes occurring within the device.

Nomenclature

PROBLEMS

10.1 Show that the attachment point model equation (10.10),

$$\cos\theta = \frac{3}{2}\left[\tanh\frac{\sigma_e y_A}{s_6 + s_o} - \frac{1}{3}\tanh^3\frac{\sigma_e y_A}{s_6 + s_o}\right]$$

has the solution

$$\tanh\frac{\sigma_e y_A}{s_6 + s_o} = 2\cos\left[\frac{\sigma + \pi}{3}\right]$$

when $0 < \theta < \pi/2$.

10.2 Show that the entrained flow given in Eq. (10.27),

$$\frac{q_{ent}}{q_s} = \frac{1}{2}\left[\left(\frac{s_6 + \sigma_e b/3}{b\sigma_e/3}\right)^{1/2} - 1\right]$$

can also be obtained by integrating the transverse velocity v [Eq. (4.85b) modified],

$$q_{ent} = h\int_0^{s_6} v\,dx.$$

10.3 A bistable switch has an aspect ratio of 2 and control and supply ports 0.5 mm wide. The other geometric parameters are $B = 0.5$ mm, $\alpha = 15$ degrees, $x_v = 5.0$ mm, $l_s = 7.5$ mm, and $c_{dc} = 0.70$. The power jet pressure is 20 kN/m².

(a) Find the input characteristic.
(b) Find the output characteristic.
(c) Find the fan-out.

10.4 Repeat Problem **10.3** for the case when the splitter distance $l_s = 4.0$ mm.

10.5 Show that the fan-out of the bistable switch with equal control and power nozzle widths can be approximated by

$$n_F = \frac{A_o}{A_s}\left[\frac{p_{ob}/p_s}{[0.30 - 0.16(B/b - 1)^2]^2} - \frac{0.490}{c_{dc}^2}\right]^{1/2}$$

for offsets between 0 and 1.

NOMENCLATURE

A_c Control area
b Power nozzle width
b_c Control nozzle width
b_o Output width [Eq. (10.36)]
B Offset distance from nozzle to side wall
c_{dc} Control nozzle discharge coefficient
c_p Pressure coefficient, $= (p_b - p_a)/(p_s - p_a)$
D Secondary throat width

h	Distance between top and bottom plates
J	Power jet momentum flux per unit height
J_1	Momentum flux per unit height moving downstream
J_2	Momentum flux per unit height into the bubble
J_3	Momentum flux per unit height moving upstream at the distance s_2
$\dot{m}$	Mass flow
$\dot{m}_f$	Reference mass flow
n_F	Fan out
N_R	Reynolds number
p_a	Ambient pressure
p_b	Mean bubble pressure
p_{b1}	Pressure along base wall
p_c	Control pressure
p_g	Pressure at control when control is blocked
p_o	Output pressure
p_{ob}	Blocked output pressure
p_s	Supply pressure
q_1	Flow contained between streamlines at $y = 0$ and $y = y_1$
q_c	Control flow
q_{ent}	Entrained flow
q_o	Output flow
q_r	Returned flow
q_s	Power jet supply flow
r	Radial distance
R	Radius of curvature of jet
s	Distance along curved jet centerline
s_o	Distance to virtual origin
s_6	Distance along curved jet centerline to attachment wall, $\overline{c_1 c_6}$ of Fig. 10.8
t_a	Streamline parameter associated with y_a, ($=\tanh[\sigma_e y_a/(s + s_o)]$)
u	$= u(s, y)$, Axial velocity
u_c	Centerline velocity
x_R	Distance along wall to attachment point
y	Dimension perpendicular to jet axis
y_a	Value of y corresponding to attachment streamline
y_f	Factor reducing attachment distance due to boundary layers
α	Angle of inclination of the attachment wall
β	Angle of jet deflection at the nozzle
δ_1	Displacement thickness
δ_2	Momentum thickness
θ	Angle jet centerline extended makes with attachment wall
ν	Kinematic viscosity
ρ	Density
σ_e	Spread coefficient
ϕ_1	Angle for sinusoidal curve
ϕ_m	Maximum angle of attachment

REFERENCES

1. M. Kadosch, F. G. Paris, J. LeFoll, and J. Bertin, "Jet Propulsion Units," U.S. Patent 2,825,204, Issued 4 Mar. 1958.
2. M. Kadosch, Deviation des jets par adherence a une paroi convexe. *J. Phys. Radium, Suppl. 4* **19** (1958).

References

3. M. Kadosch, The curved wall effect. *Proc. Cranfield Conf.*, *2nd* Paper A4 (3–5 Jan. 1967).
4. T. Young, Outlines of experiments and inquiries respecting sound and light (16 Jan. 1800), cited by J. L. Pritchard in the Dawn of Aerodynamics, *J. Roy. Aero. Soc.* (March 1957).
5. A. Lefay, Contribution a l'etude de L'etude de l'effet Chilowsky. *Memorial Artillerie Franc.* **8**, 385–392 (1929).
6. E. Ravelli, Etude pour la theorie de frein de bouche. *Mem. Artillerie franc.* **9**, 488–500 (1930); also *Riv. Artig. Genio* Nov.–Dec. 1928, Jan. 1929.
7. H. Bouasse, "Tourbillons," pp. 341–347. Delagrave, Paris. 1931.
8. Henri Coanda, "Device for Deflecting a Stream of Elastic Fluid Projected into an Elastic Fluid," filed April 19, 1935, U.S. patent issued September 1, 1936, patent #2,052,869, application in France made October 8, 1934.
9. A. Metral and F. Zerner, "L'effet Coanda—Publications Scientifiques et techniques du Ministere de l'Air," no. 218, 1 a5 (1948).
10. J. A. Giles, A. P. Hays, and R. A. Sawyer, Turbulent wall jets on a logarithmic spiral surface. *Aeronaut. Quart.* **17** (1966).
11. G. Mon, "Two-Dimensional, Incompressible, Laminar and Turbulent Curved-Wall-Jet," HDL, TM-72-34, Sept. 1972.
12. J. A. Bronchart, "Study of a Fluid Bistable Amplifier." Thesis submitted to the Catholic Univ. of Lourain, Belgium, 1967.
13. A. W. Rechten, Flow stability in bi-stable fluid elements. *Cranfield Conf. 2nd* (3–5 Jan. 1967).
14. B. B. Beeken, "A Theoretical and Experimental Study of a Coanda Curved Wall Attachment Device," paper #68 WA/FE-27, ASME, 1–5 Dec. 1968.
15. T. Sarpkaya and J. M. Kirshner, The comparative performance characteristics of vented and unvented, cusped, and straight and curved-walled bistable amplifiers. *Cranfield Fluidics Conf., 3rd* paper F3, May 1968.
16. T. Sarpkaya, The performance characteristics of geometrically similar bistable amplifiers. *ASME J. Basic Eng.* 257 (June 1969).
17. C. Bourque and B. G. Newman, Reattachment of a two dimensional incompressible jet to an adjacent flat plate. *Aeronaut. Quart.* **Xi**, 201ff (August 1960).
18. R. A. Sawyer, The flow due to a two-dimensional jet issuing parallel to a flat plate. *J. Fluid Mech.* **9**, 543ff (1960).
19. I. J. Dodds, Ph.D. Thesis, Univ. of Cambridge, 1960.
20. R. A. Sawyer, Two-dimensional reattaching jet flows including the effect of curvature on entrainment. *J. Fluid Mech.* **17**, Part 4, 481 (1963).
21. T. Wada and A. Shimizu, Experimental study of attaching jet flow on inclined flat plate with small offset. *IFAC Symp. Fluidics, 2nd*, 28 June 1971; *Fluidics Quart.* **14** 4 (1972).
22. The theoretical curve for $\sigma_e = 10$ is not given in Bourque and Newman [17] but is given in a subsequent paper by C. Bourque entitled, Reattachment of a two-dimensional jet to an adjacent flat plate, *Advan. Fluidics* 192 ASME.
23. H. R. Chaplin, "Effect of Jet Mixing on the Annular Jet." David Taylor Model Basin Rep. #1375, February 1959.
24. C. C. Perry, Two-Dimensional Jet Attachment. *Advan. Fluidics* 205 (*1967 Fluidics Symp. Proc.*) ASME.
25. M. Kimura and K. Fukuzawa, Corrected control volume model of wall attachment jet, *Proc. Cranfield Conf., 4th*, Paper X7, 17–20 Mar. 1970.
26. S. G. Levin and F. M. Manion, "Fluid Amplification 5. Jet Attachment Distance as a Function of Adjacent Wall Offset and Angles," 31 December 1962, TR-1087. Harry Diamond Lab.
27. R. F. Boucher, Incompressible jet reattachment analysis using a good free jet model. *Cranfield Fluidics Conf., 3rd* Paper F1, 8–10 May 1968.
28. C. Bourque, Reattachment of a two-dimensional jet to an adjacent flat plate. *Advan. Fluidics ASME* p. 192ff (May 1967).

29. G. Rodrigue, "Recollement d'un Jet Incompressible a une Paroi Adjacente avec Injection dans la Bulle de Separation." Thesis, Laval Univ., Quebec, September 1966.
30. T. Wada and A. Shimizu, Discussion of experimental study of attaching jet flow on inclined flat plate with small offset. *Fluidics Quart.* **4**, No. 1 (1972).
31. R. A. Sawyer, Reply to Dr. Begg on Paper A2. *Cranfield Conf.*, 2nd 3–5 Jan. 1967.
32. R. D. Trapani, "Fluerics 22. An Experimental Study of Bounded and Confined Jets," AD-644737, Harry Diamond Lab., Nov. 1966.
33. V. C. Raju and S. Kar, Studies on bounded jets. *Proc. Int. JSME Symp. Fluid Machinery Fluidics*, 2nd **3**, 27 (1972).
34. C. Bourque, Effect of nozzle boundary layers on the position of reattachment of a two-dimensional jet to an adjacent flat plate. *Fluidics Quart.* **3**, No. 3, 1–19 (1971).
35. D. I. McRee and H. L. Moses, The effect of aspect ratio and offset on nozzle flow and jet reattachment. *Advan. Fluidics* 142ff. (1967) ASME.
36. D. I. McRee and J. A. Edwards, "Three-Dimensional Turbulent Jet Reattachment," Paper #70-WA/Flcs-5, *ASME Winter Annu. Meeting, Nov. 29–Dec. 3, 1970.*
37. F. T. Brown, "Pneumatic Pulse Transmission With Bistable-Jet-Relay Reception and Amplification," p. 130ff, Sc.D. Thesis, MIT, May 1962.
38. N. C. Sher, Jet attachment and switching in bistable fluid amplifiers, *Fluids Eng. Conf.* Paper #64-FE-19 18–21 May 1964, ASME.
39. J. R. Keto, Transient behavior of bistable fluid elements. *HDL Fluid Amplification Symp.* **3** (May 1964).
40. J. M. Kirshner (ed.), "Fluid Amplifiers," p. 192ff. McGraw-Hill, New York, 1966.
41. M. P. Wilson Jr., "The Switching Process in Bistable Fluid Amplifiers," Paper #69, Flcs 23, ASME, 16–18 June 1969.
42. R. W. Warren, Personal communication.
43. T. Drzewiecki, Personal communication, 1972.
44. J. M. Goto and T. M. Drzewiecki, Fluerics 32. "An Analytical Model for the Response of Flueric Wall Attachment Amplifiers," HDL TR-1598, June 1972. (See also *Fluidics Quart.* **5**, No. 1, 63–66 (1973) (18).)
45. P. A. Lush, Investigation of the switching mechanism in a large scale model of a turbulent Reattachment amplifier, *Cranfield Conf.*, 2nd Paper A1 (3–5 Jan. 1967).
46. P. A. Lush, A theoretical and experimental investigation of the switching mechanism in a wall attachment fluid amplifier. *Proc. IFAC Symp. Fluidics* (Nov. 1968); P. A. Lush, "The Development of a Theoretical Model for the Switching Mechanism of a Wall Attachment Amplifier." Ph.D. Thesis, Univ. of Bristol, Sept. 1968.
47. H. L. Moses and D. I. McRee, "Switching in Digital Fluid Amplifiers," ASME paper 69-FLCS-31.
48. M. Kimura, Switching in wall attachment device. *IFAC Symp. Fluidics*, 2nd (28 June 1971); *Fluidics Quart.* 14 **4**, No. 1, 1–12 (1972).
49. M. R. Ogzu and A. H. Stenning, "Switching Dynamics of Bistable Fluid Amplifiers." Tech. Rep. #3 Contract #N00014-69-0417, Lehigh Univ., March 1971. See also *J. Dynam. Syst. Measurement Contr. Ser. G* **94**, No. 1, 21 (1972).
50. M. Epstein, Theoretical investigation of the switching mechanism in a bistable wall attachment fluid amplifier. Paper #70-Flcs-3, *Trans. ASME Ser. D (J.) Basic Eng.* 55–62 (March 1971).
51. T. Drzewiecki, The prediction of the dynamic and quasi-static performance characteristics of flueric wall attachment amplifiers. *Fluidics Quart.* Vol. **5**, No. 2, 19 (1973).
52. W. L. Shields and K. Karamcheti, "An Experimental Investigation of the Edgetone Flow Field," SUDAAR, No. 304, Stanford Univ., Feb. 1967.

Chapter 11

THE TRANSITION NOR

11.1 INTRODUCTION

The use of laminar jet transition to turbulence in fluid control devices is over eighty years old [1]. In the first applications, acoustic signals controlled transition and the output was a mechanical displacement [1, 2]. One arrangement for these devices featured a laminar jet passing through a hole in a downstream diaphragm. An appropriate acoustic signal caused the jet to break into turbulence and spread considerably. The resulting larger turbulent jet exerted a force on the diaphragm and moved it. In another arrangement, a jet directed at a flexible diaphragm was used as a hydroacoustic voice amplifier [1, 3]. With the introduction of the "turbulence amplifier" [4] in 1962, the transition principle was applied for the first time to a no-moving part fluid logic device.

Figure 11.1 shows schematic drawings of the original transition NOR element (the "turbulence amplifier") configuration [4]. The device consists of round supply, collector, and control tubes mounted within a larger vented cylindrical enclosure. The supply and collector tubes lie on a common axis. The axes of the control tubes intersect the supply–collector axis at various angles from about 75 to 105 degrees. In operation, a relatively low Reynolds number flow (N_R from 1200 to 1800) in the supply tube produces a laminar jet within the cylindrical enclosure. The vent in the enclosure maintains the static pressure of the jet at atmospheric pressure. Thus the energy in the jet is totally in the form of kinetic energy.

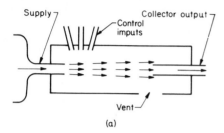

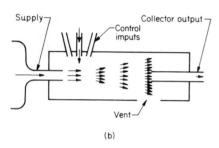

Fig. 11.1 The transition NOR ("turbulence amplifier"): (a) without control flow, (b) with control flow.

In the absence of flow in the control tubes (Fig. 11.1a) the jet reaches the output collector with a large kinetic energy distribution near the jet axis. When a sufficient flow passes through any control tube (Fig. 11.1b) the laminar jet becomes turbulent upstream of the collector tube entrance. Since the turbulent jet has increased spread there is a redistribution of the kinetic energy that continues downstream. As a result the kinetic energy available at the collector entrance is significantly reduced.

The transition device performs the logic NOT when only one control port is used and the logic NOR when more than one control port is used. In this presentation we consider the device as a NOR element. The first step is the definition of performance requirements for a workable NOR and the relation of this performance to geometric and circuit signal variables (Section 11.2). After this we give specific details and data for the characteristic relations that affect performance (Sections 11.3–11.5). The combination of the individual characteristics is demonstrated analytically in Section 11.6 and graphically in Section 11.7. Section 11.8 considers the dynamic response of transition devices and Section 11.9 indicates some other configurations that operate on the transition principle.

11.2 PERFORMANCE CRITERIA

To determine appropriate performance criteria for the transition element requires a clear understanding of the connection of NOR elements in circuits. Figure 11.2 shows a typical NOR element fan-out circuit in which the output of a source element (designated as Source 1) provides control signals to a number n

11.2 Performance Criteria

of identical load elements (designated as Load 2). For proper operation, all the load elements must be in the OFF or turbulent state when the source unit is in the ON or laminar state. Conversely, an OFF source element must result in ON load elements. The fan-out n_F is the maximum number of load elements that can operate from a single source element of identical design. To reduce logic circuit

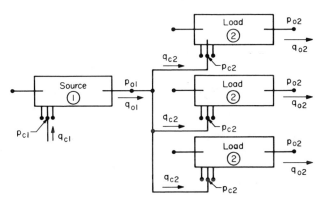

Fig. 11.2 Transition NOR fan-out circuit.

power requirements, a source element with large fan-out is desirable. For this reason, fan-out is a good indication of NOR element performance. Our objective, therefore, is to express fan-out in terms of the device geometry and characteristics.

In Chapter 6 we characterized fluidic components in terms of a two-terminal pair model. We showed that components with adequate venting ports could be described by determining the characteristics given in Eqs. (6.6a) and (6.6b). In matrix form these equations are

$$\begin{bmatrix} \Delta p_o \\ \Delta p_c \end{bmatrix} = \begin{bmatrix} Z_o & T_{oc} \\ 0 & Z_c \end{bmatrix} \begin{bmatrix} \Delta q_o \\ \Delta q_c \end{bmatrix} \quad (11.1)$$

where the output source characteristic yields the output impedance $Z_o \, (= \partial p_o / \partial q_o)$, the transfer characteristic yields $T_{oc} \, (= \partial p_o / \partial q_c)$, and the input characteristic yields the input impedance $Z_c \, (= \partial p_c / \partial q_c)$.

Matrix (11.1) applies also to the transition NOR element (the turbulence amplifier). We must remember, however, that this formulation presupposes that the power jet pressure and flow are invariant. This is, of course, true once we have the device and wish to use it in a circuit. However, if we want to design a turbulence amplifier we must have more information. Bell [5] proposes a matrix which follows Verhelst's [6] presentation of transition NOR characteristics and is

$$\begin{bmatrix} \Delta p_s \\ \Delta p_o \\ \Delta p_c \end{bmatrix} = \begin{bmatrix} Z_s & 0 & 0 \\ T_{os} & Z_o & T_{oc} \\ 0 & 0 & Z_c \end{bmatrix} \begin{bmatrix} \Delta q_s \\ \Delta q_o \\ \Delta q_c \end{bmatrix} \quad (11.2)$$

where the subscript s refers to supply conditions, the supply impedance Z_s ($=\partial p_s/\partial q_s$), and the pressure recovery characteristic T_{os} ($=\partial p_o/\partial q_s$). Note that the four lower right matrix elements of (11.2) are the same as those of matrix (11.1).

Now each matrix element represents the differentiation of a specific characteristic of the transition NOR. In this chapter we indicate how these characteristics are related to the geometry. First, however, let us work with the characteristics to determine the fan-out capability of the device.

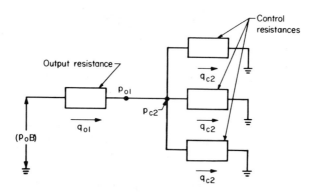

Fig. 11.3 Static equivalent impedance circuit for transition elements.

Figure 11.3 shows the static equivalent fan-out circuit. The total pressure supplied to this circuit is the pressure recovered by the collector when it is blocked ($p_{oB} = T_{os}q_s$). This pressure comes from the kinetic energy of the jet at the entrance to the collector tube. We assume, for purposes of circuit analysis, that the total pressure p_{oB} remains constant at either the laminar level p_L or the turbulent level p_T, regardless of the output flow. The magnitude of the output and control resistances depend on the geometry of the tubes. Usually they more nearly follow a square law relation than a linear relation. The continuity equation applied to the equivalent circuit yields

$$p_{o1} = p_{c2} \tag{11.3a}$$

$$q_{o1} = nq_{c2} \tag{11.3b}$$

Note that n in Eq. (11.3b) is the number of loads and not the fan-out.

This number becomes the fan-out only after we apply the special conditions required for a NOR element. The impedance relations have the approximate form

$$p_{oB} - p_{o1} = (1 + K_{Lo})\rho q_{o1}^2/2A_o^2 \tag{11.4a}$$

$$p_{c2} = (1 + K_{Lc})\rho q_{c2}^2/2A_c^2 \tag{11.4b}$$

where K_{Lo} and K_{Lc} are the total loss coefficients for the outlet and control resistance and A_o and A_c are the minimum cross-sectional area of the outlet and control tubes. Since p_{oB} is almost independent of q_{o1}, the output resistance

11.2 Performance Criteria

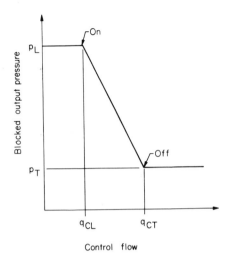

Fig. 11.4 Idealized transition NOR transfer characteristic.

$\partial p_o/\partial q_o$ is approximately equal to the outlet resistance. Experimental data for the input and output characteristics [Eq. (11.4)] are given in Section 11.5.

Figure 11.4 shows an idealized transfer characteristic (p_{oB} versus q_c) for a turbulence amplifier. When the control flow is less than q_{cL} the amplifier is in the laminar state and $p_{oB} = p_L$. On the other hand, control flows greater than q_{cT} cause turbulence and $p_{oB} = p_T$. Thus, the transfer characteristic provides the condition that

$$p_{oB} = p_L, \qquad q_c \leq q_{cL} \tag{11.5a}$$

$$p_{oB} = p_T, \qquad q_c \geq q_{cT} \tag{11.5b}$$

Control flows between q_{cL} and q_{cT} are not permitted for proper operation.

Now in a fan-out circuit the source and load amplifiers must be in opposite states. We may, therefore, combine Eq. (11.3b) and inequality (11.5) to obtain

$$(q_{o1})_T/(q_{c2})_L \leq n_F \leq (q_{o1})_L/(q_{c2})_T \tag{11.6}$$

where q_{o1} may have any value between 0 and $A_o \cdot (2p_{oB}/\rho[1 + K_{Lo}])^{1/2}$ [Eq. (11.4a)] but the control flows are constrained by inequality (11.5). The number of loads n now represents the fan-out because the NOR conditions have been included. Inequality (11.6) indicates that the fan-out for a workable NOR component has a lower limit (often called cascade check) as well as an upper limit. The establishment of a lower limit prevents a source amplifier in the OFF condition from providing control flows that are large enough to cause turbulence in the succeeding stage load amplifiers. If we may sometimes want to connect a source amplifier to a single-load amplifier, one performance requirement is that

$$(q_{o1})_T/(q_{c2})_L \leq 1 \tag{11.7}$$

When this requirement is violated we can still obtain a workable system by connecting the output to a dummy load. In the usual case, however, only the upper limit of fan-out is of interest. The upper limit of fan-out occurs when n_F equals the flow ratio $(q_{o1})_L/(q_{c2})_T$. From the combination of Eqs. (11.4) and (11.6) the upper fan-out limit is

$$n_F = \frac{A_o}{A_c}\left[\frac{(1+K_{Lc})}{(1+K_{Lo})}\left(\frac{2p_L A_c^2}{(1+K_{Lc})\rho q_{cT}^2}-1\right)\right]^{1/2} \quad (11.8)$$

If inequality (11.7) is satisfied we may operate a circuit correctly with any number of load amplifiers less than the number specified by Eq. (11.8). Now the upper limit given in Eq. (11.8) represents the fan-out for a specific geometry. To obtain an optimum fan-out we must adjust the geometric parameters to maximize Eq. (11.8).

In Sections 11.3–11.5 that follow, we consider the geometric and supply flow variables that affect p_L, q_{cT}, K_{Lc}, and K_{Lo}.

These latter quantities reflect information from the matrix elements T_{os}, T_{oc}, Z_c, and Z_o, respectively.

In Sections 11.6 and 11.7 we demonstrate how variations in these quantities affect fan-out.

11.3 SUPPLY CHARACTERISTICS

To design a transition element we need to know two factors about the supply jet:

(1) the laminar length
(2) the downstream pressure recovery

The first factor determines the distance between supply and collector tubes. The second factor indicates the blocked output pressure recovered in the collector. We treat these factors separately in the following subsections.

11.3.1 The Laminar Length

There are four distinct types of jet flow that may occur within a particular fixed-length enclosure [7]:

(a) dissipated laminar jet ($N_R < 300$)
(b) laminar jet throughout ($300 < N_R < 1000$)
(c) laminar–turbulent transition jet ($1000 < N_R < 3000$)
(d) turbulent jet throughout ($N_R > 3000$)

The first three types have quite variable characteristics which depend on the length of the enclosure. The transition NOR operates with the laminar–turbulent transition jet. Figure 11.5 shows a schematic drawing of the laminar length L for an enclosed laminar–turbulent jet. The parameters that may affect the laminar

11.3 Supply Characteristics

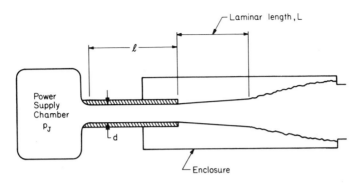

Fig. 11.5 Schematic drawing of laminar length.

length are the length and diameter of the supply tube (l and d), the Reynolds number based on tube diameter (N_R), and the dimensions of the enclosure. To obtain laminar length information we rely on experimental results.

Unfortunately measurements of laminar jet length often disagree by several orders of magnitude [7–10]. Perhaps the reason for the large discrepancies is that the laminar length depends on the rate of growth of disturbances within the jet. The probable origin of these disturbances according to Marsters [8] is in the supply chamber. Thus, the difference in supply chamber configurations used in each experiment may explain the large disparity in experimental results. To add substance to this view, experiments performed with supply tube lengths from 0 to 426 nozzle diameters ($0 < l/d < 426$) show conclusively that the laminar length does not depend on the length of the supply tube. However, rounding off the exit of the supply chamber at the entrance to the supply tube more than doubled the laminar length. Apparently, disturbances generated in the supply chamber are never completely damped out even in a long supply tube. Additional experimental results [8] confirm the independence of laminar length and supply tube length.

Somewhat contradictory evidence still exists about the effects of tube diameter on the laminar length. In some investigations by McNaughton and Sinclair [7] and Marsters [8], changes in tube diameter did not change the laminar length ratio L/d. Experiments by Werbow [9], on the other hand, show that the L/d ratio increases as tube diameter decreases. However, many of these latter tests were made at Reynolds numbers less than 1000, which is near the lower limit for the transition jet. The results become less dependent on tube diameter when the Reynolds number exceeds 1000.

McNaughton and Sinclair [7] find that the enclosure length has a considerable effect on the laminar length. A decrease in enclosure length causes a decrease in laminar length. Here again, however, the effect is most pronounced at low Reynolds numbers.

The most significant factor in the determination of the laminar length seems to be the conditions and geometry of the supply chamber. As mentioned previously, the results of some investigations differ by a factor of 1000. Thus, design

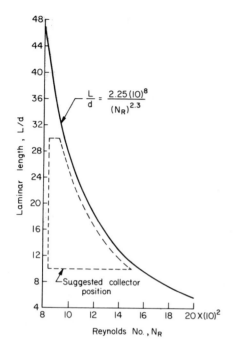

Fig. 11.6 The effect of Reynolds number on laminar length.

guidelines for the spacing between the collector and the supply tube are difficult to establish. Despite this, the total of the data seems to follow the empirical formula proposed by Marsters [8]:

$$L/d = A/N_R^B \qquad (11.9)$$

where A and B are experimentally determined constants. The value of A varies widely among different investigators ($10^7 < A < 7 \times 10^{10}$), but the value of B does not change appreciably. For a sharp entry from supply chamber to supply tube, $A = 2.25 \times 10^8$ and $B = 2.3$ are suggested [8]. Figure 11.6 shows the laminar length ratio plotted against Reynolds number.

The distance between the supply tube and the output collector must be less than the laminar length. However, this distance should always be more than 10 nozzle diameters so that p_T and $(q_{o1})_T$ will be small and inequality (11.7) will be satisfied. For most practical "turbulence amplifiers" the supply and collector tubes are spaced between 10 and 30 diameters apart. The dashed line area on Fig. 11.6 represents the region where transition elements usually operate.

11.3.2 Pressure Recovery

The fan-out [as expressed in Eq. (11.8)] increases with an increase in the blocked laminar state output pressure p_L. Let us consider, then, the factors that contribute to the pressure recovered from a laminar jet.

11.3 Supply Characteristics

Figure 11.7 shows a schematic drawing of the supply and collector geometry. On the supply side there is a plenum chamber at pressure p_s, and a tube of length l and diameter d. Downstream, at a distance L_c from the supply tube, is an axially aligned collector tube. The magnitude of the blocked output pressure

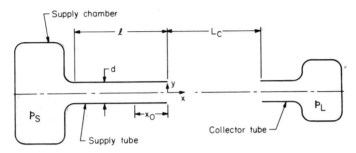

Fig. 11.7 Schematic of collector pressure recovery.

depends on the distance L_c, the diameter of the collector tube, and the shape of the velocity profile at the exit of the supply tube. More pressure is recovered from a parabolic velocity profile than from a uniform velocity profile at the same Reynolds number ($\bar{u}d/v$). Thus, although a long supply tube does not affect the laminar length, it is necessary as a means of shaping the velocity profile. Of course, profile shaping does increase power consumption; however, this increase is usually not particularly important since the transition NOR is inherently a low-power device. For example, a typical commercially available turbulence amplifier with a long supply tube requires only about 50 mW to operate.

The jet dynamic pressure recovered by an axially located collector is related essentially to the centerline velocity of a free jet in the absence of the collector [11]. The centerline velocity u_c of an axisymmetric laminar jet emanating from a point source is given in Eq. (4.62). If we include the apparent point of emanation, Eq. (4.62) becomes

$$u_c = 3J/\rho/8\pi v(x + x_o) \qquad (11.10)$$

where J is the momentum flux, ρ is the density, v is the kinematic viscosity, and x_o is the apparent point of emanation upstream of the nozzle exit. To bring this similarity solution [Eq. (11.10)] and the jet exit profile into correspondence, we match the jet momentum and the centerline velocity at the nozzle exit ($x = 0$). For a parabolic exit profile the momentum flux is

$$J = \tfrac{1}{3}\pi\rho\bar{u}^2 d^2 \qquad (11.11)$$

where $\bar{u}$ is the average velocity at the nozzle exit. Since momentum flux is conserved, we may substitute Eq. (11.11) into Eq. (11.10) to obtain

$$u_c = \bar{u}N_R/8(x/d + x_o/d) \qquad (11.12)$$

Now if we apply the additional condition that $u_c = 2\bar{u}$ when $x/d = 0$, the apparent point of emanation is expressed as

$$x_o/d = N_R/16 \qquad (11.13)$$

Thus, the centerline velocity downstream of the nozzle exit, from Eqs. (11.12) and (11.13), is

$$u_c = \frac{2\bar{u}(N_R/16)}{(x/d) + (N_R/16)} \qquad (11.14)$$

Figure 11.8 shows the centerline velocity ratio $u_c/\bar{u}$ plotted against the Reynolds number for various values of x/d in the range used by "turbulence amplifiers". Experimental hot-wire measurements of centerline velocity [12] are in good agreement with Eq. (11.14). To demonstrate where transition might occur on

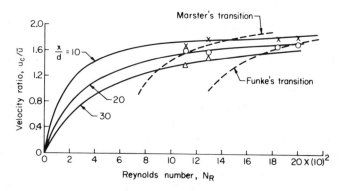

Fig. 11.8 Laminar jet centerline velocity: ———, theory; ×, Funke's data, $x/d = 10$; ○, Funke's data, $x/d = 20$; △, Funke's data, $x/d = 30$.

these curves, we superimpose transition curves [8, 12]. Thus, for example, at $x/d = 20$ the centerline velocity would increase until $u_c/\bar{u} = 1.54$ and $N_R = 1100$ for the jet in one investigation [8] and until $u_c/\bar{u} = 1.70$ and $N_R = 1970$ in the other [12]. However, the small change in velocity ratio is a bit deceiving because the dynamic pressure recovered p_L is approximately equal to $\rho u_c^2/2$, and this is considerably larger at the higher Reynolds numbers. From Eq. (11.14) the recovered pressure may be written as

$$\frac{p_L}{\rho \bar{u}^2/2} = \left[\frac{2(N_R/16)/(x/d)}{1 + (N_R/16)/(x/d)}\right]^2 \qquad (11.15)$$

Equation (11.15) describes the pressure recovery for a small collector tube that does not alter the jet flow field. In practice the collector tube diameter is larger than the momentum flux diameter of the jet, and as a result the actual pressure recovery is less than that indicated in Eq. (11.15). The result is not improved by integrating over the similarity profile of the free jet because, as

11.4 Transfer Characteristics

Beatty and Markland [10] show, stagnation at the probe causes a considerable change in flow pattern. If we assume that the blocked output pressure is proportional to the centerline value, Eq. (11.15) becomes

$$\frac{p_L}{\rho \bar{u}^2/2} = K \left[\frac{2(N_R/16)/(x/d)}{1 + (N_R/16)/(x/d)} \right]^2 \quad (11.16)$$

where K is a proportionality constant that takes into account the diameter of the collector tube. Figure 11.9 compares the pressure recovered by a probe of

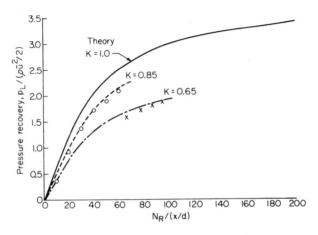

Fig. 11.9 Laminar jet pressure recovery: ○, Beatty and Markland; ×, present tests on turbulence amplifier.

diameter equal to the supply tube diameter and the pressure calculated from Eq. (11.16). For data taken on a free laminar jet by Beatty and Markland [10], the proportionality constant is 0.85. In a "turbulence amplifier", however, the jet is confined within a cylindrical enclosure. Tests on a commercially available element ($l/d = 90$, $L_c/d = 18$, $d = 0.788$ mm) produce a proportionality constant of about 0.65.

11.4 TRANSFER CHARACTERISTICS

The relation between the blocked output pressure and the control input signal is the transfer characteristic of the element. The control signal, which may come from any control tube (Fig. 11.1), causes the laminar jet to break down into turbulence. However, the exact nature of the turbulence triggering mechanism in the control signal is not well established. Verhelst [6] attributes transition to the volume flow or velocity of the control signal. Drazan [13] assumes that the transition is determined by the power of the control signal. We will examine

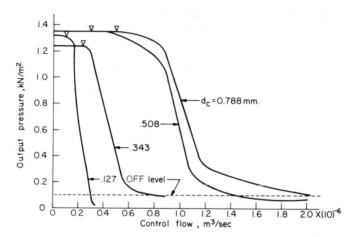

Fig. 11.10 Actual transition NOR transfer characteristics: ▽, ON; d_c, control diameter.

these possibilities in greater detail by introducing test results from a "turbulence amplifier" ($l/d = 90$, $L_c/d = 18$, $d = 0.788$ mm, $p_s = 3.2$ kN/m²) with variable control nozzle diameter.

Figure 11.10 shows some typical "turbulence amplifier" transfer characteristics (output pressure versus control volume flow) for several different control nozzle sizes. With little or no control flow the output pressure is steady at its maximum value. In this device the maximum occurred between 1.25 and 1.35 kN/m². The application of control flow reduces the output pressure to a rather low value. Between these extremes the output pressure is unstable and exhibits erratic behavior. We define OFF as the low level at which the output reaches a stable state. For the current tests OFF was about 0.1 kN/m². Similarly, upon gradual

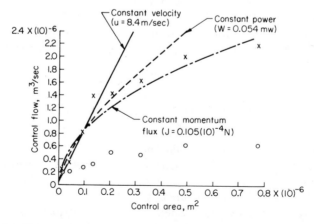

Fig. 11.11 The effect of control area on the control flow that causes transition: ×, OFF; ○, ON.

removal of the control signal, ON represents a return to the stable maximum pressure. The small triangles on Fig. 11.10 indicate ON. In this connection, we observe that the control flows, which cause both OFF and ON, increase as the control diameter increases.

The relation between the control flows (OFF and ON) and the control area is shown more clearly in Fig. 11.11. Eight different control diameters (from 0.127 to 0.991 mm) are represented. Neither OFF nor ON occurs at constant flow. Lines of constant control velocity, control power, and control momentum flux are also plotted on Fig. 11.11 for constant values that most nearly match the OFF condition. In the control area range tested, the results lie between constant power and constant momentum flux. When the control area becomes large the constant momentum flux representation is clearly preferable. The constant velocity assumption is not appropriate at large or small control areas. In any event the relation between q_{cT} and A_c may be approximated in three different ways as

$$q_{cT} = \bar{u}_{iT} A_c \quad \text{(constant velocity)} \tag{11.17a}$$

$$q_{cT} = [2W/\rho]^{1/3}(A_c)^{2/3} \quad \text{(constant power)} \tag{11.17b}$$

$$q_{cT} = [J_c/\rho]^{1/2}(A_c)^{1/2} \quad \text{(constant momentum flow)} \tag{11.17c}$$

where $\bar{u}_{iT}$ is the input (control) velocity, W is the control power, and J_c is the control momentum flux required to reach the OFF state. In Section 11.6 we will use these relations in the analytical optimization of fan-out.

Bell [14] presents considerable data for q_{cT} versus control diameter d_c for various values of L. His data show a trend similar to that given in Fig. 11.11 when replotted as q_{cT} vs A_c. However, Bell goes a step further: he applied signals through two control ports simultaneously and shows that they do not sum. In other words, the q_{cT} at each control is independent of whether other controls are active. Thus, although the data from one control fit the momentum flux assumption best, this cannot be the true transition-causing mechanism.

11.5 INPUT AND OUTPUT CHARACTERISTICS

In Section 11.2 we assumed that Eqs. (11.4a) and (11.4b) were appropriate forms for the input and output characteristics. These equations contained the empirical total loss coefficients K_{Lc} and K_{Lo}. Now we compare the formulation with experimental results and determine approximate values for the loss coefficients.

The input characteristic is an orifice characteristic and should agree well with Eq. (11.4a). Figure 11.12 shows the data from four control nozzles that were used in the transfer characteristic tests. The data are plotted as p_c vs $\rho q_c^2/2A_c^2$ to provide evidence of the accuracy of the form as well as to evaluate K_{Lc}. The slope of the straight line through the data, presented in this manner, is equal to

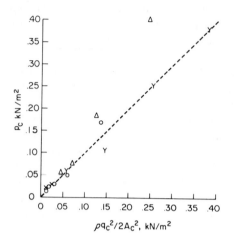

Fig. 11.12 Accuracy of square law approximation for the input characteristic. Diameter: **Y**, 0.343; △, 0.407; ○, 0.508; ×, 0.635 mm.

$1 + K_{Lc}$. We observe that there is considerable scatter and that the loss coefficient depends on the particular orifice. Despite this, the trend of the data conforms most nearly when $K_{Lc} = 0$.

Figure 11.13 shows the output characteristic for the transition element tested. This device had an output tube that was about 40 nozzle diameters long ($d = 0.788$ mm). The solid lines represent the laminar and turbulent state output characteristics. The dashed line indicates Eq. (11.4b) with $K_{Lo} = 6.7$. Although the fit to Eq. (11.4b) is not good, it is adequate for use in calculating the amplifier performance analytically.

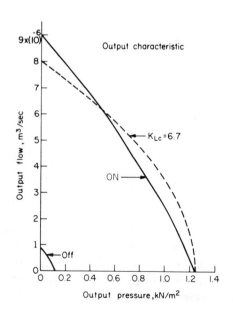

Fig. 11.13 Transition NOR output characteristics.

11.6 PERFORMANCE OPTIMIZATION

Equation (11.8) expresses the upper limit of fan-out as a function of the geometrical and circuit variables. The control flow variable at OFF (q_{cT}), however, is also a function of geometry. That is, the flow at OFF depends on the control area. In the "turbulence amplifier" the precise functional relation between q_{cT} and A_c is not known. Therefore we have postulated the three relations given in Eq. (11.17): (a) constant velocity, (b) constant power, and (c) constant momentum flux. The substitution of each of these into Eq. (11.8) provides three different expressions for the fan-out.

Figure 11.14 shows the fan-out plotted against the control-power diameter ratio for the three assumptions. The required values for the quantities in Eq. (11.8) (K_{Lc}, K_{Lo}, p_L) were taken from the experimental results previously described in Sections 11.4 and 11.5. We observe that the constant momentum flux and constant power curves have finite maximum values of fan-out, whereas the constant velocity curve increases without limit as the control diameter approaches zero. When the control diameter equals the supply diameter, the fan-out for all cases is approximately 2.5. In most practical devices, d_c/d is in the neighborhood of $\frac{1}{3}$. This corresponds to fan-out values of 7, 12, and 18 for the momentum, power, and velocity assumptions, respectively. The actual fan-out of the "turbulence amplifier" lies between the constant momentum flux and the constant power curves.

To find the maximum fan-out (n_M) we use Eq. (11.17) in conjunction with Eq. (11.8) and obtain

$(n_M)_U \to \infty$, $\quad A_c/A_o \to 0 \quad$ (constant velocity) $\quad$ (11.18a)

$(n_M)_W = (2/B)(D/3)^{3/2}$, $\quad A_c/A_o = (3B/2D)^{3/2} \quad$ (constant power) $\quad$ (11.18b)

$(n_M)_J = F/2B^{1/2}$, $\quad A_c/A_o = 2B/F \quad$ (constant momentum flux)

$\qquad$ (11.18c)

where

$$B = (1 + K_{Lc})/(1 + K_{Lo}), \qquad D = [p_L/(1 + K_{Lo})](2A_o^2/\rho W^2)^{1/3}$$

and

$$F = 2p_L A_o/(1 + K_{Lo})J_c$$

For the experimental conditions $p_L = 1.245$ kN/m², $K_{Lc} = 0$, $K_{Lo} = 6.7$, $A_o = 0.487 \times 10^{-6}$ m², $W = 0.054$ mW, and $J_c = 0.105 \times 10^{-4}$ N, the maximum fan-out for constant power is 69 and occurs when $d_c/d = 0.06$. When we apply the constant momentum flux assumption the maximum fan-out is 10 at $d_c/d = 0.187$. We will see in the following section that this latter assumption is in closer agreement with experimental results.

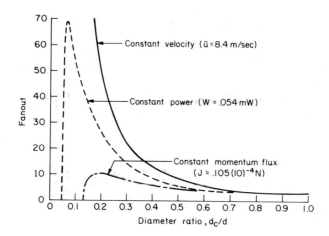

Fig. 11.14 Analytical fan-out as a function of control diameter.

11.7 MODIFIED VERHELST DIAGRAM

Since the analytical expressions used in Section 11.6 are only approximations and are not generally applicable in advance, the performance of transition elements is usually obtained from characteristic curves. Verhelst [6] introduced a compact and useful presentation of these curves. There are five curves, one for each element in matrix (11.2). The " Verhelst diagram " contains characteristics in each of the four quadrants, which are arranged as follows:

Quadrant 1. Transfer characteristics (T_{oc} matrix element)
Quadrant 2. Output impedance characteristics (Z_o matrix element) and output recovery characteristics (T_{os} matrix element)
Quadrant 3. Supply impedance characteristics (Z_s matrix element)
Quadrant 4. Input control characteristics (Z_c matrix element)

Each point of the supply impedance characteristic and the output recovery characteristic leads to a unique output source impedance characteristic. The representation of many supply conditions thus tends to crowd the second quadrant and the performance is more difficult to determine. As a result the Verhelst diagram usually presents the output characteristic for only a single supply condition. Under these circumstances the supply and recovery curves are unnecessary in the determination of performance.

Figure 11.15 shows a two-quadrant modified Verhelst diagram. Quadrant 1 contains the transfer characteristic (p_o vs q_c) and the input characteristic (p_c vs q_c). The second quadrant has the laminar (ON) and turbulent (OFF) output characteristics. To determine the upper limit of fan-out we begin at the point marked OFF in quadrant 1. The OFF control flow q_{cT} is equal to 0.88×10^{-6} m³/sec.

11.8 Dynamic Response

The control pressure required to produce this flow is the ordinate of point a on the input characteristic. Now from Eq. (11.3a) we know that the output pressure of one stage must equal the control pressure of the succeeding stage. Thus, a constant pressure line through a intersects the output characteristic at point b. The abscissa of point b represents the flow available from the output $(q_{o1})_L$. In this case, $(q_{o1})_L = 6.0 \times 10^{-6}$ m³/sec. The upper limit of fan-out [from Eq. (11.6)] is $(q_{o1})_L/q_{cT}$ or 6.82. Thus, the amplifier whose characteristics are shown has a fan-out of six identical elements.

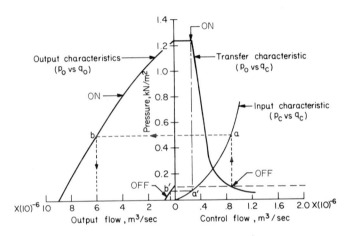

Fig. 11.15 Modified Verhelst diagram.

To determine the lower fan-out limit we begin at the point marked ON on the transfer characteristic and follow this flow ($q_{cL} = 0.26 \times 10^{-6}$ m³/sec) to point a' on the input characteristic. The constant pressure line through a' intersects the OFF output characteristic at point b'. This point supplies an output flow $(q_{o1})_T = 0.35 \times 10^{-6}$ m³/sec. Now from Eq. (11.4) the fan-out must exceed $(q_{o1})_T/q_{cL}$. We find, therefore, that the lower fan-out limit is 1.35. This indicates that, if this output was connected to only one load amplifier, the load amplifier would always be in the turbulent state. Thus, either a dummy input or two inputs from the load amplifier must be used.

11.8 DYNAMIC RESPONSE

Response time measurements on "turbulence amplifiers" often exhibit a large scatter. The variability occurs mainly during the switch to ON because the random quality of the turbulent state provides no unique flow field for the laminar return. In this section we consider a theoretical average value for the dynamic characteristics.

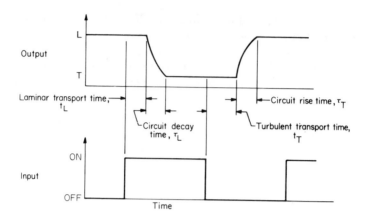

Fig. 11.16 Transition NOR time delay schematic.

Figure 11.16 shows a schematic drawing of the idealized amplifier response to an input square wave. Hayes [15] separates the output signal into four distinct time periods:

(1) the laminar transport delay time t_L
(2) the circuit decay time τ_L
(3) the turbulent transport delay time t_T
(4) the circuit rise time τ_T

Initially the input signal is OFF and the output is in the laminar state. Then, the input goes ON. There is no output change until the portion of the laminar jet that was downstream of the input disturbance passes the entrance to the collector. This time period is the laminar transport delay t_L and is expressed in a similar manner to Eq. (5.6) approximately by

$$t_L = \int_0^{L_c} \frac{dx}{u_c} \tag{11.19}$$

If the centerline velocity of the laminar jet given in Eq. (11.14) is substituted into Eq. (11.19), the integration results in

$$t_L = \frac{d^2}{\nu} \left[4 \left(\frac{L_c/d}{N_R} \right)^2 + \frac{1}{2} \left(\frac{L_c/d}{N_R} \right) \right] \tag{11.20}$$

Now, after the remaining laminar portion of the jet has passed through the amplifier, the output begins to decrease towards the turbulent steady state condition. O'Keefe [16] shows that the decay time is a function of the output circuit impedance and the shape of the transition signal at the collector entrance. There are several ways to represent the shape of the transition signal. A ramp-type transition signal can be obtained by calculating the transport time of the slower-moving particles of the laminar jet velocity profile as Hayes [15] demon-

11.8 Dynamic Response

strated. Here we consider that the transition signal is effectively a step function decrease in pressure. We define the circuit decay time as the time for the output signal to reach 90% of its final value. When the circuit components are linear the decay time is

$$\tau_L = 2.30 R_o n C \tag{11.21}$$

where R_o is the output resistance of the source amplifier and C is the capacitance of one load amplifier. The dependence of decay time on the number of loads is confirmed by O'Keefe's experiments [16].

When the amplifier is in the turbulent steady state and the input is turned OFF there is another transport delay. This delay is sometimes also represented by the laminar centerline velocity and Eq. (11.19) [15] because the laminar velocity exceeds the turbulent velocity and will presumably overtake it. However, the presence of any residual turbulence impedes the return to the laminar state. We therefore instead define turbulent transport delay t_T as

$$t_T = \int_0^{L_c} \frac{dx}{u_{cT}} \tag{11.22}$$

where u_{cT} is the centerline velocity of the turbulent jet. The relation between u_{cT} and the downstream distance x can be found for an initial parabolic velocity profile by the superposition method (see Section 4.43) and is

$$u_{cT} = 2\bar{u}[1 + F\{\exp(-1/F) - 1\}] \tag{11.23}$$

where $F = 4C_2^2 x^2/d^2$ and C_2 is the jet spread parameter. We can compute the turbulent transport delay by substituting Eq. (11.23) into Eq. (11.22) and performing a numerical integration.

Following the passage of turbulent flow we again represent the transition signal as a step function. The subsequent increase in output pressure takes place in a rise time τ_T that is assumed equal to the decay time τ_L [Eq. (11.21)].

The transition from laminar to turbulent state requires the sum of the laminar transport delay time and the circuit decay time ($t_L + \tau_L$). Similarly the turbulent to laminar transition time is the sum of the turbulent transport time and circuit rise time ($t_T + \tau_T$). Since the turbulent time exceeds the laminar time, the largest possible operating frequency will depend on the turbulent time. Therefore this frequency is

$$f = 1/2(t_T + \tau_T) \tag{11.24}$$

To ascertain the inherent response capability of the "turbulence amplifier" we estimate the response times. Figure 11.17, for example, shows the laminar transport delay [Eq. (11.20)] plotted against the Reynolds number parameter $N_R/(L_c/d)$ for two standard amplifier diameters. The transport time decreases as the Reynolds number parameter increases. Since "turbulence amplifiers" usually operate with $N_R/(L_c/d)$ between 60 and 100, the laminar transport time ranges from 0.150 to 0.400 msec. Transport response time reduces when L_c/d decreases.

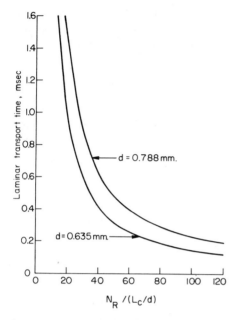

Fig. 11.17 Laminar transport time.

However, as the collector moves closer to the supply tube the turbulent state output pressure increases. Since this impairs the operation of the device, the value of L_c/d and the transport time are limited. Hayes [15] shows that the experimental results are in good agreement with the magnitude of the theoretical laminar transport delay.

Figure 11.18 shows a comparison of the laminar and turbulent transport delays for a parabolic entrance profile and the geometry $d = 0.788$ mm and $L_c/d = 22.4$. In the usual operating region ($1200 < N_R < 1800$) the turbulent transport delay

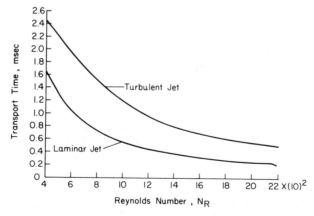

Fig. 11.18 Comparison of laminar and turbulent transport times: $d = 0.788$ mm, $L_c = 22.4$, $C_s = 0.073$.

is more than twice as long as the laminar transport delay. This, also, agrees well with experimental data [15]. We note that the use of transition elements at low Reynolds numbers, to conserve power, sacrifices response time significantly.

The circuit response decay and rise time are difficult to estimate since they depend so much on the physical size of the interconnections between amplifiers. Typical values, however, might be expected to range from 0.500 to 1.000 msec per load [16]. Thus, a "turbulence amplifier" connected to four loads would have a frequency response capability [Eq. (11.24)] from about 90 to 180 Hz.

11.9 ALTERNATE TRANSITION ELEMENT CONFIGURATIONS

Up to now we have been considering a "conventional" tubular axisymmetric "turbulence amplifier." Much of the previous discussion is general and applies equally well to alternate amplifier configurations. However, there are some planar and other axisymmetric types with desirable features that are worth mentioning.

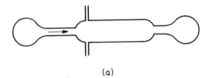

(a)

Fig. 11.19 Planar transition devices: (a) two input NOR [12], (b) four input NOR [15].

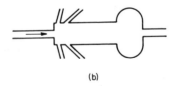

(b)

Figure 11.19 shows two planar transition element configurations. These devices offer significant advantages in fabrication, interconnection, and packaging. The planar unit in Fig. 11.19a has a channel aspect ratio of 1.5 and an interaction region that is vented perpendicular to the flow direction [6]. The device has a fan-out of 4, a supply width of 0.2 mm, and operates on 7 mW of power. Figure 11.19b represents a planar element that uses the turbulent wall attachment effect in addition to jet transition [10]. The control input triggers turbulence, but the side walls are close enough for the turbulent jet to attach. As a result, the turbulent state output is smaller than it is in devices without wall attachment. The interaction cavity is closed top and bottom except for the output vents on each side of the output. The channels have square cross sections of about

0.5 mm on a side. The device requires 50 mW to operate and the fan-out is between 4 and 6.

Figure 11.20 shows some other axisymmetric configurations of transition devices. The element depicted in Fig. 11.20a has another tube on the supply–collector axis [17]. This tube lies between the supply and collector tubes but is closer to the supply tube. The purpose of the intermediate conduit is to permit operation at higher power supply pressures without moving the transition point in front of the output collector and to improve response time. In effect, the tube delays transition. As a result, a laminar state output pressure recovery of up to 9 kN/m^2 is attainable, and this signal is of sufficient magnitude to actuate moving-part power valves. However, the output pressure in the turbulent state is also larger and this acts as a disadvantage.

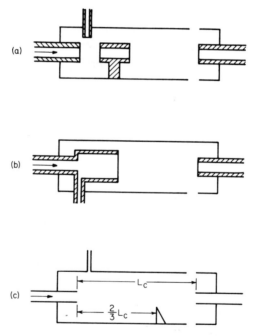

Fig. 11.20 Modifications in tubular transition devices: (a) Howie amplifier [17], (b) Mott's amplifier [18], (c) Siwoff's amplifier [19].

The device shown in Fig. 11.20b has an axisymmetric shroud at the exit of the supply tube [18]. The control input tube enters the upstream side of the shroud. This device has some bistable features. A control input causes transition to turbulence but, when the control signal is removed and the control is vented to atmosphere, the jet remains turbulent. The reason for this is that the turbulent jet lowers the ambient pressure within the shroud. Thus, when the control signal is replaced by atmospheric pressure, a control flow is still entrained into the shroud. The device does not return to the laminar state until the control is blocked mechanically or a negative pressure is applied to the control. In addition, the device has proportional features when the control signals are negative pressures. This configuration produces a special-purpose device which is generally unsuited to logic function implementation.

Figure 11.20c shows a "turbulence amplifier" developed by Siwoff [19] with an edge projecting into the interaction chamber between the supply and collector tubes. Normally the edge is $\frac{2}{3}L_c$ from the supply tube exit. The edge provides a feedback mechanism for the turbulent jet. In the laminar state the edge does not affect the pressure recovery. When the jet begins to undergo transition due to a control input, the edge acts on the distributed jet and hastens the arrival of the turbulent steady state. In addition to the advantage of faster response the device also has a low turbulent state output pressure recovery.

PROBLEMS

11.1 The output and input characteristics of a particular transition NOR have been measured as

$$p_o = 30 - q_o^2, \qquad p_c = 2q_c^2$$

(a) Find the fan-out if the control flow required for transition q_{cT} equals 0.5.
(b) Find the fan-out if $q_{cT} = 2.0$.

11.2 A "turbulence amplifier" has a fan-out of 8 and $q_{cT} = 1.0 \text{ cc/sec}$, $p_{cT} = 0.5 \text{ kN/m}^2$. The output resistance is linear and equal to $0.25 \times 10^6 \text{ kN-sec/m}^5$. Find the pressure which this device must recover in the blocked output condition when $p_c = q_c = 0$.

11.3 The supply tube of a "turbulence amplifier" is 0.75 mm in diameter and 75 mm long. The supply pressure is 3.5 kN/m².

(a) Calculate the supply flow and Reynolds number. [Hint: Refer back to Chapter 2, Eq. (2.27).]
(b) Calculate the pressure that could be recovered on the centerline 12 mm downstream of the supply tube.

SUGGESTED RESEARCH PROJECTS

11.1 Determine the mechanism in the control jet flow that causes transition of the power jet flow.

11.2 Find the properties that determine the laminar length.

NOMENCLATURE

A_c	Control area at minimum control cross section
A_o	Output collector area at minimum output cross section
C	Fluid capacitance
C_2	Turbulent jet spread parameter
d	Supply tube diameter
d_c	Control tube diameter
J_c	Control momentum flux
K	Proportionality constant
K_{Lo}	Output total loss coefficient
K_{Lc}	Control total loss coefficient
l	Supply tube length
L	Laminar length of jet
L_c	Distance from supply tube exit to collector tube entrance
n	Number of loads
n_F	Fan-out
n_M	Maximum fan-out
N_R	Reynolds number ($\bar{u}d/\nu$)
p	Total pressure
p_c	Control pressure
p_s	Plenum chamber supply pressure
p_L	Blocked laminar output pressure
p_T	Blocked turbulent output pressure
p_o	Output pressure
q	Volume flow
q_c	Control volume flow
q_{cL}	TURN-ON control volume flow
q_{cT}	TURN-OFF control volume flow
R_o	Output resistance
t	Transport delay time
$\bar{u}$	Average supply tube velocity
u_c	Centerline velocity
u_i	Input control velocity
W	Control power
x	Axial dimension from supply tube exit
x_o	Apparent jet origin-distance upstream of supply tube exit
ν	Kinematic viscosity
ρ	Fluid density
τ	Circuit response time

REFERENCES

1. C. A. Bell, On the sympathetic vibrations of jets. *Phil. Trans.* **177**, Part 2, 383–422 (1886).
2. R. E. Hall, "Method of and Means for Translating Sounds," U.S. Patent 1,205,530, November 1916.
3. C. L. Stong, The amateur scientist. *Sci. Amer.* 177 (April 1961).
4. R. N. Auger, Turbulence amplifier design and application. *Proc. Fluid Amplification Symp., DOFL* **1**, 357 (October 1962).

5. A. C. Bell, An analytical and empirical basis for the design of turbulence amplifiers, Part I: Analysis and experimental confirmation. *J. Dynam. Syst. Measurements ASME Ser. G* (June 1973).
6. H. A. M. Verhelst, On the design characteristics and production of turbulence amplifiers. *Cranfield Fluidics Conf.*, 2nd Paper F-2 (January 1967).
7. K. J. McNaughton and C. G. Sinclair, Submerged jets in short cylindrical flow vessels. *J. Fluid Mech.* **25**, Part 2 (June 1966).
8. G. F. Marsters, Some observations of the transition to turbulence in small unconfined free jets. *Fluidics Quart.* **3**, 12 (July 1971).
9. A. R. Werbow, "Effects of Reynolds Number and Nozzle Geometry on the Non-Turbulent Length of Free Axi-Symmetric and Rectangular Jets." Masters Thesis, Virginia Polytechnic Inst., December 1966.
10. E. K. Beatty and E. Markland, Feasibility study of laminar jet deflection fluidic elements. *Cranfield Fluidics Conf.*, 3rd Paper H1 (May 1968).
11. K. Reid, "Static and Dynamic Interaction of a Fluid Jet and a Receiver-Diffuser." Sc.D. Thesis, M.I.T., Sept. 1964.
12. M. Funke, Personal communication, 1966.
13. P. Drazan, Optional design of the control jet of a fluid amplifier. *Cranfield Fluidics Conf.*, 2nd Paper B-2 (Jan. 1967).
14. A. C. Bell, An analytical and empirical basis for the design of turbulence amplifiers, Part II: Empirical relationships and design procedure. *J. Dynam. Syst. Measurements ASME Ser. G* (June 1973).
15. W. Hayes, The dynamic response of fluidic turbulence amplifiers. *Cranfield Fluidics Conf.*, 4th Paper A-1 (March 1970).
16. R. F. O'Keefe, "Development and Application of a New Jet Interaction Fluid Amplifier," Instrum. Soc. of Amer., Preprint No. 15-2-MCID-67, September 1967.
17. K. Howie, "Turbulence Amplifier," U.S. Patent No. 3,428,068, February 18, 1969.
18. R. C. Mott, "Fluid Amplifier," U.S. Patent No. 3,429,323, February 25, 1969.
19. F. Siwoff, Improvement of the static and dynamic behavior of the turbulence amplifier by inbuilding of an edge over the distance between the emitter and the collector. *Cranfield Fluidics Conf.*, 3rd Paper H-2 (May 1968).

Appendix A

Circular Transmission Line Characteristics for Air

ω/ω_v	$\mathrm{Re}(Z_c/Z_{ca})$	$\mathrm{Im}(Z_c/Z_{ca})$	$\mathrm{Re}(Y_c/Y_{ca})$	$\mathrm{Im}(Y_c/Y_{ca})$	$\mathrm{Re}(\phi)$	$\mathrm{Im}(\phi)$
0.5	1.2043	−0.5858	0.6715	0.3266	0.4635	0.7775
0.6	1.1675	−0.5017	0.7230	0.3107	0.4895	0.8929
0.7	1.1433	−0.4378	0.7628	0.2921	0.5116	1.0080
0.8	1.1269	−0.3878	0.7935	0.2730	0.5311	1.1230
0.9	1.1154	−0.3478	0.8171	0.2547	0.5486	1.2378
1.0	1.1073	−0.3152	0.8354	0.2378	0.5647	1.3526
2.0	1.0810	−0.1702	0.9027	0.1421	0.6909	2.4940
3.0	1.0708	−0.1254	0.9212	0.1079	0.7961	3.6164
4.0	1.0628	−0.1027	0.9322	0.0901	0.8907	4.7201
5.0	1.0567	−0.0883	0.9398	0.0786	0.9759	5.8101
6.0	1.0520	−0.0782	0.9453	0.0703	1.0535	6.8907
7.0	1.0483	−0.0706	0.9496	0.0640	1.1251	7.9644
8.0	1.0453	−0.0647	0.9530	0.0590	1.1918	9.0328
9.0	1.0428	−0.0600	0.9558	0.0550	1.2545	10.0970
10.0	1.0407	−0.0561	0.9581	0.0516	1.3139	11.1575
12.0	1.0372	−0.0500	0.9619	0.0463	1.4245	13.2700
14.0	1.0345	−0.0454	0.9648	0.0423	1.5263	15.3734
16.0	1.0324	−0.0418	0.9671	0.0392	1.6212	17.4695
18.0	1.0305	−0.0389	0.9690	0.0366	1.7104	19.5596
20.0	1.0290	−0.0365	0.9706	0.0344	1.7948	21.6449
22.0	1.0277	−0.0345	0.9720	0.0326	1.8751	23.7259
24.0	1.0265	−0.0327	0.9732	0.0310	1.9518	25.8033
26.0	1.0255	−0.0312	0.9743	0.0297	2.0255	27.8775
28.0	1.0245	−0.0299	0.9752	0.0284	2.0964	29.9489
30.0	1.0237	−0.0287	0.9761	0.0274	2.1648	32.0177
35.0	1.0220	−0.0262	0.9779	0.0251	2.3265	37.1804
40.0	1.0206	−0.0243	0.9793	0.0233	2.4771	42.3317
45.0	1.0194	−0.0227	0.9805	0.0218	2.6185	47.4738
50.0	1.0184	−0.0214	0.9815	0.0206	2.7524	52.6081
55.0	1.0176	−0.0202	0.9824	0.0195	2.8797	57.7359

ω/ω_v	$\text{Re}(Z_c/Z_{ca})$	$\text{Im}(Z_c/Z_{ca})$	$\text{Re}(Y_c/Y_{ca})$	$\text{Im}(Y_c/Y_{ca})$	$\text{Re}(\phi)$	$\text{Im}(\phi)$
60.0	1.0168	−0.0193	0.9831	0.0186	3.0014	62.8579
65.0	1.0162	−0.0184	0.9838	0.0178	3.1181	67.9750
70.0	1.0156	−0.0177	0.9844	0.0171	3.2304	73.0876
75.0	1.0150	−0.0170	0.9849	0.0165	3.3387	78.1963
80.0	1.0146	−0.0164	0.9854	0.0159	3.4436	83.3013
85.0	1.0141	−0.0159	0.9858	0.0154	3.5452	88.4032
90.0	1.0137	−0.0154	0.9862	0.0150	3.6439	93.5020
95.0	1.0134	−0.0149	0.9866	0.0145	3.7398	98.5982
100.0	1.0130	−0.0145	0.9869	0.0141	3.8333	103.6919
105.0	1.0127	−0.0141	0.9872	0.0138	3.9245	108.7832
110.0	1.0124	−0.0138	0.9875	0.0134	4.0135	113.8724
120.0	1.0119	−0.0131	0.9881	0.0128	4.1857	124.0448
130.0	1.0114	−0.0126	0.9885	0.0123	4.3509	134.2103
140.0	1.0110	−0.0121	0.9890	0.0118	4.5098	144.3694
150.0	1.0106	−0.0116	0.9893	0.0114	4.6632	154.5230
160.0	1.0103	−0.0112	0.9897	0.0110	4.8115	164.6715
170.0	1.0100	−0.0109	0.9900	0.0106	4.9553	174.8154
180.0	1.0097	−0.0105	0.9903	0.0103	5.0949	184.9552
190.0	1.0095	−0.0102	0.9905	0.0100	5.2307	195.0911
200.0	1.0092	−0.0099	0.9908	0.0098	5.3629	205.2235

Appendix B

Rectangular Transmission Line Characteristics for Air

σ	ω/ω_v	Re(Z_c/Z_{ca})	Im(Z_c/Z_{ca})	Re(Y_c/Y_{ca})	Im(Y_c/Y_{ca})	Re(ϕ)	Im(ϕ)
1.0	0.5	2.3823	-2.1325	0.2330	0.2086	1.5146	1.6451
1.0	0.6	2.1987	-1.9252	0.2574	0.2254	1.6459	1.8173
1.0	0.7	2.0580	-1.7628	0.2803	0.2401	1.7636	1.9794
1.0	0.8	1.9462	-1.6308	0.3019	0.2529	1.8705	2.1339
1.0	0.9	1.8549	-1.5207	0.3224	0.2643	1.9684	2.2824
1.0	1.0	1.7789	-1.4268	0.3421	0.2744	2.0588	2.4260
1.0	2.0	1.3973	-0.9059	0.5039	0.3267	2.7051	3.7246
1.0	3.0	1.2582	-0.6684	0.6198	0.3293	3.1065	4.9283
1.0	4.0	1.1917	-0.5277	0.7016	0.3107	3.3954	6.1089
1.0	5.0	1.1558	-0.4346	0.7581	0.2850	3.6241	7.2830
1.0	6.0	1.1349	-0.3690	0.7969	0.2591	3.8176	8.4547
1.0	7.0	1.1219	-0.3209	0.8239	0.2357	3.9891	9.6246
1.0	8.0	1.1135	-0.2845	0.8431	0.2154	4.1462	10.7923
1.0	9.0	1.1075	-0.2563	0.8570	0.1983	4.2933	11.9575
1.0	10.0	1.1031	-0.2339	0.8675	0.1839	4.4334	13.1197
1.0	12.0	1.0967	-0.2009	0.8822	0.1616	4.6993	15.4338
1.0	14.0	1.0917	-0.1779	0.8923	0.1454	4.9519	17.7325
1.0	16.0	1.0874	-0.1610	0.8999	0.1332	5.1948	20.0153
1.0	18.0	1.0836	-0.1478	0.9060	0.1236	5.4291	22.2825
1.0	20.0	1.0801	-0.1373	0.9111	0.1158	5.6554	24.5351
1.0	22.0	1.0769	-0.1285	0.9156	0.1093	5.8740	26.7745
1.0	24.0	1.0739	-0.1211	0.9195	0.1037	6.0850	29.0020
1.0	26.0	1.0712	-0.1147	0.9230	0.0988	6.2889	31.2191
1.0	28.0	1.0687	-0.1091	0.9260	0.0945	6.4860	33.4269
1.0	30.0	1.0665	-0.1041	0.9288	0.0907	6.6767	35.6266
1.0	35.0	1.0617	-0.0939	0.9346	0.0827	7.1283	41.0955
1.0	40.0	1.0578	-0.0859	0.9392	0.0763	7.5489	46.5295
1.0	45.0	1.0545	-0.0794	0.9429	0.0710	7.9436	51.9359
1.0	50.0	1.0518	-0.0741	0.9461	0.0667	8.3168	57.3196
1.0	55.0	1.0494	-0.0696	0.9487	0.0630	8.6714	62.6843

Appendix B

σ	ω/ω_v	$\mathrm{Re}(Z_c/Z_{ca})$	$\mathrm{Im}(Z_c/Z_{ca})$	$\mathrm{Re}(Y_c/Y_{ca})$	$\mathrm{Im}(Y_c/Y_{ca})$	$\mathrm{Re}(\phi)$	$\mathrm{Im}(\phi)$
1.0	60.0	1.0474	−0.0658	0.9510	0.0598	9.0101	68.0326
1.0	65.0	1.0456	−0.0625	0.9530	0.0570	9.3348	73.3666
1.0	70.0	1.0439	−0.0596	0.9548	0.0545	9.6470	78.6879
1.0	75.0	1.0425	−0.0571	0.9564	0.0523	9.9481	83.9979
1.0	80.0	1.0411	−0.0548	0.9578	0.0504	10.2391	89.2977
1.0	85.0	1.0399	−0.0527	0.9591	0.0486	10.5209	94.5882
1.0	90.0	1.0388	−0.0508	0.9603	0.0470	10.7944	99.8703
1.0	95.0	1.0378	−0.0492	0.9614	0.0455	11.0601	105.1446
1.0	100.0	1.0369	−0.0476	0.9624	0.0442	11.3187	110.4118
1.0	105.0	1.0360	−0.0462	0.9633	0.0429	11.5706	115.6725
1.0	110.0	1.0352	−0.0449	0.9642	0.0418	11.8163	120.9270
1.0	120.0	1.0337	−0.0425	0.9658	0.0397	12.2907	131.4193
1.0	130.0	1.0324	−0.0404	0.9671	0.0379	12.7445	141.8917
1.0	140.0	1.0312	−0.0386	0.9683	0.0363	13.1801	152.3465
1.0	150.0	1.0302	−0.0370	0.9694	0.0348	13.5992	162.7855
1.0	160.0	1.0293	−0.0356	0.9704	0.0336	14.0034	173.2104
1.0	170.0	1.0284	−0.0343	0.9713	0.0324	14.3939	183.6226
1.0	180.0	1.0276	−0.0331	0.9721	0.0313	14.7720	194.0231
1.0	190.0	1.0269	−0.0320	0.9729	0.0303	15.1385	204.4130
1.0	200.0	1.0262	−0.0310	0.9736	0.0294	15.4943	214.7931
2.0	0.5	2.6123	−2.3913	0.2083	0.1907	1.6937	1.8084
2.0	0.6	2.4057	−2.1637	0.2298	0.2067	1.8435	1.9942
2.0	0.7	2.2468	−1.9856	0.2499	0.2208	1.9785	2.1683
2.0	0.8	2.1202	−1.8410	0.2689	0.2335	2.1018	2.3334
2.0	0.9	2.0164	−1.7204	0.2870	0.2449	2.2152	2.4915
2.0	1.0	1.9296	−1.6178	0.3043	0.2551	2.3205	2.6437
2.0	2.0	1.4858	−1.0486	0.4493	0.3171	3.0900	3.9931
2.0	3.0	1.3156	−0.7874	0.5596	0.3349	3.5832	5.2107
2.0	4.0	1.2294	−0.6300	0.6442	0.3302	3.9410	6.3875
2.0	5.0	1.1800	−0.5236	0.7080	0.3142	4.2216	7.5495
2.0	6.0	1.1497	−0.4468	0.7557	0.2937	4.4541	8.7055
2.0	7.0	1.1301	−0.3891	0.7911	0.2724	4.6549	9.8582
2.0	8.0	1.1170	−0.3444	0.8175	0.2521	4.8336	11.0088
2.0	9.0	1.1079	−0.3090	0.8375	0.2336	4.9962	12.1574
2.0	10.0	1.1014	−0.2804	0.8527	0.2171	5.1465	13.3041
2.0	12.0	1.0928	−0.2375	0.8738	0.1899	5.4209	15.5923
2.0	14.0	1.0874	−0.2071	0.8875	0.1690	5.6709	17.8732
2.0	16.0	1.0835	−0.1846	0.8969	0.1528	5.9046	20.1466
2.0	18.0	1.0805	−0.1674	0.9038	0.1400	6.1268	22.4123
2.0	20.0	1.0779	−0.1538	0.9092	0.1297	6.3406	24.6701
2.0	22.0	1.0756	−0.1428	0.9136	0.1213	6.5477	26.9199
2.0	24.0	1.0735	−0.1337	0.9173	0.1142	6.7493	29.1618
2.0	26.0	1.0716	−0.1261	0.9205	0.1083	6.9462	31.3957
2.0	28.0	1.0697	−0.1195	0.9233	0.1032	7.1388	33.6221
2.0	30.0	1.0680	−0.1139	0.9258	0.0987	7.3273	35.8411
2.0	35.0	1.0639	−0.1024	0.9313	0.0897	7.7821	41.3585
2.0	40.0	1.0604	−0.0937	0.9357	0.0827	8.2145	46.8377
2.0	45.0	1.0573	−0.0867	0.9395	0.0770	8.6259	52.2844
2.0	50.0	1.0545	−0.0809	0.9428	0.0723	9.0177	57.7037
2.0	55.0	1.0521	−0.0760	0.9456	0.0683	9.3915	63.0998
2.0	60.0	1.0499	−0.0718	0.9480	0.0648	9.7488	68.4763
2.0	65.0	1.0480	−0.0681	0.9501	0.0617	10.0913	73.8359
2.0	70.0	1.0463	−0.0649	0.9520	0.0590	10.4202	79.1810
2.0	75.0	1.0448	−0.0620	0.9537	0.0566	10.7370	84.5133
2.0	80.0	1.0434	−0.0595	0.9553	0.0544	11.0426	89.8343

Rectangular Transmission Line Characteristics for Air

σ	ω/ω_v	$\text{Re}(Z_c/Z_{ca})$	$\text{Im}(Z_c/Z_{ca})$	$\text{Re}(Y_c/Y_{ca})$	$\text{Im}(Y_c/Y_{ca})$	$\text{Re}(\phi)$	$\text{Im}(\phi)$
2.0	85.0	1.0422	-0.0572	0.9566	0.0525	11.3381	95.1453
2.0	90.0	1.0410	-0.0551	0.9579	0.0507	11.6244	100.4472
2.0	95.0	1.0400	-0.0532	0.9591	0.0491	11.9021	105.7408
2.0	100.0	1.0390	-0.0515	0.9601	0.0476	12.1720	111.0268
2.0	105.0	1.0381	-0.0499	0.9611	0.0462	12.4347	116.3060
2.0	110.0	1.0373	-0.0484	0.9620	0.0449	12.6905	121.5787
2.0	120.0	1.0357	-0.0458	0.9636	0.0426	13.1837	132.1067
2.0	130.0	1.0344	-0.0435	0.9651	0.0406	13.6547	142.6140
2.0	140.0	1.0332	-0.0415	0.9663	0.0388	14.1057	153.1029
2.0	150.0	1.0321	-0.0397	0.9675	0.0372	14.5389	163.5757
2.0	160.0	1.0311	-0.0381	0.9685	0.0358	14.9558	174.0340
2.0	170.0	1.0303	-0.0367	0.9694	0.0345	15.3579	184.4792
2.0	180.0	1.0294	-0.0354	0.9702	0.0333	15.7464	194.9125
2.0	190.0	1.0287	-0.0342	0.9710	0.0323	16.1224	205.3352
2.0	200.0	1.0280	-0.0331	0.9717	0.0313	16.4867	215.7480
3.0	0.5	2.9491	-2.7598	0.1808	0.1692	1.9493	2.0467
3.0	0.6	2.7099	-2.5027	0.1992	0.1839	2.1251	2.2531
3.0	0.7	2.5255	-2.3017	0.2163	0.1971	2.2844	2.4457
3.0	0.8	2.3780	-2.1387	0.2325	0.2091	2.4305	2.6274
3.0	0.9	2.2568	-2.0031	0.2479	0.2200	2.5657	2.8006
3.0	1.0	2.1550	-1.8877	0.2626	0.2300	2.6917	2.9667
3.0	2.0	1.6257	-1.2495	0.3867	0.2972	3.6350	4.4074
3.0	3.0	1.4134	-0.9562	0.4854	0.3284	4.2628	5.6657
3.0	4.0	1.3001	-0.7776	0.5665	0.3388	4.7278	6.8561
3.0	5.0	1.2316	-0.6547	0.6331	0.3366	5.0937	8.0167
3.0	6.0	1.1871	-0.5642	0.6871	0.3266	5.3944	9.1632
3.0	7.0	1.1570	-0.4947	0.7308	0.3124	5.6498	10.3026
3.0	8.0	1.1358	-0.4396	0.7657	0.2963	5.8726	11.4381
3.0	9.0	1.1206	-0.3950	0.7937	0.2798	6.0708	12.5713
3.0	10.0	1.1095	-0.3583	0.8162	0.2635	6.2503	13.7028
3.0	12.0	1.0950	-0.3017	0.8488	0.2339	6.5673	15.9624
3.0	14.0	1.0864	-0.2606	0.8704	0.2088	6.8445	18.2187
3.0	16.0	1.0810	-0.2297	0.8851	0.1881	7.0940	20.4723
3.0	18.0	1.0775	-0.2058	0.8954	0.1711	7.3239	22.7233
3.0	20.0	1.0749	-0.1869	0.9030	0.1570	7.5394	24.9718
3.0	22.0	1.0729	-0.1717	0.9087	0.1454	7.7443	27.2173
3.0	24.0	1.0713	-0.1592	0.9132	0.1357	7.9412	29.4595
3.0	26.0	1.0700	-0.1488	0.9169	0.1275	8.1319	31.6981
3.0	28.0	1.0687	-0.1400	0.9199	0.1205	8.3178	33.9328
3.0	30.0	1.0675	-0.1324	0.9225	0.1144	8.4999	36.1634
3.0	35.0	1.0648	-0.1176	0.9278	0.1025	8.9415	41.7205
3.0	40.0	1.0623	-0.1067	0.9320	0.0936	9.3680	47.2492
3.0	45.0	1.0599	-0.0982	0.9355	0.0867	9.7812	52.7499
3.0	50.0	1.0576	-0.0914	0.9385	0.0811	10.1819	58.2240
3.0	55.0	1.0555	-0.0858	0.9412	0.0765	10.5701	63.6735
3.0	60.0	1.0535	-0.0810	0.9436	0.0725	10.9459	69.1009
3.0	65.0	1.0517	-0.0768	0.9458	0.0691	11.3096	74.5084
3.0	70.0	1.0500	-0.0732	0.9478	0.0661	11.6614	79.8981
3.0	75.0	1.0484	-0.0699	0.9496	0.0634	12.0018	85.2721
3.0	80.0	1.0470	-0.0670	0.9512	0.0609	12.3313	90.6322
3.0	85.0	1.0457	-0.0644	0.9527	0.0587	12.6504	95.9797
3.0	90.0	1.0444	-0.0621	0.9541	0.0567	12.9599	101.3162
3.0	95.0	1.0433	-0.0599	0.9553	0.0548	13.2602	106.6426
3.0	100.0	1.0423	-0.0579	0.9565	0.0531	13.5519	111.9601
3.0	105.0	1.0413	-0.0561	0.9576	0.0516	13.8356	117.2694

σ	ω/ω_v	$\mathrm{Re}(Z_c/Z_{ca})$	$\mathrm{Im}(Z_c/Z_{ca})$	$\mathrm{Re}(Y_c/Y_{ca})$	$\mathrm{Im}(Y_c/Y_{ca})$	$\mathrm{Re}(\phi)$	$\mathrm{Im}(\phi)$
3.0	110.0	1.0404	-0.0544	0.9585	0.0501	14.1117	122.5713
3.0	120.0	1.0388	-0.0513	0.9603	0.0475	14.6432	133.1553
3.0	130.0	1.0374	-0.0487	0.9619	0.0452	15.1496	143.7161
3.0	140.0	1.0361	-0.0464	0.9632	0.0431	15.6338	154.2571
3.0	150.0	1.0350	-0.0443	0.9644	0.0413	16.0981	164.7805
3.0	160.0	1.0340	-0.0425	0.9655	0.0397	16.5445	175.2886
3.0	170.0	1.0330	-0.0408	0.9665	0.0382	16.9747	185.7829
3.0	180.0	1.0322	-0.0394	0.9674	0.0369	17.3900	196.2649
3.0	190.0	1.0314	-0.0380	0.9682	0.0357	17.7918	206.7356
3.0	200.0	1.0307	-0.0367	0.9690	0.0345	18.1810	217.1961
4.0	0.5	3.2746	-3.1080	0.1607	0.1525	2.1912	-2.2764
4.0	0.6	3.0048	-2.8224	0.1768	0.1661	2.3913	-2.5033
4.0	0.7	2.7964	-2.5994	0.1918	0.1783	2.5731	-2.7143
4.0	0.8	2.6294	-2.4188	0.2060	0.1895	2.7405	-2.9130
4.0	0.9	2.4919	-2.2685	0.2194	0.1998	2.8959	-3.1016
4.0	1.0	2.3763	-2.1409	0.2323	0.2093	3.0411	3.2820
4.0	2.0	1.7687	-1.4368	0.3406	0.2767	4.1457	4.8241
4.0	3.0	1.5186	-1.1139	0.4281	0.3141	4.9015	6.1389
4.0	4.0	1.3810	-0.9168	0.5026	0.3337	5.4731	7.3602
4.0	5.0	1.2950	-0.7802	0.5666	0.3413	5.9285	8.5357
4.0	6.0	1.2373	-0.6785	0.6214	0.3407	6.3044	9.6870
4.0	7.0	1.1967	-0.5995	0.6680	0.3346	6.6233	10.8249
4.0	8.0	1.1673	-0.5360	0.7075	0.3249	6.8998	11.9552
4.0	9.0	1.1454	-0.4840	0.7408	0.3130	7.1439	13.0810
4.0	10.0	1.1289	-0.4405	0.7686	0.3000	7.3626	14.2040
4.0	12.0	1.1064	-0.3724	0.8119	0.2732	7.7433	16.4454
4.0	14.0	1.0927	-0.3216	0.8422	0.2479	8.0690	18.6832
4.0	16.0	1.0839	-0.2827	0.8638	0.2253	8.3556	20.9192
4.0	18.0	1.0782	-0.2522	0.8794	0.2057	8.6136	23.1541
4.0	20.0	1.0744	-0.2277	0.8908	0.1888	8.8499	25.3882
4.0	22.0	1.0717	-0.2078	0.8993	0.1744	9.0697	27.6216
4.0	24.0	1.0697	-0.1914	0.9058	0.1620	9.2766	29.8542
4.0	26.0	1.0682	-0.1776	0.9109	0.1514	9.4736	32.0858
4.0	28.0	1.0671	-0.1660	0.9150	0.1423	9.6626	34.3162
4.0	30.0	1.0661	-0.1560	0.9183	0.1344	9.8454	36.5451
4.0	35.0	1.0641	-0.1366	0.9245	0.1187	10.2825	42.1087
4.0	40.0	1.0624	-0.1224	0.9289	0.1071	10.7004	47.6574
4.0	45.0	1.0607	-0.1117	0.9324	0.0982	11.1054	53.1883
4.0	50.0	1.0591	-0.1033	0.9353	0.0912	11.5006	58.7000
4.0	55.0	1.0575	-0.0964	0.9378	0.0855	11.8871	64.1921
4.0	60.0	1.0560	-0.0907	0.9401	0.0808	12.2655	69.6648
4.0	65.0	1.0544	-0.0859	0.9421	0.0768	12.6356	75.1186
4.0	70.0	1.0530	-0.0817	0.9440	0.0733	12.9975	80.5548
4.0	75.0	1.0516	-0.0781	0.9458	0.0702	13.3509	85.9744
4.0	80.0	1.0502	-0.0748	0.9474	0.0675	13.6958	91.3787
4.0	85.0	1.0490	-0.0719	0.9489	0.0650	14.0322	96.7689
4.0	90.0	1.0478	-0.0692	0.9503	0.0628	14.3602	102.1463
4.0	95.0	1.0466	-0.0668	0.9516	0.0607	14.6800	107.5119
4.0	100.0	1.0456	-0.0646	0.9528	0.0588	14.9918	112.8668
4.0	105.0	1.0446	-0.0625	0.9539	0.0571	15.2958	118.2120
4.0	110.0	1.0436	-0.0606	0.9550	0.0555	15.5923	123.5482
4.0	120.0	1.0419	-0.0572	0.9569	0.0526	16.1642	134.1970
4.0	130.0	1.0404	-0.0543	0.9585	0.0500	16.7098	144.8183
4.0	140.0	1.0391	-0.0516	0.9600	0.0477	17.2317	155.4163
4.0	150.0	1.0379	-0.0493	0.9613	0.0457	17.7321	165.9944

Rectangular Transmission Line Characteristics for Air

σ	ω/ω_v	$\text{Re}(Z_c/Z_{ca})$	$\text{Im}(Z_c/Z_{ca})$	$\text{Re}(Y_c/Y_{ca})$	$\text{Im}(Y_c/Y_{ca})$	$\text{Re}(\phi)$	$\text{Im}(\phi)$
4.0	160.0	1.0368	−0.0472	0.9625	0.0439	18.2130	176.5550
4.0	170.0	1.0358	−0.0454	0.9636	0.0422	18.6762	187.1004
4.0	180.0	1.0349	−0.0437	0.9645	0.0407	19.1233	197.6322
4.0	190.0	1.0341	−0.0421	0.9654	0.0393	19.5557	208.1519
4.0	200.0	1.0334	−0.0407	0.9662	0.0381	19.9747	218.6607
5.0	0.5	3.5781	−3.4279	0.1457	0.1396	2.4137	2.4903
5.0	0.6	3.2803	−3.1158	0.1603	0.1522	2.6360	2.7366
5.0	0.7	3.0499	−2.8723	0.1738	0.1636	2.8384	2.9651
5.0	0.8	2.8651	−2.6753	0.1865	0.1741	3.0250	3.1799
5.0	0.9	2.7128	−2.5114	0.1985	0.1838	3.1986	3.3834
5.0	1.0	2.5846	−2.3723	0.2100	0.1927	3.3613	3.5777
5.0	2.0	1.9065	−1.6069	0.3067	0.2585	4.6116	5.2217
5.0	3.0	1.6228	−1.2571	0.3851	0.2983	5.4840	6.5996
5.0	4.0	1.4639	−1.0434	0.4530	0.3229	6.1550	7.8615
5.0	5.0	1.3626	−0.8950	0.5127	0.3368	6.6965	9.0625
5.0	6.0	1.2931	−0.7841	0.5654	0.3429	7.1475	10.2289
5.0	7.0	1.2431	−0.6973	0.6119	0.3433	7.5317	11.3746
5.0	8.0	1.2059	−0.6272	0.6527	0.3395	7.8654	12.5078
5.0	9.0	1.1776	−0.5692	0.6884	0.3327	8.1596	13.6332
5.0	10.0	1.1558	−0.5203	0.7194	0.3239	8.4225	14.7536
5.0	12.0	1.1250	−0.4427	0.7697	0.3029	8.8769	16.9859
5.0	14.0	1.1052	−0.3839	0.8074	0.2804	9.2615	19.2123
5.0	16.0	1.0922	−0.3381	0.8355	0.2586	9.5961	21.4360
5.0	18.0	1.0834	−0.3015	0.8566	0.2384	9.8934	23.6584
5.0	20.0	1.0774	−0.2719	0.8726	0.2202	10.1620	25.8802
5.0	22.0	1.0732	−0.2475	0.8847	0.2041	10.4082	28.1018
5.0	24.0	1.0702	−0.2272	0.8941	0.1898	10.6365	30.3236
5.0	26.0	1.0680	−0.2101	0.9014	0.1773	10.8504	32.5454
5.0	28.0	1.0664	−0.1955	0.9072	0.1664	11.0527	34.7672
5.0	30.0	1.0652	−0.1830	0.9119	0.1567	11.2456	36.9890
5.0	35.0	1.0631	−0.1585	0.9202	0.1372	11.6971	42.5418
5.0	40.0	1.0617	−0.1406	0.9257	0.1226	12.1186	48.0896
5.0	45.0	1.0605	−0.1271	0.9296	0.1114	12.5212	53.6293
5.0	50.0	1.0594	−0.1166	0.9327	0.1026	12.9112	59.1581
5.0	55.0	1.0583	−0.1082	0.9352	0.0956	13.2921	64.6743
5.0	60.0	1.0571	−0.1013	0.9373	0.0898	13.6659	70.1764
5.0	65.0	1.0560	−0.0955	0.9393	0.0849	14.0334	75.6639
5.0	70.0	1.0549	−0.0906	0.9411	0.0808	14.3950	81.1366
5.0	75.0	1.0537	−0.0863	0.9427	0.0772	14.7508	86.5947
5.0	80.0	1.0526	−0.0825	0.9442	0.0740	15.1006	92.0385
5.0	85.0	1.0515	−0.0792	0.9456	0.0712	15.4443	97.4686
5.0	90.0	1.0504	−0.0762	0.9470	0.0687	15.7818	102.8858
5.0	95.0	1.0494	−0.0735	0.9483	0.0665	16.1129	108.2907
5.0	100.0	1.0484	−0.0711	0.9494	0.0644	16.4376	113.6842
5.0	105.0	1.0475	−0.0688	0.9506	0.0625	16.7559	119.0670
5.0	110.0	1.0466	−0.0667	0.9516	0.0607	17.0678	124.4398
5.0	120.0	1.0449	−0.0630	0.9536	0.0575	17.6727	135.1584
5.0	130.0	1.0433	−0.0597	0.9553	0.0547	18.2533	145.8453
5.0	140.0	1.0419	−0.0569	0.9569	0.0522	18.8109	156.5049
5.0	150.0	1.0407	−0.0543	0.9583	0.0500	19.3471	167.1409
5.0	160.0	1.0396	−0.0520	0.9595	0.0480	19.8634	177.7566
5.0	170.0	1.0385	−0.0499	0.9607	0.0462	20.3613	188.3545
5.0	180.0	1.0376	−0.0481	0.9617	0.0446	20.8424	198.9367
5.0	190.0	1.0367	−0.0464	0.9627	0.0430	21.3079	209.5050
5.0	200.0	1.0359	−0.0448	0.9635	0.0417	21.7592	220.0609

Appendix C

Weighting Factors for Circular Sections

IMPEDANCE WEIGHTING FACTOR (CIRCULAR)

The impedance formulation given in Eq. (3.123a) is

$$Z(s) = R + L_a s + Z_1(s) \qquad \text{(C-1)}$$

The transmission line impedance derived by Brown (Chapter 3 [11]) is

$$Z(s) = \frac{L_a s}{1 - 2J_1(G)/GJ_0(G)} \qquad \text{(C-2)}$$

where $G = jr_w(s/v)^{1/2}$, r_w is the radius, and v is the kinematic viscosity. Thus, from Eqs. (C-1) and (C-2) the time-dependent impedance $Z(s)$ is

$$Z_1(s) = \frac{L_a s}{1 - 2J_1(G)/GJ_0(G)} - L_a s - R \qquad \text{(C-3)}$$

The impedance weighting function $W_z(s) \equiv Z_1(s)/L_a s$ and Eq. (C-3) may be written as

$$W_z(s) = \frac{1}{1 - 2J_1(G)/GJ_0(G)} - 1 + \frac{8}{G^2} \qquad \text{(C-4)}$$

since $R/L_a s = -8/G^2$. Now from the Bessel function identity that $G[J_0(G) + J_2(G)] = 2J_1(G)$ and algebraic manipulation we obtain

$$W_z(s) = \frac{[8 - G^2]J_2(G) - G^2 J_0(G)}{G^2 J_2(G)} \qquad \text{(C-5)}$$

To invert Eq. (C-5) the residue method is used with the result that

$$W_z(t) = \sum_{i=1}^{\infty} \frac{\exp(s_i t)}{(\partial/\partial s)[1/W_z(s)]_{s=s_i}} \tag{C-6}$$

The roots s_i are the values of the zeros of $J_2(G)$. The derivative in the denominator is

$$\frac{\partial}{\partial s}\left[\frac{1}{W_z(s)}\right]_{J_2=0} = -\frac{G}{2}\frac{\partial G}{\partial s} = \frac{r_w^2}{4v} \tag{C-7}$$

and therefore Eq. (C-6) becomes

$$W_z(t) = \frac{4v}{r_w^2}\sum_{i=1}^{\infty}\exp(s_i t) \tag{C-8}$$

where

$$s_i = -C_i^2 v/r_w^2$$

and C_i are the zeros of $J_2(x)$. From this and the normalized time $t' = ta/l$, Eq. (C-8) takes the normalized form

$$W_z(t') = (R/K_s L_a)4\left[\sum_{i=1}^{\infty}\exp(-C_i^2 t'/N_k K_s)\right] \tag{C-9}$$

where $K_s = 8$ is the circular section shape factor. From Eq. (3.116a), $W(t') = (R/K_s L_a)\phi_z(t'/N_k K_s)$,

$$\phi_z(t'/N_k K_s) = 4\sum_{i=1}^{\infty}\exp(-C_i^2 t'/N_k K_s) \tag{C-10}$$

The first 20 values of C_i and C_i^2 are given in Table C.1. Since the convergence of (C-10) is relatively slow, Brown (Chapter 3 [32]) uses various approximations instead of Eq. (C-10) to obtain values of time. The plot of ϕ_z shown in Fig. 3 37a, however, is based on Eq. (C-10) with 20 terms.

ADMITTANCE WEIGHTING FACTOR (CIRCULAR)

The procedure for the determination of the admittance weighting factor is the same as for the impedance weighting factor. We begin with the admittance formulation of Eq. (3.123b):

$$Y(s) = C_a s + Y_1(s) \tag{C-11}$$

From Brown's (Chapter 3 [11]) theory the admittance is

$$Y(s) = C_a s + C_a s(2)(\gamma - 1)J_1(F)/F J_0(F) \tag{C-12}$$

where

$$F = jr_w(N_p s/v)^{1/2}$$

TABLE C.1
Zeros

	Impedance ($J_2(x)$)		Admittance ($J_0(x)$)	
i	C_i	C_i^2	K_i	K_i^2
1	5.136	26.378	2.405	5.784
2	8.417	70.846	5.520	30.470
3	11.620	135.024	8.654	74.892
4	14.796	218.922	11.792	139.051
5	17.960	322.562	14.931	222.935
6	21.117	445.928	18.071	326.561
7	24.270	589.033	21.212	449.949
8	27.421	751.911	24.352	593.020
9	30.569	934.464	27.493	755.865
10	33.717	1136.836	30.635	938.503
11	36.863	1358.881	33.776	1140.818
12	40.008	1600.640	36.917	1362.864
13	43.153	1862.181	40.058	1604.643
14	46.298	2143.505	43.200	1866.240
15	49.442	2444.511	46.341	2147.488
16	52.586	2765.287	49.483	2448.567
17	55.730	3105.833	52.624	2769.285
18	58.873	3466.030	55.766	3109.847
19	62.016	3845.984	58.907	3470.035
20	65.159	4245.695	62.048	3849.954

The time-dependent admittance $Y_1(s)$ is therefore

$$Y_1(s) = C_a(s)[2(\gamma - 1)]J_1(F)/F J_0(F) \tag{C-13}$$

The definition of the admittance weighting factor $W_y(s) \equiv Y_1(s)/C_a s$, so that

$$W_y(s) = \frac{2(\gamma - 1)J_1(F)}{F J_0(F)} \tag{C-14}$$

Once again the residue method is used to put the weighting factor into the time domain, so that Eq. (C-14) transforms to

$$W_y(t) = 2(\gamma - 1) \sum_{i=1}^{\infty} \frac{\exp(s_i t)}{(\partial/\partial s)(1/W_y(s))_{s=s_i}} \tag{C-15}$$

where in this case the roots s_i are the zeros of $J_0(F)$. Now, if we evaluate the derivative in the denominator of Eq. (C-15),

$$\frac{\partial}{\partial s}\left(\frac{1}{W_y(s)}\right)_{J_0=0} = -F\frac{\partial F}{\partial s} = \frac{N_p r_w^2}{2\nu} \tag{C-16}$$

and Eq. (C-15) then becomes

$$W_y(t) = \frac{4(\gamma - 1)\nu}{N_p r_w^2} \sum_{i=1}^{\infty} \exp(s_i t) \tag{C-17}$$

where $s_i = -K_i^2 v/(N_p r_w^2)$ and K_i are the zeros of $J_0(x)$. The normalized form of Eq. (C-17) is

$$W_y(t') = [(\gamma - 1)R/N_p K_s L_a]\left[4\sum_{i=1}^{\infty} \exp(-K_i^2 t'/N_p K_s N_k)\right] \quad \text{(C-18)}$$

Now, since $W_y(t') = [(\gamma - 1)R/N_p K_s L_a] \phi_y(t'/N_p K_s N_k)$, the normalized weighting factor is

$$\phi_y(t'/N_p N_k K_s) = 4\sum_{i=1}^{\infty} \exp(-K_i^2 t'/N_p N_k K_s) \quad \text{(C-19)}$$

Table C.1 also gives the first 20 values of K_i and K_i^2. Equation (C-19) with the first 20 terms is plotted in Fig. 3.37a for the case of $N_p = 0.7$.

Appendix D

Weighting Factors for Rectangular Sections

IMPEDANCE WEIGHTING FACTOR (RECTANGULAR)

The impedance formulation [Eq. (3.123a)] is the starting point of the determination of impedance weighting function:

$$Z(s) = R + L_a s + Z_1(s) \tag{D-1}$$

The impedance of a rectangular transmission line from Eq. (3.72a) is

$$Z(s) = L_a s/[1 - \tanh(H)/H] \tag{D-2}$$

where $H = \{sA/(4[\sigma + 1/\sigma]v)\}^{1/2}$, A is the area of the cross section, and σ is the aspect ratio. The combination of Eqs. (D-1) and (D-2) yields the time-dependent impedance as

$$Z_1(s) = \frac{L_a s}{1 - \tanh(H)/H} - L_a s - R \tag{D-3}$$

and, from the definition of weighting factor, Eq. (D-3) leads to

$$W_z(s) = \frac{1}{1 - \tanh H/H} - 1 - \frac{3}{H^2} \tag{D-4}$$

where

$$R/L_a s = 12v(\sigma + 1/\sigma)/As = 3/H^2$$

As H approaches infinity $\tanh H$ approaches unity, so that for small values of time we may approximate Eq. (D-4),

$$W_z(s) = 1/(H - 1) \tag{D-5}$$

or, in terms of s,

$$W_z(s) = \frac{1}{[As/4(\sigma + 1/\sigma)v]^{1/2} - 1} \tag{D-6}$$

The inversion of Eq. (D-6) into the time domain yields

$$W_z(t) = 4B[1/(4\pi t/B)^{1/2} - e^{4t/B} \operatorname{erfc}(4t/B)^{1/2}] \tag{D-7}$$

where

$$B = v(\sigma + 1/\sigma)/A = R/K_s L_a$$

and $K_s = 12$ for a rectangular section. In normalized form, Eq. (D-7) is

$$W_z(t') = \frac{R}{K_s L_a} \left[4 \left(\frac{1}{[4\pi t'/(N_k K_s)]^{1/2}} - \exp(4t'/N_k K_s) \operatorname{erfc}(4t'/N_k K_s)^{1/2} \right) \right] \tag{D-8}$$

and thus the normalized impedance weighting factor for the rectangular line is

$$\phi_z\left(\frac{t'}{N_k K_s}\right) = 4\left[\frac{1}{[4\pi t'/(N_k K_s)]^{1/2}} - \exp(4t'/N_k K_s) \operatorname{erfc}(4t'/(N_k K_s))^{1/2} \right] \tag{D-9}$$

Equation (D-9) is shown plotted in Fig. 3.37b.

ADMITTANCE WEIGHTING FACTOR (RECTANGULAR)

The admittance is given in Eq. (3.123b) as

$$Y(s) = C_a(s) + Y_1(s) \tag{D-10}$$

For rectangular sections, Eq. (3.72b) gives

$$Y(s) = C_a s + \frac{C_a(s)(\gamma - 1) \tanh(E)}{E} \tag{D-11}$$

where

$$E = \{sN_p A/(4v[\sigma + 1/\sigma])\}^{1/2}$$

A comparison of Eqs. (D-10) and (D-11) shows that the time-dependent admittance is

$$Y_1(s) = C_a(s)(\gamma - 1) \tanh(E)/E \tag{D-12}$$

so that the admittance weighting factor in this case is

$$W_y(s) = (\gamma - 1) \tanh(E)/E \tag{D-13}$$

If we again use the approximation that tanh E approaches unity for large values of s, the inverse transform of Eq. (D-13) becomes

$$W_y(t) = \frac{2(\gamma - 1)R}{N_p K_s L_a} \left[\frac{1}{(\pi t R/N_p K_s L_a)^{1/2}} \right] \quad \text{(D-14)}$$

and, in normalized form,

$$W_y(t') = \frac{(\gamma - 1)R}{N_p K_s L_a} \left\{ \frac{2}{[\pi t'/(N_p N_k K_s)]^{1/2}} \right\} \quad \text{(D-15)}$$

The normalized admittance weighting factor for the rectangular line is [see Eq. (3.131b)]

$$\phi_y\left(\frac{t'}{N_p N_k K_s}\right) = \frac{2}{[\pi t'/(N_p N_k K_s)]^{1/2}} \quad \text{(D-16)}$$

Figure 3.37b shows the normalized admittance weighting factor when $N_p = 0.7$.

Appendix E

Computer Programs

**SMALL-AMPLITUDE METHOD OF CHARACTERISTICS:
RESPONSE OF BLOCKED TUBE**

```
C     THIS PROGRAM GIVES THE RESULTS SHOWN IN FIGURES 3.32 AND 3.33
C
C
C     THERE ARE 250 INITIAL PRESSURE VALUES , E(TIME)
C        THE MESH SIZE IS 25 INCREMENTS
      DOUBLE PRECISION P(75),Q(75),E(250)
      DT=1./25.
C        ENTER INPUT - IN THIS CASE A RECTANGULAR PULSE
      DO 1 I=1,50
      E(I)=1.0
    1 CONTINUE
      DO 2 J=51,250
      E(J)=0.0
    2 CONTINUE
C     SELECT VALUES OF NK=ENK
      DO 3 K=1,3
      ENK=-.2+.4*FLOAT(K)
      EMM=ENK-DT/4.
      EMP=ENK+DT/4.
      EQ=E(1)/ENK
C     SET INITIAL CONDITIONS
      DO 4 L=1,75
      P(L)=0.0
      Q(L)=0.0
    4 CONTINUE
C     BEGIN CALCULATIONS
C
C
```

```
      DO 5 M=2,250
      T=DT*FLOAT(M-1)
C     LEFT BOUNDARY
      P(26)=.5*(E(M-1)+P(1)+EMM*(EQ-Q(1)))
      Q(26)=.5*(E(M-1)-P(1)+EMM*(EQ+Q(1)))/EMP
      EQ=(EMM*Q(26)+E(M)-P(26))/EMP
C     INTERIOR POINTS
      DO 6 N=27,50
      P(N)=.5*(P(N-26)+P(N-25)+EMM*(Q(N-26)-Q(N-25)))
      Q(N)=.5*(P(N-26)-P(N-25)+EMM*(Q(N-26)+Q(N-25)))/EMP
    6 CONTINUE
      DO 7 NN=51,74
      P(NN)=.5*(P(NN-25)+P(NN-24)+EMM*(Q(NN-25)-Q(NN-24)))
      Q(NN)=.5*(P(NN-25)-P(NN-24)+EMM*(Q(NN-25)+Q(NN-24)))/EMP
    7 CONTINUE
C     RIGHT BOUNDARY FOR A BLOCKED TUBE
      Q(75)=0.0
      P(75)=P(50)+EMM*Q(50)
      WRITE (6,8) ENK,T,P(75)
    8 FORMAT (2F10.3,F15.5)
C
C     UPDATE INITIAL CONDITIONS
      DO 9 IE=1,25
      P(IE)=P(IE+50)
      Q(IE)=Q(IE+50)
    9 CONTINUE
    5 CONTINUE
    3 CONTINUE
      STOP
      END
```

SMALL-AMPLITUDE METHOD OF CHARACTERISTICS WITH THROUGHFLOW

```
C     THIS PROGRAM GIVES THE RESULTS SHOWN IN FIGURE 3.36
C     SET MESH SIZE =10
      DOUBLE PRECISION P(32),Q(32),E(101)
      DX=.1
C     SET RATIO OF TERMINIATING RESISTANCE TO LINE RESISTANCE , R
      R=3.
C     SET CHARACTERISTIC NUMBER NK=ENK
      ENK=1.
C     SET VELOCITY OF THROUGHFLOW U/A=UA
      DO 99 JT=1,3
      UA=.1*FLOAT(JT-1)
      UP=1.+UA
      UM=1.-UA
      DT=DX/UP
      AD=ENK+DT/4.
      AM=ENK-DT/4.
      DTL=DX/UM
      ADL=ENK+DTL/4.
      AML=ENK-DTL/4.
      DN=1.-UA*UA
      EM=1.-2.*UA/DN
      EP=2.*UA/DN
C
C     THE INPUT SIGNAL HAS 101 TIME INCREMENTS
C     ENTER INPUT SIGNAL - IN THIS CASE A STEP FUNCTION
      DO 1 I=1,101
      E(I)=1.0
    1 CONTINUE
C
C     SET INITIAL CONDITIONS
      DO 2 J=1,32
      P(J)=0.0
      Q(J)=0.0
    2 CONTINUE
C
```

```
C     INITALIZE LEFT BOUNDARY
      P(1)=E(1)
      Q(1)=P(1)/ENK
C
C     BEGIN CALCULATIONS OF INTERIOR POINTS
      DO 3 K=1,99
      T=.1*FLOAT(K)/UP
      DO 4 LX=1,19
      IF (LX-10) 5,5,6
    5 NX=LX+11
      GO TO 7
    6 NX=LX+12
    7 QD=(Q(NX-10)+UA*Q(NX-11))/UP
      PD=(P(NX-10)+UA*P(NX-11))/UP
      P(NX)=.5*(P(NX-11)+PD+AM*(Q(NX-11)-QD))
      Q(NX)=.5*(P(NX-11)-PD+AM*(Q(NX-11)+QD))/AD
    4 CONTINUE
C
C     CALCULATE RIGHT BOUNDARY
      P(32)=(P(21)+AM*Q(21))/(1.+AD/R)
      Q(32)=P(32)/R
      WRITE (6,20) ENK,R,UA,T,P(32)
   20 FORMAT (4F10.3,F15.4)
C
C     UPDATE LEFT BOUNDARY
      PE=EM*E(K+1)+EP*E(K+2)
      QE=(PE-P(12)+AML*Q(12))/ADL
      Q(22)=(Q(1)*(-UA)+QE)/UM
C
C     UPDATE INITIAL CONDITIONS
      DO 9 JX=2,11
      P(JX)=P(JX+21)
      Q(JX)=Q(JX+21)
    9 CONTINUE
      Q(1)=Q(22)
      P(1)=E(K+1)
    3 CONTINUE
   99 CONTINUE
      STOP
      END
C     END OF THROUGHFLOW PROGRAM
```

SMALL-AMPLITUDE QUASI METHOD OF CHARACTERISTICS

```
C     THIS PROGRAM GIVES THE RESULTS SHOWN IN FIGURES 3.39,3.40,AND 3.41
C     SET MESH SIZE = 10
      DOUBLE PRECISION P(101,21),Q(101,21),WY(101),WZ(101)
C     SET RATIO OF TERMINIATING RESISTANCE TO DC LINE RESISTANCE , R
C     R=1000 IS CONSIDERED THE SAME AS A BLOCKED LINE
      R=1000.
C     SET CHARACTERISTIC NUMBER NK=ENK
      ENK=1.0
C     SET SHAPE FACTOR KS
C     KS=8 FOR A CIRCULAR LINE
      AKS=8.
C     SET PRANDTL NUMBER NP=PR
      PR=.7
C     SET RATIO OF SPECIFIC HEATS , GAM
      GAM=1.4
      A=1./AKS
      B=(GAM-1.)/(PR*ENK*AKS)
      DT=.1
      AQ=ENK-DT/2.
      TZ2=DT/(2.*AKS*ENK)
      TY2=TZ2/PR
```

```
C
C      THE INPUT SIGNAL HAS 101 TIME INCREMENTS , P(TIME,1)
C      CALCULATE AND STORE WEIGHTING FUNCTION VALUES FOR ALL TIMES
       DO 1 I=1,101
       V=FLOAT(2*I-1)
       TZ=TZ2*V
       TY=TY2*V
       WY(I)=WTY(TY)
       WZ(I)=WTZ(TZ)
     1 CONTINUE
       DP=B*DT*WY(1)
       DQ=A*DT*WZ(1)
C
C      SET INITIAL CONDITIONS
       DO 2 LA=1,101
       DO 3 JA=1,21
       P(LA,JA)=0.0
       Q(LA,JA)=0.0
     3 CONTINUE
     2 CONTINUE
C
C      SET INPUT SIGNAL
C      IN THIS CASE THE INPUT IS A PULSE
       DO 4 LB=1,21
       P(LB,1)=1.0
     4 CONTINUE
C
C      INITIAL LEFT BOUNDARY FLOW
       Q(1,1)=P(1,1)/ENK
C
C      BEGIN CALCULATION OF INTERIOR POINTS
       DO 5 LT=2,101
       T=.1*FLOAT(LT-1)
       IT=LT-1
       IB=LT-2
       DO 6 JX=1,19
       IF (JX-10) 7,7,8
     7 IX=2*JX
       LO=IT
       SP=DP*P(LO,IX)
       SQ=DQ*Q(LO,IX)
       GO TO 9
     8 IX=2*JX-19
       LO=LT
       SP=DP*P(LO-1,IX)
       SQ=DQ*Q(LO-1,IX)
     9 HP=AQ*(Q(LO,IX-1)-Q(LO,IX+1))+P(LO,IX-1)+P(LO,IX+1)+SP
       HQ=AQ*(Q(LO,IX-1)+Q(LO,IX+1))+P(LO,IX-1)-P(LO,IX+1)+SQ
       SMP=0.0
       SMQ=0.0
       IF (IB) 10,10,11
    11 DO 12 LD=1,IB
       LF=LT-LD
       SGP=B*DT*WY(LD+1)*(P(LF,IX)-P(LF-1,IX))
       SGQ=A*DT*WZ(LD+1)*(Q(LF,IX)-Q(LF-1,IX))
       SMP=SMP+SGP
       SMQ=SMQ+SGQ
    12 CONTINUE
    10 P(LT,IX)=(HP-SMP)/(2.+DP)
       Q(LT,IX)=(HQ-SMQ)/(2.*ENK+DQ)
     6 CONTINUE
C
C      CALCULATE RIGHT BOUNDARY
       HPE=AQ*Q(LT,20)+P(LT,20)+.5*DQ*Q(LT-1,21)+.5*DP*P(LT-1,21)
       SME=0.0
       IF (IB) 13,13,14
    14 DO 15 LG=1,IB
       LH=LT-LG
       SEP=.5*B*DT*WY(LG+1)*(P(LH,21)-P(LH-1,21))
       SEQ=.5*A*DT*WZ(LG+1)*(Q(LH,21)-Q(LH-1,21))
       SME=SME+SEP+SEQ
    15 CONTINUE
```

```
   13 DN=(ENK+.5*DQ)/R+1.+.5*DP
      P(LT,21)=(HPE-SME)/DN
      Q(LT,21)=P(LT,21)/R
C
C     CALCULATE LEFT BOUNDARY FLOW
      HQI=AQ*Q(LT,2)-P(LT,2)+.5*DQ*Q(LT-1,1)+.5*DP*(P(LT,1)-P(LT-1,1))+P(LT,1)
     1(LT,1)
      SMI=0.0
      IF (IB) 16,16,17
   17 DO 18 LK=1,IB
      LM=LT-LK
      SIP=.5*B*DT*WY(LK+1)*(P(LM,1)-P(LM-1,1))
      SIQ=.5*A*DT*WZ(LK+1)*(Q(LM,1)-Q(LM-1,1))
      SMI=SMI+SIQ-SIP
   18 CONTINUE
   16 Q(LT,1)=(HQI-SMI)/(ENK+.5*DQ)
      WRITE (6,19) ENK,R,T,P(LT,21)
   19 FORMAT (3F10.1,F15.4)
    5 CONTINUE
      STOP
      END
C
      DOUBLE PRECISION FUNCTION WTY(TAU)
      IF (TAU-.02) 49,49,50
   50 TA=2.40482*2.40482*TAU
      TB=30.4712*TAU
      TC=74.8869*TAU
      TD=139.04*TAU
      TE=222.932*TAU
      TF=326.563*TAU
      W1=DEXP(-TA)+DEXP(-TB)+DEXP(-TC)+DEXP(-TD)+DEXP(-TE)+DEXP(-TF)
      WTY=4.*W1
      GO TO 51
   49 TS=DSQRT(TAU)
      WTY=1.128/TS-1.-.282*TS
   51 RETURN
      END
C
      DOUBLE PRECISION FUNCTION WTZ(TAU)
      IF (TAU-.01) 59,59,60
   60 SA=26.3746*TAU
      SB=70.8499*TAU
      SC=135.021*TAU
      SD=218.920*TAU
      SE=322.555*TAU
      SF=445.928*TAU
      SG=589.038*TAU
      W2=DEXP(-SA)+DEXP(-SB)+DEXP(-SC)+DEXP(-SD)
      W3=DEXP(-SE)+DEXP(-SF)+DEXP(-SG)
      WTZ=4.*(W2+W3)
      GO TO 61
   59 ST=DSQRT(TAU)
      WTZ=4.*(.2821/ST-1.25+1.058*ST+.94*TAU)
   61 RETURN
      END
C     END OF QUASI-METHOD OF CHARACTERISTICS PROGRAM
```

LARGE-AMPLITUDE METHOD OF CHARACTERISTICS FOR STEP INPUT

```
C     THIS PROGRAM GIVES THE RESULTS SHOWN IN FIGURES 3.43,3.44,AND 3.45
C
      DOUBLE PRECISION U(2,41),A(2,41)
      DOUBLE PRECISION DD,C2,TRU,TRA
      DOUBLE PRECISION GR,DR,U1,A1,FTR,AS
      DOUBLE PRECISION G1,G2,D1,D2,AA1,UU1,AA2,UU2,F1,F2
      DOUBLE PRECISION GL,DN,U2,A2,FTL,C1
      COMMON DD,C2
```

```
C
C     SET MESH SIZE HERE AND CHANGE FIRST DOUBLE PRECISION CARD
C     TO MATCH
      MESH=41
      M=MESH-1
      DX=1./FLOAT(MESH-1)
C
C     SET MAGNITUDE OF STEP INPUT
      AE=1.075
      AS=AE*AE
C
C     SET THETA HERE. ADJUST FOR LARGEST VALUE THAT PRODUCES
C     A STABLE SOLUTION
      TH=.7
      DT=TH*DX
C
C ENTER FRICTION FL/2D AS FR
      DO 90 KK=1,2
      FR=.5*FLOAT(KK)-.5
      FC=DT*FR
C
C     INITIAL CONDITIONS
      DO 1 I=1,2
      DO 2 J=1,MESH
      U(I,J)=0.0
      A(I,J)=1.0
    2 CONTINUE
    1 CONTINUE
      A(1,1)=(5.+DSQRT(6.*AS-5.))/6.
      U(1,1)=5.*(A(1,1)-1.)
C BEGIN TIME PROGRESSION
      DO 3 JT=1,600
      T=DT*FLOAT(JT)
C LEFT BOUNDARY POINT
      GL=U(1,2)*A(1,1)-U(1,1)*A(1,2)
      DN=1.+TH*(A(1,1)-U(1,1)+U(1,2)-A(1,2))
      U2=(U(1,1)+TH*GL)/DN
      A2=(A(1,1)+TH*GL)/DN
      FTL=FC*U2*DABS(U2)
      C1=U2-5.*A2-FTL
      A(2,1)=(-C1+DSQRT(6.*AS-.2*C1*C1))/6.
      U(2,1)=5.*A(2,1)+C1
C INTERIOR POINTS
      DO 4 LX=2,M
      G1=U(1,LX-1)*A(1,LX)-U(1,LX)*A(1,LX-1)
      G2=U(1,LX+1)*A(1,LX)-U(1,LX)*A(1,LX+1)
      D1=1.+TH*(U(1,LX)+A(1,LX)-U(1,LX-1)-A(1,LX-1))
      D2=1.+TH*(A(1,LX)-U(1,LX)+U(1,LX+1)-A(1,LX+1))
      AA1=(A(1,LX)-TH*G1)/D1
      UU1=(U(1,LX)+TH*G1)/D1
      AA2=(A(1,LX)+TH*G2)/D2
      UU2=(U(1,LX)+TH*G2)/D2
      F1=FC*UU1*DABS(UU1)
      F2=FC*UU2*DABS(UU2)
      A(2,LX)=.1*(UU1-UU2+5.*(AA1+AA2)+F2-F1)
      U(2,LX)=.5*(UU1+UU2+5.*(AA1-AA2)-F1-F2)
    4 CONTINUE
C RIGHT BOUNDARY FOR ORIFICE TERMINATION
      GR=U(1,M)*A(1,MESH)-U(1,MESH)*A(1,M)
      DR=1.+TH*(U(1,MESH)+A(1,MESH)-U(1,M)-A(1,M))
      U1=(U(1,MESH)+TH*GR)/DR
      A1=(A(1,MESH)-TH*GR)/DR
      FTR=FC*U1*DABS(U1)
      C2=U1+5.*A1-FTR
C
C     SET ORIFICE SIZE DO/U=DD
      DD=.5
      TRU=0.0
      TRA=C2/5.
C
C     FOR BLOCKED LINE REMOVE NEXT 2 STATEMENTS AND GO TO 11
      IF (TRA-1.) 11,11,16
   16 CALL ROOT(TRU,TRA,DD*DD)
```

```
   11 A(2,MESH)=TRA
      U(2,MESH)=TRU
C
C  TRANSFER OF GRID
C     UPDATE INITIAL CONDITIONS
      DO 5 N=1,MESH
      A(1,N)=A(2,N)
      U(1,N)=U(2,N)
      A(2,N)=0.0
      U(2,N)=0.0
    5 CONTINUE
      P=A(1,MESH)**7.
      WRITE (6,10) AE,FR,T,A(1,MESH),U(1,MESH),P
   10 FORMAT (3F10.3,3F15.5)
    3 CONTINUE
   90 CONTINUE
      STOP
      END
C
C     SIMULTANEOUS SOLUTION OF EQUATIONS 43(A) AND 46
      SUBROUTINE ROOT(U,A,DDD)
      DOUBLE PRECISION U,A,DDD,DD,C2,Z,ZP,UP,R
      COMMON DD,C2
      Z=A-1.D0
      ZP=Z
      UP=U
      N=0
      R=1.D0
    5 Z=ZP+R*(UP**2-5.D0*DDD*ZP*(ZP+2.D0)/((ZP+1.D0)**10-DDD))
      N=N+1
      IF (N.GE.500) STOP
      U=U-R*( UP+5.D0*Z-(C2-5.D0))
      IF (U.LT.0.D0) GO TO 15
      IF (DABS(1.D0-UP/U)+DABS(1.D0-ZP/Z).LT.1.D-10) GO TO 10
      ZP=Z
      UP=U
      GO TO 5
   15 R=.25D0
      ZP=0.D0
      UP=0.D0
      GO TO 5
   10 A=Z+1.D0
      RETURN
      END
C
C     END OF LARGE AMPLITUDE PROGRAM
```

INDEX

Page numbers for definitions are in italics.

A

Abramovich, G. N., 180, 191
Acoustic effects on a jet, 185
Acoustic equations, 76
Albertson, M. S., 174, 180, 191
  jet velocity profile, 174
Alexander, L. G., 191
Amplifier
  beam deflection, see Beam deflection amplifier
  bistable, see Bistable switch
  turbulence, see Transition NOR
  vortex, see Vortex triode
Analog
  current, *10, 11*
  voltage, *10, 11*
AND unit, 249
Anderson, W. W., 304, 307, 308, 309, 314
Andrade, E. N. daC., 160, 191
Apparent loss coefficient, *16*
Aspect ratio, *3*
Auger, R. N., 448
Axisymmetric laminar jet profile, 155-161
  equations, 158
Ayers, B. O., 61

B

Baker, P. J., 35, 61
Baron, T., 191
Bauer, A. B., 216, 220, 225
Baumeister, T., 60
Beam deflection amplifier
  aspect ratio, effect of, 360
  bandwidth, 372
  control bias pressure, *335*
  control duct impedance, 368
  control edge impedance, 368
  dynamic analysis, 361-370
  dynamic range, *371*
  figures, 3
  flow gain, 375
  gain optimization, 356-358
  general discussion, 315-320
  input characteristics, 340-348
  input impedance, 362, *363,* 364, 374-375
  input region, 326-348
  input resistance, 333
  input resistance, negative, 348
  introduction, 3
  jet impedance, 369
  momentum controlled, 323-324, 333
  negative input resistance, 348
  outlet impedance, 369
  output characteristics, 358-360

Bean deflection amplifier (cont'd)
  output region, 353-358
  power gain, 375
  pressure controlled, 320-323, 333
  pressure gain, 375
  pressure recovery, 359-360
  Reynolds number effects, 373
  splitter, 354
  staging, 371-376
  transfer characteristics, 354
  transfer function, 364-367
  vent impedance, 366
  vents, 358-352
Beatty, E. K., 435, 449
Beeken, B. B., 423
Beij, K. H., 61
Belforte, G., 61
Bell, A. C., 437, 449
Bell, C., 448
Bellman, R. H., 380
Belsterling, C. A., 361, 369, 380
Benson, R. S., 106
Bermel, T. W., 61
Bertia, J., 61, 423
Biagi, F., 61
Bias level, *335*
Bichara, R. T., 291, 292, 293, 295, 297, 299, 313
Bickley, W., 151, 160, 191
Binder, R. C., 60
Bistable amplifier, *see* Bistable switch
Bistable switch
  attachment, 383-384
  attachment point theory, 392-395
  characteristics, 240-246, 409-416
  control volume assumption, 396-397
  controls, effect of, 385
  curved walls, 388-389
  discharge coefficient, effect of, 405
  dynamic modeling, 417-420
  fan-out ratio, *389*
  figures, 4, 385
  figures of merit, 389
  introduction, 4
  outlet, blocked, 385-386
  pressure recovery, *389*
  splitter, 387-388
  vents, 386-387
Bjornsen, B. G., 255, 275
Blackburn, J. F., 275
Bleeds, *see* Vents
Blosser, R. L., 61
Boothe, W. A., 361, 373, 380, 381

Borda-Carnot relation, 27
Bouasse, H., 383, 423
Boucher, R. F., 400, 423
Bourque, C., 182, 191, 380, 390, 392, 395, 396, 397, 399, 400, 402, 403, 405, 408, 423, 424
Bouyoucos, J. V., 208, 210, 215, 226
Bowles, R. E., 278, 313, 382
Branch resistance, 37-39
Bronchart, J. A., 423
Brown, F. T., 62, 64, 79, 83, 107, 118, 180, 191, 225, 320, 323, 326, 349, 361, 362, 379, 380, 406, 408, 424
Brown, G. B., 208, 210, 211, 215, 225, 226

C

Camarata, F. J., 352, 380
Capacitance, *43*
  jet, 55, 206
  massless piston, 44-45
  variable volume (compliance), 51-55
  variation with frequency, 45-49
Carlson, R. J., 64, 89
Carriere, A., 383
Cascade, *see* Staging
Cascade diode, *see* Diode, cascade
Center dump, 3, 354
Chaplin, H. R., 423
Characteristic frequency, 12, *81*
Characteristic impedance, *66*
Characteristics, *see* Static characteristics
Characteristics, method of
  fluidic line circuits, 136-140
  large amplitude signals, 125-130
  quasi method of, 118-124
  small amplitude signals, 107-118
Circuit
  branch, 42
  regimes, 11
  variables, 10-11
Coanda, H., 382, 383, 389, 423
Coanda effect, 382-420
Cochran, W. L., 380
Comings, E. W., 191
Compliance, *43*, *see also* Capacitance, variable volume
Controls, input, 3
Cook, R. K., 61
Cornish, R. J., 88
Courant, R., 106
Crandall, J. B., 63, 79, 80
Critical Reynolds number, 19
Curle, N., 208, 226

Index

D

Dagan, J., 62
Dai, V. B., 191
Daniels, F. B., 61, 63
Dean number, *23*
Decay factor, 259, *260,* 261-268
Dexter, E. M., 348, 380
Diameter, equivalent, *19*
Diaphragm capacitor, 52
Diode
  cascade, 34
  flow, *32*
  fluid, 31-37
  momentum interaction, 35
  pressure, *32*
  scroll, 35
  Tesla, 35
  vented, 32, 41-42
  vortex, *see* Vortex diode
Diodicity, 8, *31*
Dockery, R. J., 254
Dodds, I, J., 390, 423
Douglas, J. F., 323, 379
Drazan, P., 435, 449
Drzewiecki, T., 413, 415, 419, 424
Duct, resistance, 16
Dump, 4
  center, 3
Dynamic range, *371*

E

Eastman, J. M., 380
Edgetones, 207-220
Edwards, J. A., 405, 424
Energy distribution factor, *19*
Eutrance profile, 20
Entrance resistance, 18
Epstein, M., 418, 424
Equivalent (hydraulic) diameter, *19*
Evans, R. A., 330, 380
Exponent, polytropic
  variation with frequency, 45-49

F

Fan-out, *4, 389,* 427
Fill time, *303*
Fluid circuit regimes, 11
Fluid diode, 31-37
Fluid transmission line, *see* Transmission line
Foss, J. F., 191, 326, 327, 380
Foster, K., 254
Franke, M. E., 64, 83, 97

Frequency, characteristic, 12, *81*
Fukuzawa, K., 397, 423
Funke, M., 449

G

Gebben, V. D., 284, 313
Gerber, H., 61, 62
Gibson, A. H., 61
Giles, J. A., 423
Goertler, H., 162, 191
Goertler equations, 163
Goldschmied, F. R., 81, 145
Goodson, R. E., 64, 74, 76
Goto, J. M., 326, 379, 413, 415, 419, 424
Gray, W. E., 278, 313
Griffin, W. S., 349, 355, 380
Grogan, E. C., 64

H

Hagen-Poiseuille resistance, 31, 63, 368
Hall, R. E., 449
Han, L. S., 61
Hartman, R. B., 61
Hastie, E., 62
Hatch, R. W., 61
Hayes, A. P., 423
Hayes, W., 442, 444, 449
Healey, A. J., 64, 89, 361, 380
Heim, R., 61, 277, 279, 313
Hixon, C. W., 61
Horton, B. M., 278, 313, 315, 317, 319, 320, 348, 379, 382
Howard, J. H., 62
Howie, K., 449
Howland, G. R., 328, 380
Humphrey, R. A., 62, 225, 349, 362, 380
Hydraulic (equivalent) diameter, *19*
Hyer, J. M., 380

I

Iberall, A. S., 64, 79
Impact modulator
  annular control pressure effect, 265
  figure, 7
  general discussion, 255-259
  impact region, 7, 255
  introduction, 6
  pressure gain, 269-273
  source flow modulation, 268-269
  transverse control pressure effect, 266-268

Impedance
  characteristic, *see* Characteristic impedance
  input, terminated line, 67
Inductance, *56*
Inertance, *56*
Inverter, *4, 5*
Iseman, J. M., 254

J

Jacobs, B. E. A., 35, 61
Jensen, R. A., 191
Jet
  bounded, 187-189
  capacitance, 55, 206
  in constant pressure gradient, 197-199
  impedance, 205-207
  profile, *see* Profile, jet velocity
  response to an impulse, 195-197
  spread, 164
  spread parameter, 164
  stability, 185
  steamlines, 164
  velocity distribution, *see* Profile, jet velocity
  width, 160, 172
Jones, J. B., 191

K

Kadosch, M., 278, 313, 382, 423
Kallevig, J. A., 352, 353, 380
Kantola, R., 64, 105
Kantrowitz, A., 106
Kar, S., 188, 191, 404, 424
Karam, J. T., 64, 83, 105
Karamcheti, K., 211, 215, 216, 220, 221, 223, 225, 226, 420, 424
Katz, S., 61, 62, 191, 254, 256, 260, 262, 263, 268, 275, 361, 379, 380
Kelley, L. R., 373, 381
Kesavan, H. K., 60
Keshock, E. G., 280, 313
Keto, J. R., 408, 424
Keulegan, G. H., 61
Kimura, M., 397, 413, 423, 424
King, R. W. P., 63
Kirshner, J. M., 60, 107, 191, 225, 266, 268, 275, 379, 380, 423, 424
Kline, S. J., 61
Koenig, H. E., 60
Kuo, B. C., 111
Kwok, C. C. K., 61, 62

L

Langhaar, H. L., 61
Lawley, T. J., 301, 313
Lechner, T. J., 82, 255, 275
Le Clerc, A., 260, 275
Lefay, A., 423
Le Foll, J., 278, 313, 423
Leonard, R. G., 64, 76
Levin, S. G., 397, 399, 423
Limit circle, *8*
Load, *227*
  characteristics, 230-232
Loss coefficient, *16, 19*
  apparent, *16*
  relation between apparent and total, 17
  total, *16*
Losses, major and minor, 16
Lumped lines, *12*
Lush, P. A., 413, 417, 418, 419, 424

M

Mach number relation to circuit regimes, 11
Manion, F. M., 62, 225, 325, 326, 327, 328, 349, 352, 353, 355, 361, 362, 373, 374, 379, 380, 381, 397, 399, 423
Manning, J. R., 106
Markland, E., 435, 449
Marks, L., 60
Marsters, G. F., 431, 432, 449
Mayer, E. A., 280, 289, 290, 291, 294, 313
McGinn, J. J., 61
McNaughton, K. J., 431, 449
McRee, D. I., 405, 413, 424
Method of characteristics, *see* Characteristics, method of
Metral, A., 382, 383, 423
Miller, D. E., 191
Mon, G., 225, 325, 349, 355, 373, 374, 379, 423
Moses, H. L., 405, 413, 424
Mott, R. C., 449

N

Neve, R. S., 323, 379
Newman, B. G., 182, 191, 380, 390, 392, 395, 396, 397, 399, 400, 408, 423
Nichols, N. B., 64, 79, 81, 83
NOR, *5*
  beam deflection, 5
  impact modulator, 6
  transition, 5

Norwood, R. E., 254
Nozzle-baffle, 24-26
Nozzle, yower jet, 3
Number
  Dean, *23*
  Prandtl, *46*
  Reynolds, *see* Reynolds number
  Strouhal, *224*
Nyborg, W. S., 208, 209, 210, 211, 212, 215, 220, 225, 226

O

Offset, 390, *395*
Ogzu, M. R., 418, 419, 424
O'Keefe, R. F., 442, 443, 449
Oldenburger, R., 74
Olson, R. E., 352, 380
Operating point, *282*
Orner, P. A., 291, 292, 293, 295, 297, 299, 313
Outlets, 3
Owczarek, J., 106

P

Palmisano, R. R., 320
Pan, C. H. T., 327, 380
Parker, G. A., 254
Paul, F. W., 61
Pavlin, C., 328, 380
Peperone, S. J., 317, 323, 352, 379
Performance ratio, diode, *see* Diodicity
Perry, C. C., 423
Peterman, F., 61
Peters, H., 61
Plate decay factor, 259, *260,* 261-268
Polytropic exponent variation with frequency, 45-49
Powell, A., 208, 209, 211, 225, 325, 379
Prandtl number, *46*
Prandtl's new mixing theory, 161
Pressure gain, *3,* 375
Pressure recovery, *4,* 359-360
Price, D. C., 301, 313
Profile, jet velocity
  Albertson, 174
  axisymmetric laminar jet, 155-160
    equations, 158
  Goertler, H., 390, 391, 399, 404, 417
    equations, 163
  modified Goertler, 182
  modified Schlichting-Bickley, 182

  oscillating jet, 220-222
  Reichardt inductive theory, 167-172
  Schlichting-Abramovich, 180
  Schlichting-Bickley, 151
  Simson-Brown, 180-182
  turbulent jet from finite aperture, 173
  two-dimensional laminar jet, 147-155
  two-dimensional turbulent jet, 162-165
Propagation constant, *66*

R

Raju, V. C., 181, 191, 404, 424
Ravelli, E., 423
Rayleigh, J. W. S., 63
Reader, T., 61
Rechten, A. W., 423
Recovery pressure, *4,* 359-360
Reed, H. R., 63
Reethof, G., 275
Reichardt, H., 167, 191
Reichardt inductive theory, 167-172
Reid, K. N., 61, 191, 262, 275, 354, 380, 449
Resistance
  ac, *14*
  branch, 37-39
  calculation example, 30
  curved passage, 22
  dc, *14*
  entrance, 16
  exit, *16,* 26-30
  Hagen-Poiseuille, 31, 63, 368
  linear, 16
  nonlinear, 16
  nozzle-baffle, 25
  square law, 16
  straight passage, 20
  subsonic diffuser, 25
  sudden enlargement, 30
  variational, 16
Reynolds number, *154*
  critical flat plate, 19
  equivalent diameter, *19*
  parameter, *25*
  radial, *293, 299, 301*
  stretched, *373*
  tangential, *299*
Richardson, H. H., 289, 300, 301, 309, 313
Roark, R. J., 62
Rodrigue, G., 402, 424
Roffman, G. L., 185, 191, 320, 361, 379, 380
Rohmann, C. P., 64
Rouse, H., 61, 191

Royle, J. K., 295, 296, 303, 313
Rudinger, G., 106, 126
Rupert, J. G., 352, 354, 380

S

Sakao, F., 187, 191
Sarpkaya, T., 423
Sato, H., 187, 191
Savino, J. M., 280, 313
Sawyer, R. A., 390, 393, 394, 399, 400, 404, 423, 424
Schaedel, H., 64, 89
Schlichting, H., 151, 160, 180, 191
Schlichting-Abramovich velocity distribution, 180
Schlichting-Bickley equations, 151
Self preservation, *146*
Shapiro, A. H., 61, 106
Shavitt, C., 61
Shearer, J. L., 275
Sher, N., 408, 424
Shields, W. L., 215, 221, 223, 226, 420, 424
Shimizu, A., 395, 403, 404, 423, 424
Shunt admittance, 77, 85
Siegel, R., 61
Similarity, *146*
Simson, A. K., 180, 191, 400, 419
Simson-Brown velocity distribution, 180-182
Sinclair, C. G., 431, 449
Siwoff, F., 449
Sorenson, P. H., 255, 275
Source, *227*
　characteristics, 228-230
Sparrow, E. M., 61
Speed of wave propagation, *78, 79*
Spread factor, 164
Staging, 371-376
Static characteristics
　bistable switch, 240-246, 409-416
　NOR element, 247-249, 428-430, 435-438
　passive AND, 249-251
　passive inclusive OR, 251-252
　proportional amplifier, 235-240
　vortex triode, 282-284
Stegen, G. R., 226
Stenning, A. H., 418, 419, 424
Stern, H., 278, 313
Stiffler, A. K., 222, 226
Stong, C. L., 449
Streeter, V. L., 106
Strouhal number, 187, *224*

Switch, bistable, *see* Bistable switch
Syred, N., 295, 296, 303, 313

T

Tank, *see* Capacitance
Taplin, L. B., 303, 307, 313, 314
Tee, 38, *see also* Branch
Tesla, N., 61
Tesla diode, 35-37
Thoma, D., 51, 277, 313
Thoma counterflow brake, 33
Tippetts, J. R., 313
Toda, K., 185, 191
Tokad, Y., 60
Transition, 5
Transition NOR
　alternate configurations, 445-447
　characteristics, 247-249, 428-430, 435-438
　dynamic response, 441-445
　fan out, 427-430
　input characteristics, 437-438
　introduction, 425-426
　laminar length, 431-432
　output characteristics, 437-438
　performance criteria, 427-430
　performance optimization, 439
　pressure recovery, 433-435
　transfer characteristic, 435-437
　Verhelst diagram, 440-441
Transmission line, fluid
　characteristic impedance, *66, 82*
　frequency response, 99-104
　impulse response, 105
　input impedance, 67
　lossless line model, 77
　lumped parameter approximation, 69-75, 93-97
　matching, 130-136
　rectangular, 87-93
　series impedance, 77, 80
　shunt admittance, 77, 81
　step response, 105
　with throughflow, 115-118
Trapani, R. D., 188, 191, 404, 424
Triode, vortex, *see* Vortex triode
Tsui, K. C., 361, 369, 380
Turbulence amplifier, *see* Transition NOR
Turndown ratio, *284*
Two-dimensional geometry, *3*
Two-dimensional laminar jet profile, 147-155

Index

V

Van Tilburg, R. W., 325, 326, 360, 380
Vennard, J. K., 60
Vents, 3, 348-352, 386-388
Verhelst, H. A. M., 239, 254, 435, 440, 449
Viscous characteristic frequency, 12, *81*
Vogel, G., 61
Vortex diode, 8, 33-35, 289
Vortex triode
  ac analysis, 303-304
  characteristics, 282-284
  design chart, 300-303
  fill time, *303*
  flow angle, *286*
  flow field, 280-281
  general discussion, 276-278
  introduction, 9
  modified boundary layer coefficient, *294*
  pressure ratio figure of merit, *284*
  semi-inviscid analysis, 286-289
  time response, 304-305
  turndown ratio, *284*

W

Wada, T., 395, 403, 404, 423, 424
Wall attachment, *see* Coanda effect
Wall attachment switch, *see* Bistable switch
Ware, L. A., 63
Warren, R. W., 383, 408, 424
Werbow, A. R., 431, 449
White, C. M., 61
Wilson, M. P., 424
Wormley, D. N., 280, 289, 292, 294, 295, 300, 301, 304, 309, 313
Wright, C. P., 254
Wylie, E. B., 106

Y

Young, T., 382, 423

Z

Zalmanzon, L. A., 61
Zerner, F., 383, 423
Zielke, W., 106, 107, 118